AF881599

JAMAICA

Primary Mathematics

Second Edition

NSC Edition

Grade 4

Kerry Saadien-Raad
Jennifer Peek

hachette LEARNING

The Publishers would like to thank the following for permission to reproduce copyright material.

Photo credits

t = top, *b* = below, *l* = left, *r* = right, *c* = centre, *b/g* = background

p.7: © sumire8/stock.adobe.com; p.12: © Iryna Volina/stock.adobe.com; p.13: © stocksolutions/stock.adobe.com; p.16: © lineartestpilot/stock.adobe.com; p.18: © slava/stock.adobe.com; p.19: © Freepik/stock.adobe.com; p.20: © Octyabr/stock.adobe.com; p.22: © Gennadiy Poznyakov/stock.adobe.com; p.23: © Polina Tomtosova/stock.adobe.com; p.26: © jantima/stock.adobe.com; p.29: © Natali/stock.adobe.com; p.31: © wachiwit/stock.adobe.com; p.36: © (tl) innafoto2017/stock.adobe.com, (tr) © Alexstar/stock.adobe.com, (bl) © ChaoticDesignStudio/stock.adobe.com, (bc) © Anke van Wyk/stock.adobe.com, (br) © Alexstar/stock.adobe.com; p.40: (t) © artissp/stock.adobe.com, (b) © fotogal/stock.adobe.com; p.41: © kolonko/stock.adobe.com; p.46: © selma/stock.adobe.com; p.52: © Matthew Benoit/stock.adobe.com; p.53: © Nataliya Yakovleva/stock.adobe.com; p.59: (l) © stanproject/stock.adobe.com, (c) © Rashevskyi Media/stock.adobe.com, (r) © Valerii Evlakhov/stock.adobe.com; p.60: © hachut/stock.adobe.com; p.61: © gearstad/stock.adobe.com; p.62: (l) © Dmitriy Kazitsyn/stock.adobe.com, (cl) © Rashevskyi Media/stock.adobe.com, (cr) © weera/stock.adobe.com, (r) © butenkow/stock.adobe.com; p.63: © snaptitude/stock.adobe.com; p.64: © Iryna Volina/stock.adobe.com; p.67: © inspiredbyart/stock.adobe.com; p.69: © savvano/stock.adobe.com; p.72: (l) © muratart/stock.adobe.com, (r) © Anatoliy Karlyuk/stock.adobe.com; p.76: © gator/stock.adobe.com; p.77: (l) © misha_ru/stock.adobe.com, (r) © Winai Tepsuttinun/stock.adobe.com; p.78: © MW Photography/stock.adobe.com; p.80: © powell83/stock.adobe.com; p.82: © bigmouse108/stock.adobe.com; p.88: (t) © Monkey Business/stock.adobe.com, (b) © Scott Latham/stock.adobe.com; p.91: (l) © bohbeh/stock.adobe.com, (r) © akhenatonimages/stock.adobe.com; p.101: © Nizoli/stock.adobe.com; p.102: (t) © bankrx/stock.adobe.com, (b) © fufupix/stock.adobe.com; p.103: © Bokeh Art Photo/stock.adobe.com; p.112: © Peter/stock.adobe.com; p.115: (tl) © Drazen/stock.adobe.com, (tr) © vectorfusionart/stock.adobe.com, (bl) © WavebreakMediaMicro/stock.adobe.com, (br) © eggeeggjiew/stock.adobe.com; p.120: © Juanan Tamayo-Ramos/stock.adobe.com; p.125: © cherylvb/stock.adobe.com; p.126: © treter/stock.adobe.com; p.130: © Dixi_/stock.adobe.com; p.131: © microscience/stock.adobe.com; p.132: © spotmatikphoto/stock.adobe.com; p.137: © Tony R. Hightower/stock.adobe.com; p.140: (t) © David/stock.adobe.com, (c) © Photocreo Bednarek/stock.adobe.com, (b) © Prostock-studio/stock.adobe.com; p.146: © begun1983/stock.adobe.com; p.147: © fotografermen/stock.adobe.com; p.155: © Anatoly Tiplyashin/stock.adobe.com; p.166: © oksix/stock.adobe.com; p.167: © Iryna Volina/stock.adobe.com; p.174: (l) © boonchuay1970/stock.adobe.com, (cl) © sergio37_120/stock.adobe.com, (c) © Alexstar/stock.adobe.com, (cr) © k_samurkas/stock.adobe.com, (r) © egorxfi/stock.adobe.com; p.175: © madamsaffa/stock.adobe.com; p.177: (tl) © Eva Speshneva/stock.adobe.com, (tc) © Ortis/stock.adobe.com, (tr) © Vladimir Pospelov/stock.adobe.com, (cl) © wizdata/stock.adobe.com, (cr) © TheFarAwayKingdom/stock.adobe.com, (bl) © Valua Vitaly/stock.adobe.com, (br) © tr3gi/stock.adobe.com; p.178: © daisy.bee/stock.adobe.com; p.180: © cloverlittleworld/stock.adobe.com; p.182: © dariaren/stock.adobe.com; p.184: © dollitude/stock.adobe.com; p.190: © Photographee.eu/stock.adobe.com; p.193: © Janusz Pienkowski/Shutterstock.com; p.194: © josefkubes/stock.adobe.com; p.203: © pamela_d_mcadams/stock.adobe.com; p.204: © Юлия Буракова/stock.adobe.com; p.206: © Africa Studio/stock.adobe.com; p.207: © Julián Rovagnati/stock.adobe.com; p.208: © Lumos sp/stock.adobe.com; p.210: © Eric Isselée/stock.adobe.com; p.213: © wizdata/stock.adobe.com; p.217: © ChaoticDesignStudio/stock.adobe.com; p.218: (l) © fotofabrika/stock.adobe.com, (r) © mimagephotos/stock.adobe.com; p.220: © Debbie Ann Powell/stock.adobe.com; p.225: © Lars Christensen/stock.adobe.com; p.237: © didecs/stock.adobe.com; p.240: © Morphart/stock.adobe.com; p.243: © dana_zurki/stock.adobe.com; p.247: (t) © Stepan Popov/stock.adobe.com, (b) © PiyawatNandeenoparit/stock.adobe.com; p.249: © mast3r/stock.adobe.com; p.250: © ijp2726/stock.adobe.com; p.252: © ying/stock.adobe.com; p.255: © Nizoli/stock.adobe.com; p.257: © Irina/stock.adobe.com; p.258: © Christos Georghiou/stock.adobe.com; p.264: © Monkey Business/stock.adobe.com; p.266: © FoxyImage/stock.adobe.com; p.269: © geshas/stock.adobe.com; p.270: (t) © Peter/stock.adobe.com, (c) © icemanphotos/stock.adobe.com, (b) © JpegPhotographer/stock.adobe.com; p.271: © armo/stock.adobe.com; p.274: (t) © Krzysztof Bubel/stock.adobe.com, (b) © armo/stock.adobe.com.

To order, please visit www.hachettelearning.com or contact Customer Service at education@hachette.co.uk / +44 (0)1235 827827.

ISBN: 9781036019969

First published in 2023

This edition published in 2026 by Hachette Learning (a trading division of Hodder & Stoughton Limited)

An Hachette UK Company

Carmelite House

50 Victoria Embankment

London EC4Y 0DZ

www.hachettelearning.com

Impression number 10 9 8 7 6 5 4 3 2 1

Year 2026

Cover illustration by Oxford Designers and Illustrators Ltd

Illustrations by Magriet Brink and Leo Daly, Cathy Hughes, Rose Elphick and Andy Cooke, Peter Lubach and Aptara Ltd and Hyphen S.A.

Typeset by Hyphen S.A.

Printed and Bound in Great Britain by Bell & Bain Ltd, Glasgow

A catalogue record for this title is available from the British Library.

Contents

Unit 2: Geometry

Unit 3: Algebra

Unit 4: Statistics and probability

TERM 1

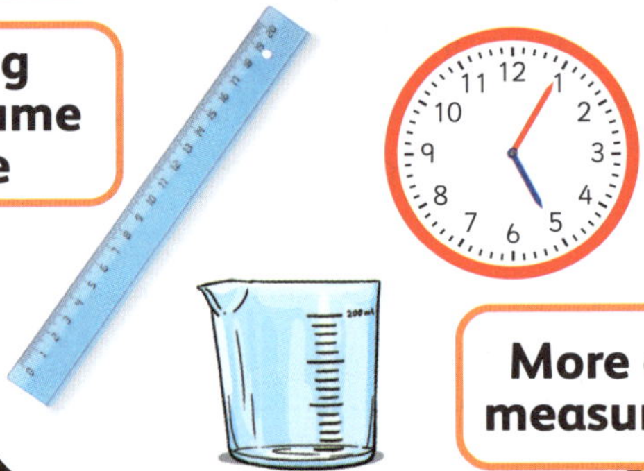

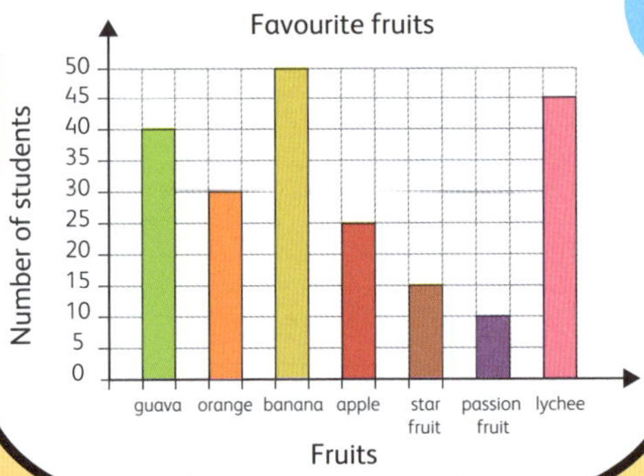

Favourite fruits
Number of students
0
5
10
15
20
25
30
35
40
45
50
guava
orange
banana
apple
star fruit
passion fruit
lychee
Fruits

Unit 1: Number

Chapter 1 Number value

In this chapter you will explore this focus question:

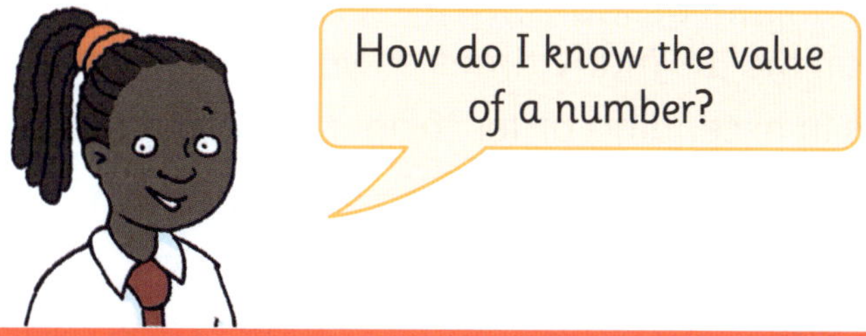

To answer this question, you will:

- work with numbers that have up to seven digits
- distinguish between value, place value and face value.

Starting point

1 Talk about the picture. What animal makes each **number**?

2 Which ten different numbers could you make if you only had the first three balloons?

3 The **face value** of the zebra is always 5. Its **value** can change depending on the position of the **digits** in a bigger number. This is called **place value**. What place value does 5 have in each of these numbers:

a 3 015 **b** 580 **c** 5 600

Key words

number
face value
value
digit
place value

4 Start a small notebook for maths definitions and examples. Draw or write definitions of value, face value and place value.

5 Use the picture of the balloon numbers on the previous page. Which animal balloons would you use to make these numbers:

a two thousand, four hundred, forty-seven

b five thousand, three hundred, six

c eight thousand nineteen.

6 a Choose four of the animal balloons. Write ten different 4-digit numbers with them. You can use each animal as many times as you like in each 4-digit number.

b Write the numbers in ascending order (from smallest value to largest value).

7 We express a part of the whole using fractions, such as one half or $\frac{1}{2}$. Use what you know about fractions to complete these sentences.

a $\frac{3}{10}$ of the balloons are _____ .

b _____ of the balloons are black-and-white.

c _____ of the balloons are orange.

d _____ of the set is brown.

8 Round the numbers you wrote in Exercise 3 off to the nearest thousand.

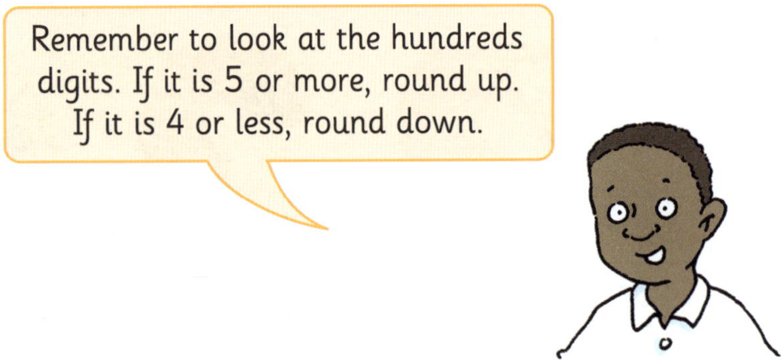

Maths detective: Find the mistake!

Work with a partner. Each person writes a 3-digit number and five different number sentences, which give that number as a sum or difference. One of your number sentences should have a mistake. Get your partner to find the mistake!

Numbers in different forms

Key maths idea

You can write numbers in different ways.

Using **standard form**: 4 627

In a place value chart:

Thousands	Hundreds	Tens	Ones
4	6	2	7

Expanded form: 4 000 + 600 + 20 + 7

Word form: four thousand, six hundred, twenty-seven

Key words

standard form
expanded form
word form

1 Are the numbers given below in standard form, expanded form or word form? Write the form and then write the number in numerals.

- **a** ten thousand one
- **b** 6 054
- **c** 7 000 + 200 + 90 + 8
- **d** 8 000 + 800 + 4
- **e** three thousand thirty-one
- **f** 740

2 Write these numbers using standard form.

- **a** ten
- **b** 10 + 5
- **c** twenty
- **d** 100 + 10 + 0
- **e** 300 + 50 + 7
- **f** four hundred seventy-five
- **g** one thousand
- **h** five thousand three hundred
- **i** 8 000 + eighty-three

3 Write the following numbers in expanded form.

- **a** five thousand fifty-five
- **b** eight thousand twenty-eight
- **c** one thousand, two hundred, nineteen

Working with bigger numbers

Key maths idea

10 × 1 000 = 10 000 We say ten thousand.
10 × 10 000 = 100 000 We say one hundred thousand.
10 ×100 000 = 1 000 000 We say one million.

When you work with bigger numbers, it is important to look at the place value. The digit furthest to the left has the **greatest** place value. That means it has the **largest** value. The digit furthest to right has the smallest, or **least**, value.

Look at this number expressed in standard form: 1 435 605

Expanded form: 1 000 000 + 400 000 + 30 000 + 5 000 + 600 + 5

Here is the place value chart:

M	Hth	Tth	Th	H	T	O
1	4	3	5	6	0	5

Word form (read the millions first): **1 million, four hundred thirty-five thousand, six hundred, five**

Key words

greatest
largest
least

1 Write the following numbers using numerals.

- **a** two hundred thousand twenty-five
- **b** five million fifty-three
- **c** three million, three hundred thousand, thirty
- **d** one million, two thousand, eight

2 Write the following numbers using standard form.

- **a** 7 000 000 + 500 000 + 40 000 + 6 000 + 10 + 9
- **b** 400 000 + 7 000 + 600 + 20 + 4
- **c** 2 000 000 + 700 000 + 60 000 + 1 000 + 500 + 10 + 9
- **d** 4 000 000 + 800 000 + 4 000 + 500 + 10

3 Write the following numbers in expanded form.

- **a** 4 678 932
- **b** 702 944
- **c** 3 659 021
- **d** 9 999 999

Maths detective: Large numbers

Case 1

1 Work in pairs. Write down ten 6-digit numbers you can make using only the digits 1 and 2.

2 Write them in order from least to greatest.

3 Compare with another pair.

Case 2

1 Look at the abacus. Make four different numbers. Draw your answers and write the numbers you made.

2 Now write the numbers in order from largest to smallest.

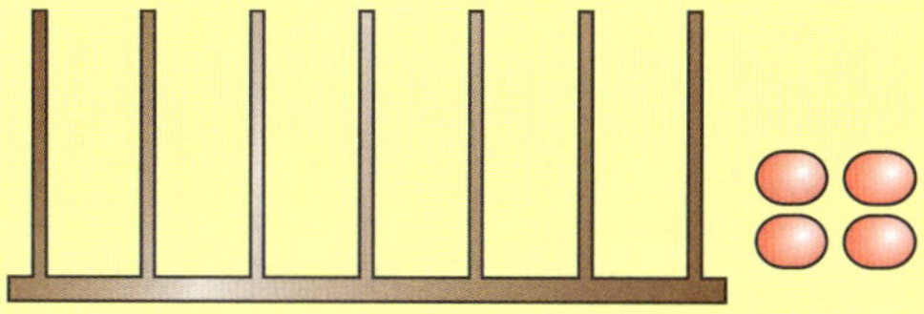

Large numbers

Key maths idea

When do we use large numbers? Use the pictures to help you.

We use large numbers to talk about the **population** of a country. The Jamaican government holds a **census** every ten years to find out the population size. Each household has to fill in a **questionnaire**. A questionnaire is a list of questions that we use to collect information on a particular topic. The census provides **information** about the population and helps the government to plan its services.

Key words

population
census
questionnaire
information

Population of Jamaica

Parish	Population (2018)	Parish	Population (2018)
1 Kingston and St Andrew	669 773	8 Hanover	70 265
2 St Thomas	94 939	9 Westmoreland	145 628
3 Portland	82 643	10 St Elizabeth	151 839
4 St Mary	114 867	11 Manchester	191 881
5 St Ann	174 203	12 Clarendon	247 702
6 Trelawny	75 981	13 St Catherine	521 249
7 St James	185 697		
Total population	**2 726 667**		

(Source: Statistical Institute of Jamaica, adapted from: https://statinja.gov.jm/Demo_SocialStats/EndofYearPopulationbyParish.aspx)

1 Use the information in the table to answer the following questions.
- **a** Which parish has the largest population?
- **b** Which parish has the smallest population?
- **c** What is the total population of Jamaica?
- **d** Choose 2 parishes that would have over 1 000 000 people if they were added together.
- **e** How many more people live in parish 3 than in parish 8? Show how you work this out.
- **f** Write the population numbers of parishes 2 to 6 in ascending order.
- **g** Add together the populations of parishes 4, 5 and 6.

2 Draw a place value chart and put the smallest population number digits in the correct columns.
- **a** Write the number in expanded form.
- **b** Write the face value for each digit.
- **c** Write the place value for each digit.

3 Draw another place value chart and put the total population number in the correct columns.
- **a** Write the number in expanded form.
- **b** Write the number value for each digit.
- **c** Write the place value for each digit.

Real-world maths: Number of students in a school

1 Why is it important to know how many children are at your school? Discuss this with your class.

2 a Make a questionnaire to find out the number of students and staff at your school, and how many children are in each class and grade.

b Use the questionnaire to collect the information.

3 List the following results:

a What is the total population of the school?

b How many of these are students and how many of these are staff?

c Which is the largest grade group?

d What is the difference in numbers between the largest and smallest grade group?

Real-world maths: Making big purchases

Choose one of the following problems to investigate.

1 Malia needs to buy a car for her cupcake business. These are her requirements:

- It should be a hatchback so that she can easily pack deliveries into the back.
- It must be no older than 3 years.
- It must have less than 20 000 km mileage.
- The budget is maximum $3 250 000.

Use a newspaper or look online (for example, *jamaicars.com*). Identify three possible cars Malia could buy that would fit her requirements.

Discuss with a partner which would be the best buy.

2 Your family wants to buy a new big-screen TV set.

- The screen must be 55 inches or larger.
- It must be a smart TV, with WiFi and Bluetooth.

Use the newspaper or internet to identify three possible products at different prices. Compare the prices and explain why each is cheaper or more expensive than the others. Identify which is the best buy and explain why you would choose that one.

Real-world maths: A museum visit

A visit to the Bob Marley Museum in Kingston costs a family of three $5 500. If they paid for it with 1 000-dollar bills, how many would they need? If they paid with 500-dollar bills, how many would they need? If they paid with 100-dollar bills, how many would they need?

Maths detective: Code crackers

Now use your skills to solve these cases!

Case 1

Chris Anderson scored over fifteen thousand runs in One Day International test cricket. The secret number is a whole number with 5 digits. The digit in the ones place is the product of 3 × 3. The digit in the hundreds place is an even number that is smaller than 8 but larger than 4. The digit in the thousands place is equal to the quotient of 21 and 3. The digit in the ten-thousand place is the numerator in every unit fraction and the digit in the tens place is a factor of 10, 15 and 20. Using all the clues you should know the amount of runs by now.

1 What is the secret number?

2 Write the secret number in word form.

Case 2

Now see if you can find this number:
I am a 4-digit number between 5 500 and 7 000.
I am odd.
I am a multiple of 5.
The digit in the thousands place is 1 more than the digit in the ones place.
The sum of my digits is 19.
The digit in the tens place is 3 more than the digit in the ones place.
Which number am I?

Case 3

Can you find this number?
I am a 4-digit number.
I am even.
The digit in the ones place is 3 times larger than the digit in the thousands.
The digit in the tens place is 5 less than the digit in the ones place.
The digit in the hundreds place is 8 more than the digit in the tens place.
The sum of my digits is 18.
Which number am I?

Case 4

Calculate a value for each of the letters of the alphabet. Then use these values to help you crack the code and read the message below.

A	the digit in the thousands place of 415 925		M	10 × 5 makes this product	
C	the place value of the 7 in the number 24 718		N	1 quarter of 1 000	
D	10 of these make 400		O	the first even number	
E	the greatest 1-digit number		P	the digit in the hundred thousands place in the number 4 352 129	
F	the only digit that is not a natural or counting number		S	the number of digits in five thousand	
G	the greatest 2-digit number		T	the smallest 3-digit number	
H	the face value of the 1 in the number 1 million		U	5 of these make one hundred	
I	the number of digits in the number 2 153 990		W	the number that multiplied by itself makes 100	
K	the number of digits in half a million		X	if you divide 4 000 by ten, you get this	
L	the digit in the thousands place in 1 982 571		Y	half of 1 000	

1	2	10		40	2		500	2	20			

100	9	5	700	1				

50	5		1	4		100	2		5

700	1	7	700	6	9	250	
							?

8	2	100	4		2	0	

9	99	99		4	5	50	3	8	9	4	
			–								!

What I have learned

Imagine that a spaceship full of aliens has arrived at your school from another planet. Use diagrams and simple examples to explain to them how we use a number system with ten numerals that can show numbers big enough to represent the distance between their planet and ours.

Practice questions

1. a. Write the smallest possible 7-digit number you can using the digits 1, 5, 7 and 2. All digits need to be used at least once.
   b. Now write the greatest possible 7-digit number using the same digits.
   c. Use one of the numbers you made to explain the difference between face value, place value and value of a number.
   d. Use one of the numbers you made to show how we write a number in expanded form.
2. Explain in your own words what we mean by:
   a. standard form
   b. expanded form.
3. Give three examples of things that cost more than $1 million.
4. Write the value of 6 in each number.
   a. 609 773 b. 94 639 c. 685 597
   d. 4 892 306 e. 6 987 032
5. Draw a place value chart. Fill the numbers in on the chart.
   a. 4 533 b. 56 709 c. 345 981
   d. 9 882 431 e. 602 f. 44 320
6. Write this set of numbers in numerals, then order from smallest to largest.

- eighty-two thousand, seven hundred, eighty-four
- two hundred forty-seven thousand, four hundred, fifteen
- six million, six hundred six thousand, eight hundred, fifty-five
- four million, seven hundred ten thousand, three hundred, one
- three million, seven hundred thousand, forty

7 Look at the abacus and answer the questions.

a What is the largest number you can make using 5 beads? Draw your answer and write the number.

b What is the smallest number you could make? Draw your answer and write the number.

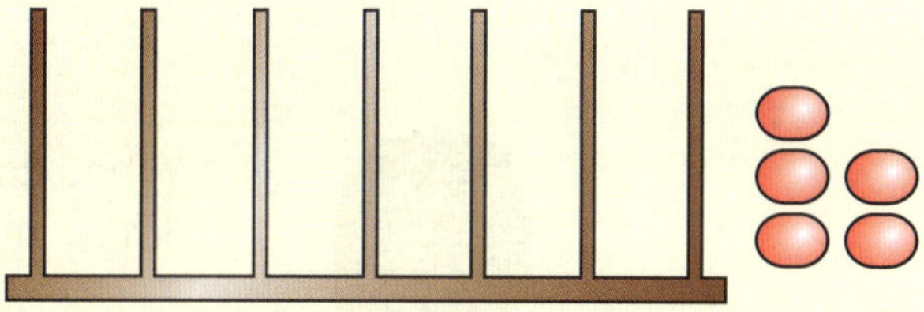

1 Self check

Check you can do this:	If you have difficulty, try this:
1 Write any 4-digit number. Give the value, place value and face value of each digit. Try it with this number: 5 847	Go to page 7 and revise what we mean by place value, face value and value.
2 Can you identify the value of numbers with 5, 6 or 7 digits? Try it with these numbers: 12 384 349 750 1 438 885	Go to page 10 and practise putting numbers into place value charts to identify the value of each digit.
3 Can you put numbers into expanded form? Try it with this number: 8 156 209	Go to page 9. Think about expanding each digit to show its full value and expressing it as an addition sum.
4 Give some examples of when we need to use 7-digit numbers. Think of two things we might buy that might have 7-digit prices.	Go back to pages 10 and 11 and refresh your memory!

Chapter 2 Sets

In this chapter you will explore this focus question:

To answer this question, you will:

- describe sets, explain what sets are
- list elements or members of a set
- name sets using curly brackets (also called braces).

Starting point

1 Look at the **set** of objects in the picture.

  a What **similarities** and **differences** do you notice?

  b How could you **group** the objects differently?

  c Equivalent sets have the same number of **members**. Which equivalent sets can you see in the picture?

2 Your teacher will give you sets of notes and coins. Work out the value of each set. Explain how you worked it out.

3 Look carefully to find and match the equivalent sets.

Set A = {2 000, 500, 20, 1}
Set B = {5, 100, 50, 1 000}
Set C = {5 000, 1, 10, 20, 50}
Set D = {20, 10, 5, 1}
Set E = {1, 10, 20, 50, 5 000}
Set F = {5, 10, 20, 1}
Set G = {500, 1, 20, 2 000}
Set H = {50, 5, 100, 1 000}

Remember: members are the objects or items in a set.

Key words
set
similarities
differences
group
members

Sets

Key maths idea

We **sort** objects into sets according to **attributes** such as size, shape, colour, texture or type of item. Each item in a set is called a **member** or **element** of the set. The picture shows a set of 10 paper clips. They are all the same shape and size, and the same type of item. There are 10 members in this set.

Set A

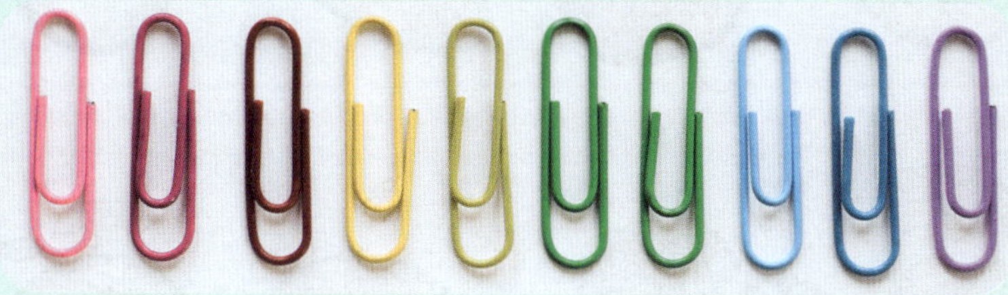

We name sets using capital letters. We list the members of a set in **curly brackets** (also called **braces**) and put a comma between each member. This is called **set notation**.
For example, Set B is a set of five different ways to travel.

Set B = {car, bus, train, aeroplane, ship}

curly brackets or braces — member or element — comma

Set C is the set of counting numbers less than 7.
Set C = {1, 2, 3, 4, 5, 6}

Key words

sort
attributes
member
element
curly brackets
braces
set notation

Remember: a set has members or elements that share the same attribute.

1 Use set notation to show these sets. The first one has been done for you.

a Set P is the set of whole numbers greater than 12 and less than 20.
Set P = {13, 14, 15, 16, 17, 18, 19}

b Set M is a set of any three kinds of flower.

c Set D is a set of five names that start with D.

d Set R is the set of days of the week.

2 Describe each set in your own words. The first one has been done for you.

a Set Y = {bananas, egg yolks, sunflowers}
Set Y is a set of things that can be yellow.

b Set E = {puppy, kitten, lamb, duckling}

c Set V = {mobile phone, computer, tablet}

3 a Group the fruits according to similar attributes to make a few different sets.
  b Write sentences to describe what makes the members in each group a set.
  c Work with a partner to find at least one different way to group the elements to form different sets.

4 Use set notation to write these sets.
  a Set W is a set of 6 wild animals.
  b Set N is a set of 5 names – each name has more than 6 letters.

5 Look around your classroom. List 15 objects you can see around you.
  a Group them according to similarities.
  b Explain to a partner what makes these objects members of a set.

6 Your teacher will show you a bag of objects. Take one object from the bag. Say which other objects you could put together with this object to make a set with shared attributes. Draw a picture of your set.

Venn diagrams

Key maths idea

A **Venn diagram** shows the members of one or more sets. It uses circles or rings. Each ring represents a set. Elements inside each ring belong in that set. Elements in the **overlapping** area belong to both sets. Elements outside of the ring are not members of the sets. Here is an example.

Set P = shapes with an odd number of sides

Set F = shapes with four sides or fewer

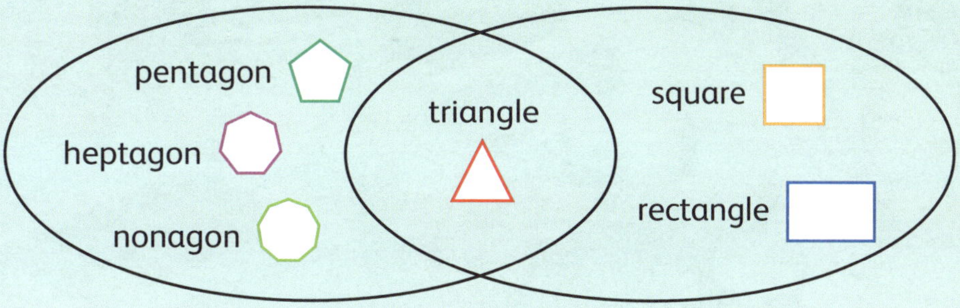

Key words

Venn diagram
overlapping

1 Talk about the sets shown above.

a How many elements are in set P?

b How many elements are in set F?

2 Into which set would each of these shapes fit?

a 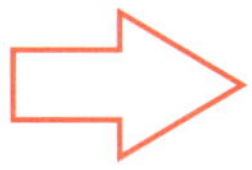

b 

c 

3 Draw another shape that could belong to set F.

Real-world maths: Make a scrapbook

Make a scrapbook of sets as pages.

1 Look at the magazines and advertising pages/leaflets/flyers that your teacher has given you. Cut out pictures of things you think you can make sets with. Glue them in their sets.

2 Name your sets. Write sentences to describe what makes the members in each group a set.

You will need:

- magazines and printed advertisements
- scissors
- glue
- blank pages.

Real-world maths: Art supplies

Imagine that your school has decided to set up an art room where children can do painting, drawing and clay modelling. You need to draw up a list of supplies.

1 Draw set diagrams for these sets:

- Set A = painting materials
- Set B = drawing materials
- Set C = useful tools for crafts
- Set D = items for keeping clean and neat
- Set E = storage containers.

In each set, write at least three elements that fit into that set.

2 Which elements would belong to:

a both set A and set B?

b both set C and set D?

Real-world maths: Games we play

1 Work in pairs or groups. Write a list of games you like to play.

2 Draw a table like this on a sheet of paper. Draw as many rows as you need. Write the games in the first column and then tick what you use to play each game.

Games	Dice	Board	Ball	Bats/Rackets	Pieces

What I have learned

Look at this diagram of two sets.

Write a paragraph that gives the following information:

- what kind of diagram this is
- what it shows
- why there can only be one element in the overlapping area.

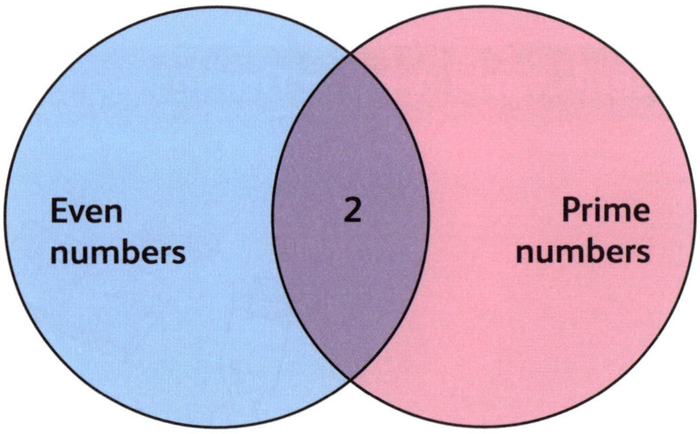

Practice questions

1 Decide whether each sentence is true or false. If it is false, correct the statement.

  a A set is a collection of objects that have something in common.
  b The objects in the set are called its parts.
  c Curly brackets (braces) are used to show that the objects written between them belong to a set.
  d Pens, pencils, sugar, erasers and books could be elements of the same set.
  e A duck, pigeon, flamingo, eagle and ostrich could be members of the same set.

2 Name each of these sets and write them in set notation.

  a even numbers between 52 and 62
  b odd numbers between 13 and 31

3 Look at these pictures of food and drink that people like. Find a way to group them into three sets. Use set notation to describe the sets.

> Remember: set notation means naming the set with a capital letter and writing the elements in curly brackets.

4 Akiel grouped these numbers in a set:
{0, 1, 2, 3, 4, 5}
 a Describe this set in words.
 b Create your own set of numbers. Remember to use curly brackets around the numbers. Swap sets with a partner and describe each other's sets.

2 Self check

Check you can do this:	If you have difficulty, try this:
1 Explain what a set is. Use the words set, element, belong.	Go to pages 18 and 19 and revise sorting objects into groups according to attributes.
2 Describe a set. Try it with this set: Set A = {0, 2, 4, 6, 8, 10}	Go to page 19. Exercise 1 describes sets for you to write in set notation. Exercise 2 gives set notation and gives you the opportunity to describe each set.
3 Name sets using curly brackets (braces) and list members of a set. Try it with this set: a set of any three animals that people keep as pets.	Go back to pages 18 and 19.

Chapter 3 Fraction ideas

In this chapter you will explore this focus question:

To answer this question, you will:

- identify whole numbers, proper fractions, improper fractions and mixed numbers
- recognise like fractions (fractions with equal denominators)
- order fractions with different denominators
- benchmark fractions using 0, $\frac{1}{2}$ and 1
- add and subtract different types of fractions
- solve problems involving fractions.

Starting point

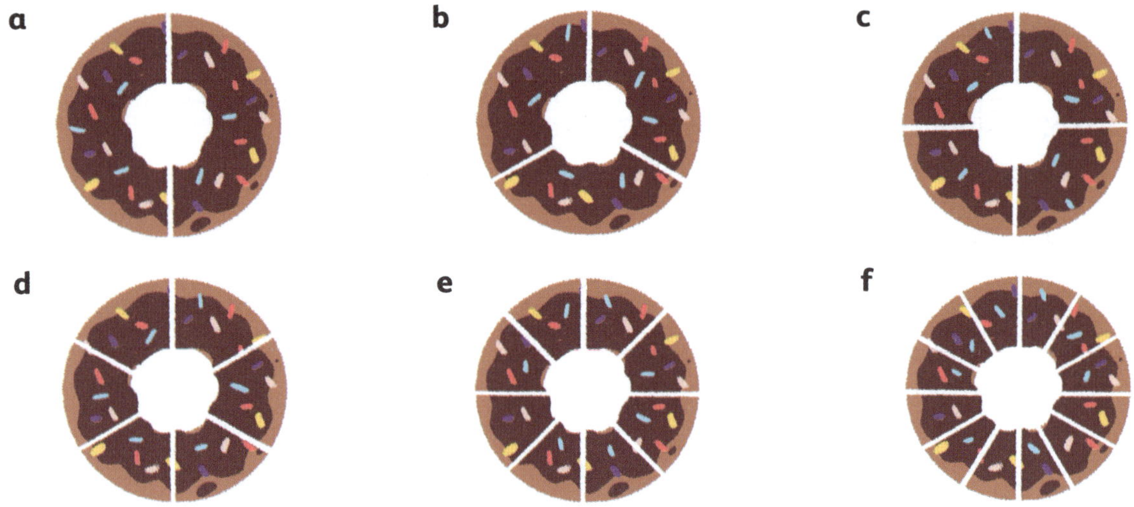

1 These six donuts have been split into fractions. Donut a is split into halves. Each part is one half or $\frac{1}{2}$.

Copy the sentence and complete it correctly for the rest of the donuts.

Donut ____ is split into ____ . Each part is ____ . We write that as ____ .

2 What do you call each part when a whole is divided into:

a five equal parts? **b** ten equal parts?

Revise what you know about fractions

Key maths idea

When you split an object or group into equal-sized parts, each part is a fraction of the whole. For example, two-thirds means two of three equal parts of one donut.

$\frac{2}{3}$

numerator – how many parts in this fraction

denominator – how many parts in the whole

A **unit fraction** has 1 as the numerator. It shows one of the equal parts.

Examples $\frac{1}{2}, \frac{1}{3}, \frac{1}{4}, \frac{1}{5}$

A **proper fraction** has a numerator that is less than the denominator.

Examples $\frac{1}{4}, \frac{3}{4}, \frac{7}{8}, \frac{2}{5}, \frac{9}{11}$

A **mixed number** has a whole number together with a proper fraction.

Examples $1\frac{1}{2}, 2\frac{3}{4}, 8\frac{9}{10}$

You can also express mixed numbers as improper fractions.

Improper fractions have a greater numerator than the denominator.

Examples $\frac{3}{2}, \frac{11}{4}, \frac{89}{10}$

Key words

numerator
denominator
unit fraction
proper fraction
mixed number

1 Look at each shape. Write the fraction that is shaded blue. Label the numerator and the denominator in each fraction.

a

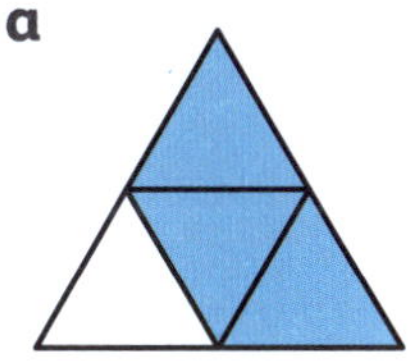

b

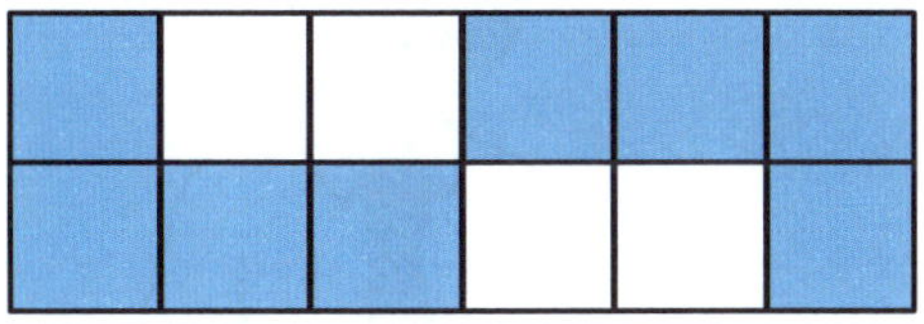

c

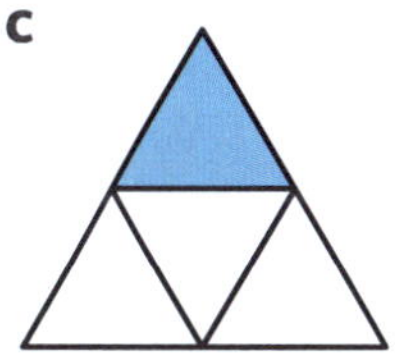

d

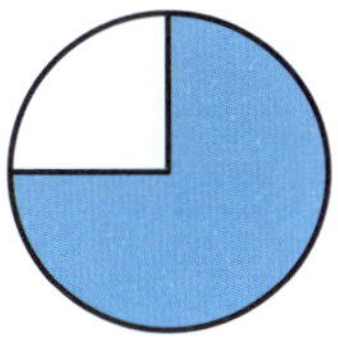

e

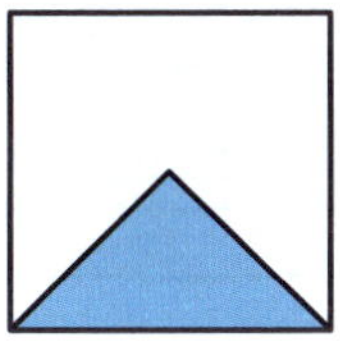

f

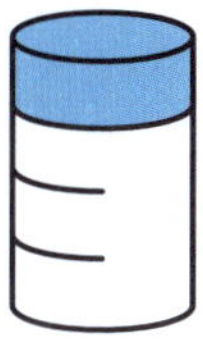

2 **a** Copy this fraction into your notebook. Label the parts of the fraction with these words: *whole number, denominator, numerator.*

$$5\frac{3}{4}$$

b Why do we call this kind of number a mixed number?

c Which numbers in this set are not mixed numbers? Give reasons for your answer.

Set C = $\{\frac{1}{2}, 1\frac{1}{2}, 15\frac{3}{4}, \frac{75}{100}\}$

3 Look again at the donuts on page 26. We say that $\frac{2}{4} = \frac{1}{2}$. The fractions are equivalent. What other fractions can you see that are equivalent to $\frac{1}{2}$?

4 Use the donuts on page 26 to help you. Write each of these fractions in their simplest form.

a $\frac{4}{6}$ **b** $\frac{9}{12}$ **c** $\frac{6}{8}$

5 Look at donuts a, c and e. Halves, quarters and eighths make a fraction family.

a Discuss with your group why you think they make a fraction family. Why are thirds not in the same family?

b What other fraction family can you see in the picture?

Talk about

Kevin says that to convert a mixed number to an improper fraction, you multiply the whole number by the denominator, then add it to the numerator. Why does this work?

Remember: look at the denominators. What similarity do you notice?

Unit fractions

Key maths idea

A **unit fraction** is simply a fraction that has 1 as its denominator. In other words, it shows 1 of the equal parts that make up the whole. When you add unit fractions that have the same denominator, you simply add the numerators together.

$\frac{1}{2} + \frac{1}{2} = \frac{2}{2}$

$\frac{2}{2}$ is the same as one whole.

$\frac{1}{2} + \frac{1}{2} = 1$

Key word

unit fraction

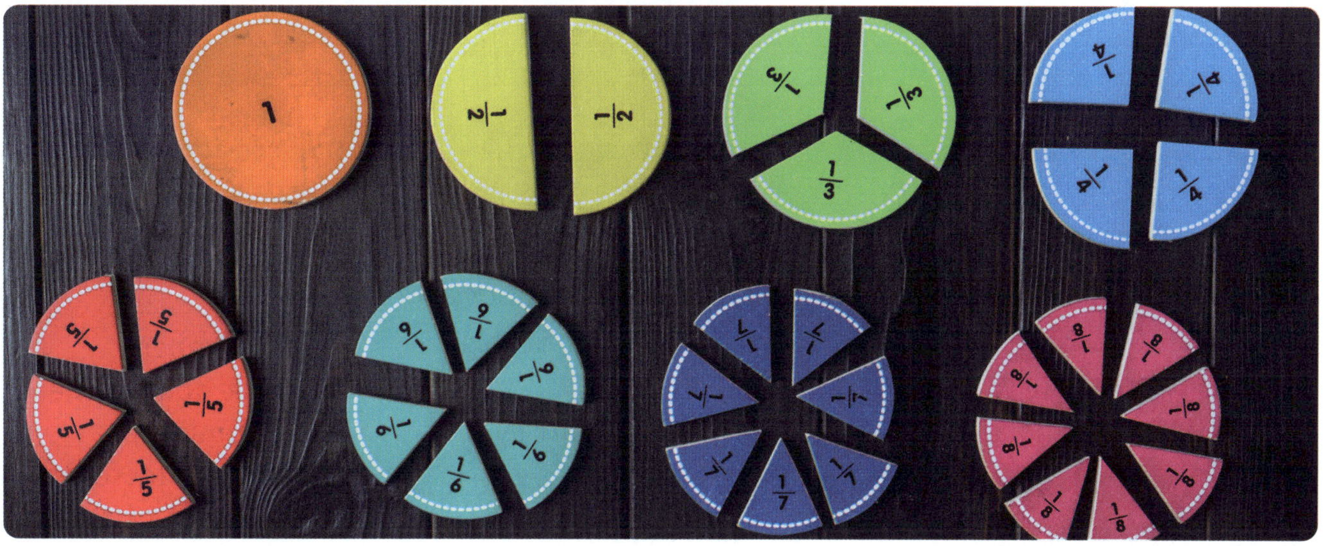

1 Which is greater:

a $\frac{1}{2}$ or $\frac{1}{3}$? **b** $\frac{1}{4}$ or $\frac{1}{5}$? **c** $\frac{1}{7}$ or $\frac{1}{6}$?

2 Add. You can use the picture to help you.

a $\frac{1}{3} + \frac{1}{3}$ **b** $\frac{1}{6} + \frac{1}{6} + \frac{1}{6}$ **c** $\frac{1}{8} + \frac{1}{8} + \frac{1}{8}$

d $\frac{3}{4} + \frac{1}{4}$ **e** $\frac{7}{8} + \frac{1}{8}$ **f** $\frac{6}{7} + \frac{1}{7}$

3 Subtract. You can use the picture to help you.

a $1 - \frac{1}{2}$ **b** $1 - \frac{1}{3}$ **c** $\frac{3}{4} - \frac{1}{4}$

d $\frac{4}{5} - \frac{1}{5}$ **e** $\frac{5}{6} - \frac{1}{6}$ **f** $\frac{7}{8} - \frac{1}{8}$

4 **a** Which different fractions can you see in the picture that are equivalent to one half?

b Which circles do not easily make a fraction equivalent to one half? What do they have in common?

Talk about

The greater the denominator, the smaller the unit fraction. Do you agree? Give reasons for your answer.

Build a fraction chart

Key maths idea

A **fraction chart** helps us to see the relationships between different fractions. It also helps us to identify **fraction families**. You can draw fraction charts using a ruler and a pencil. If you have a computer, you can use a spreadsheet. Sometimes you can draw a fraction chart by hand. Just **estimate** the relative sizes of the different fractions, like this:

<table>
<tr><td colspan="3">$\frac{1}{2}$</td><td colspan="3">$\frac{1}{2}$</td></tr>
<tr><td colspan="2">$\frac{1}{3}$</td><td colspan="2">$\frac{1}{3}$</td><td colspan="2">$\frac{1}{3}$</td></tr>
</table>

Key words

fraction chart
fraction families
estimate

<table>
<tr><th></th><th>A</th><th>B</th><th>C</th><th>D</th><th>E</th><th>F</th><th>G</th><th>H</th><th>I</th><th>J</th><th>K</th><th>L</th></tr>
<tr><td>1</td><td colspan="12">one</td></tr>
<tr><td>2</td><td colspan="6">one half</td><td colspan="6">one half</td></tr>
<tr><td>3</td><td colspan="3">one quarter</td><td colspan="3">one quarter</td><td colspan="3">one quarter</td><td colspan="3">one quarter</td></tr>
<tr><td>4</td><td colspan="4">one third</td><td colspan="4">one third</td><td colspan="4">one third</td></tr>
<tr><td>5</td><td colspan="2">one sixth</td><td colspan="2">one sixth</td><td colspan="2">one sixth</td><td colspan="2">one sixth</td><td colspan="2">one sixth</td><td colspan="2">one sixth</td></tr>
<tr><td>6</td><td></td><td></td><td></td><td></td><td></td><td></td><td></td><td></td><td></td><td></td><td></td><td></td></tr>
</table>

1 **a** Which fraction families does the fraction chart above show?
b Which other fractions could you show as part of this fraction family?
c Which fractions would not fit into this chart easily?

2 **a** Which other fractions on the chart are equal to one half?
b Which fraction is half the size of one third?
c Which fraction is half the size of one quarter?
d 'If a fraction is half the size of another fraction, its denominator is double.' Is this true? Give reasons for your answer.

Remember: think about other denominators that are factors or multiples of these denominators.

3 If you start a fraction chart divided into fifteenths, which other fractions would be easy to show as part of this fraction family?

4 Trace this fraction circle. Shade it as follows:

Shade $\frac{1}{15}$ red. Shade $\frac{1}{3}$ yellow. Shade $\frac{2}{5}$ purple.

How much is left unshaded?

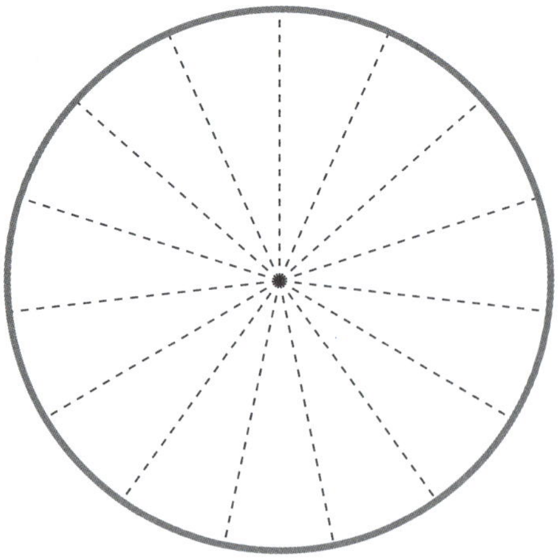

Explore equivalence

Key maths idea

You can fold paper to help you explore **equivalent fractions**. Here are some tips for making fractions using paper folding:

- Make sure to fold the paper into equal parts.
- To make neat folds, run your ruler or fingernail over the fold before you open it out.
- Check your fractions by unfolding the paper and comparing the size of the parts.

Key word

equivalent fractions

1 Look at the picture. Which fractions can you see that are equivalent to:

a $\frac{1}{3}$? **b** $\frac{2}{3}$? **c** $\frac{3}{3}$?

2 If more folds split each of the nine blocks into three smaller strips, what equivalent fractions could you make for:

a $\frac{7}{9}$? **b** $\frac{1}{9}$? **c** $\frac{8}{9}$?

3 Fold a sheet of paper to show halves, quarters, eighths and sixteenths. Use the fraction pieces you have made to work out how many:

a eighths fit into one quarter? **b** quarters fit into one half?

c quarters fit into one whole? **d** sixteenths fit into one eighth?

4 Use your folded paper from Exercise 3 to find equivalent fractions for:

a $\frac{3}{8}$ **b** $\frac{3}{4}$ **c** $\frac{7}{8}$

5 Use the shape below to help you work out which fraction is greater in each pair.

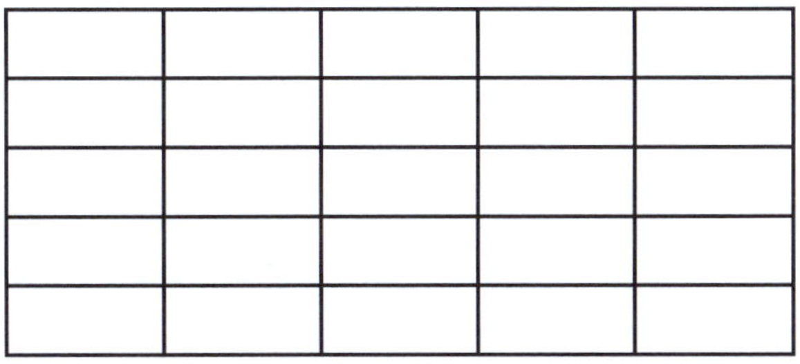

Which is greater:

a $\frac{1}{3}$ or $\frac{1}{25}$? **b** $\frac{3}{5}$ or $\frac{3}{25}$?

c $\frac{7}{25}$ or $\frac{2}{5}$? **d** $\frac{10}{25}$ or $\frac{1}{5}$?

Talk about

Is it easier to compare fractions with the same numerator or denominator? What is the same or different when they have the same denominator?

Compare and order fractions

Key maths idea

When you compare fractions, you are looking to see which has the greater value. We may use the greater than, less than and equals signs:

$<$ less than

$>$ greater than

$=$ equal to

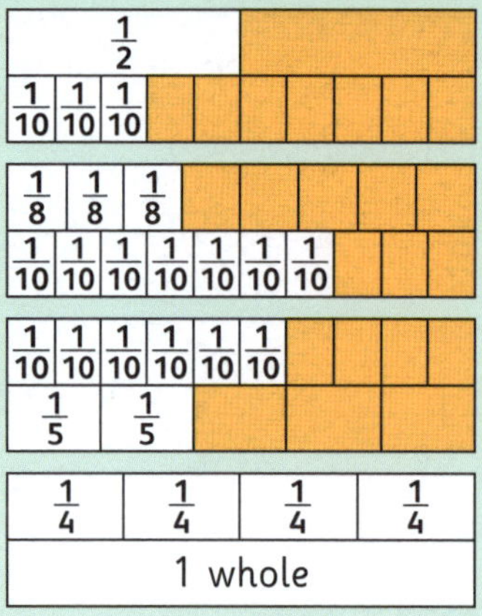

$\frac{3}{10} < \frac{1}{2}$

$\frac{3}{8} < \frac{7}{10}$

$\frac{6}{10} > \frac{2}{5}$

$\frac{4}{4} = 1$

When you compare unit fractions, you can compare the denominators. The fewer parts the whole is divided into, the greater the size of each unit fraction.

$\frac{1}{4} > \frac{1}{5}$

Fractions with the same denominator are **like fractions**. When you compare like fractions, look at the numerators. This set of fractions is in order from least to greatest:

$\frac{3}{8}, \frac{6}{8}, \frac{7}{8}$

You may also find it easy to use fraction walls to help you compare fraction families or draw sketches or diagrams to help you.

Key word

like fractions

1 Write true or false. If it is false, write it with a sign that makes it correct.

a $\frac{1}{4} > \frac{1}{10}$ **b** $\frac{1}{10} > \frac{1}{8}$ **c** $\frac{1}{2} < \frac{4}{10}$ **d** $\frac{3}{10} > \frac{1}{5}$

2 Copy the statements and fill in $<$, $>$ or $=$ to make them true.

a $\frac{3}{4} \square \frac{4}{5}$ **b** $\frac{7}{10} \square \frac{5}{8}$ **c** $\frac{3}{4} \square \frac{2}{5}$ **d** $\frac{4}{9} \square \frac{3}{5}$

3 Write each set of fractions in ascending order (from least to greatest).

a $\frac{1}{5}, \frac{1}{2}, \frac{5}{10}, \frac{1}{4}$ **b** $\frac{3}{5}, \frac{3}{9}, \frac{3}{8}, \frac{3}{4}$ **c** $\frac{3}{4}, \frac{4}{5}, \frac{7}{20}, \frac{5}{6}$

4 Use $>$, $<$ or $=$ to complete these statements.

a $\frac{1}{8} \square \frac{1}{2}$ **b** $\frac{2}{4} \square \frac{1}{2}$ **c** $\frac{3}{4} \square \frac{1}{1}$

d $\frac{1}{2} \square \frac{4}{8}$ **e** $\frac{8}{8} \square \frac{4}{4}$ **f** $\frac{3}{4} \square \frac{3}{8}$

More about equivalence

Key maths idea

You can use fraction charts to compare fractions and to find equivalent fractions. Look at the chart and see how many equivalent fractions you can find.

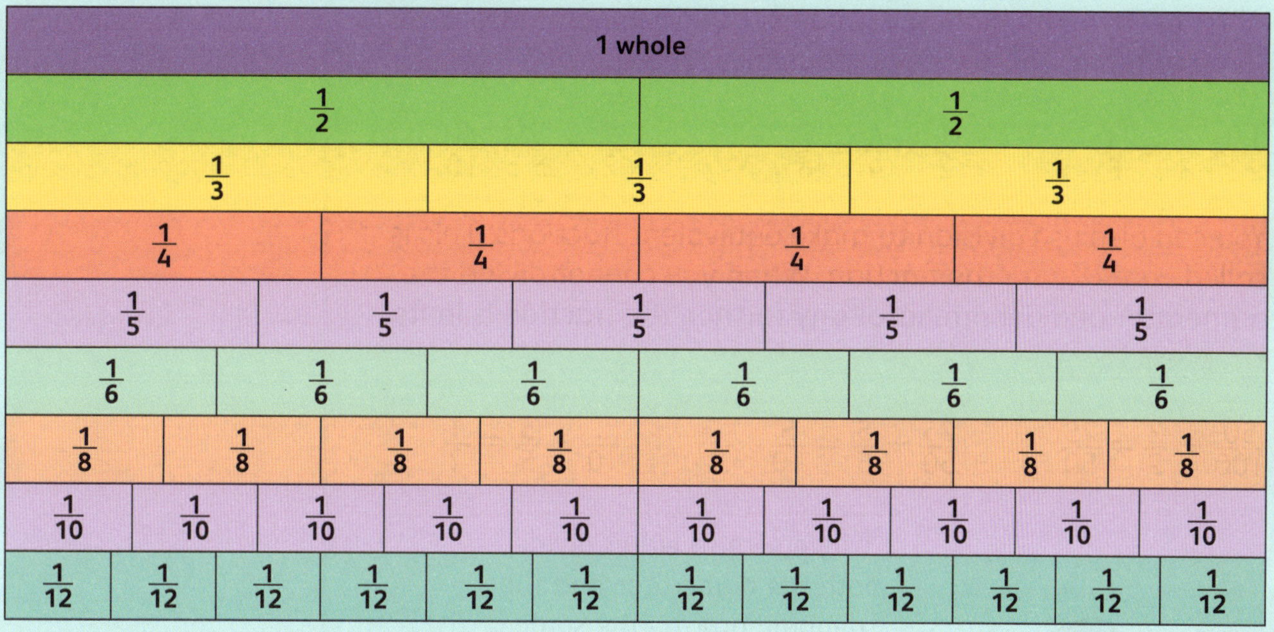

1 Use the fraction chart above to help you answer the questions.

a $\frac{1}{2} = \frac{\square}{6}$ **b** $\frac{1}{4} = \frac{\square}{8}$ **c** $\frac{1}{\square} = \frac{2}{6}$

d $\frac{2}{3} = \frac{\square}{6}$ **e** $\frac{3}{9} = \frac{1}{\square}$ **f** $\frac{6}{8} = \frac{\square}{4}$

2 Write two different equivalent fractions for each fraction.

a $\frac{2}{3}$ **b** $\frac{3}{5}$ **c** $\frac{6}{10}$ **d** $\frac{3}{6}$

3 Write all the fractions that are equal to one half.

4 Copy these fraction charts into your notebook and colour them the same way. Follow the instructions below.

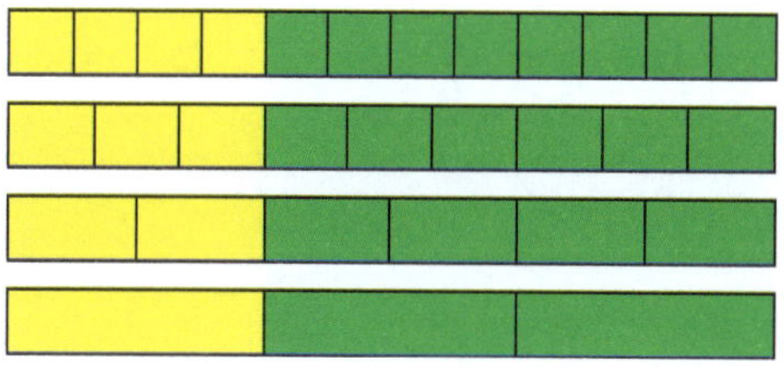

a Write a fraction into each block on the bars.

b Write a fraction for each row using the number of yellow blocks as the numerator each time. For example, row four as a fraction could be $\frac{1}{3}$.

c Which of the fractions you have written are equivalent fractions?

d Find the bar that has the fraction $\frac{4}{6}$ green. Write two other fractions that are equal to $\frac{4}{6}$.

e Four fraction families are shown on the four bars. Choose two bars and write all the fractions shown here that belong to each family.

Simplifying fractions

Key maths idea

You can use equivalent fractions to help you compare unlike fractions.

Whenever you multiply the numerator and denominator by the same number, you create an equivalent fraction.

$\frac{1}{2} \times \frac{3}{3} = \frac{3}{6}$ $\frac{1}{2} \times \frac{4}{4} = \frac{4}{8}$ $\frac{1}{2} \times \frac{5}{5} = \frac{5}{10}$

You can also use division to make equivalent fractions. This is called **simplifying** the fraction. When you cannot divide the numerator and denominator any further, the fraction is in its **simplest form**.

$\frac{50}{100} \div \frac{2}{2} = \frac{25}{50}$ $\frac{25}{50} \div \frac{5}{5} = \frac{5}{10}$ $\frac{5}{10} \div \frac{5}{5} = \frac{1}{2}$

Key words

simplifying
simplest form

1 Use multiplication to create three equivalent fractions for each fraction.
a $\frac{1}{4}$ b $\frac{3}{5}$ c $\frac{7}{10}$ d $\frac{7}{9}$

2 Simplify each fraction – write it in its simplest form.
a $\frac{4}{8}$ b $\frac{12}{15}$ c $\frac{15}{20}$ d $\frac{7}{21}$

3 Use equivalent fractions to help you work out which fraction is greater in each pair.
a $\frac{4}{5}$ or $\frac{3}{4}$ b $\frac{2}{3}$ or $\frac{3}{4}$ c $\frac{7}{8}$ or $\frac{2}{3}$

Remember: keep making equivalent fractions until both the fractions have the same denominator.

Improper fractions and mixed numbers

Key maths idea

You know that a proper fraction has a numerator that is less than its denominator. Proper fractions are always numbers with a value between 0 and 1.

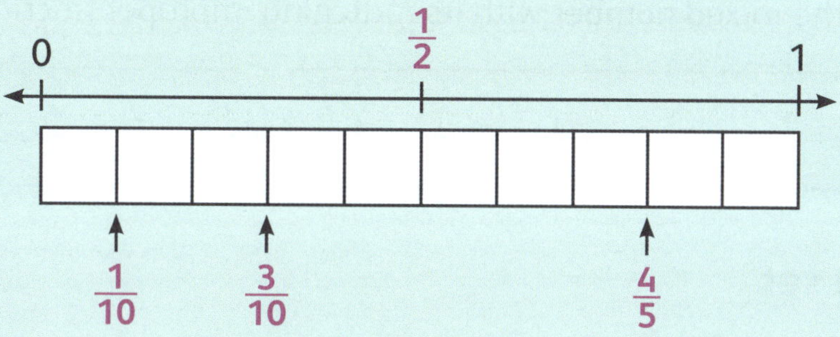

Improper fractions and mixed numbers are fractions that are greater than 1. They have a value between two whole numbers.

For example $1\frac{1}{2}$ means one whole and one half. It is the number halfway between 1 and 2. $1\frac{1}{2}$ is a mixed number. You can also write it as three halves or $\frac{3}{2}$.

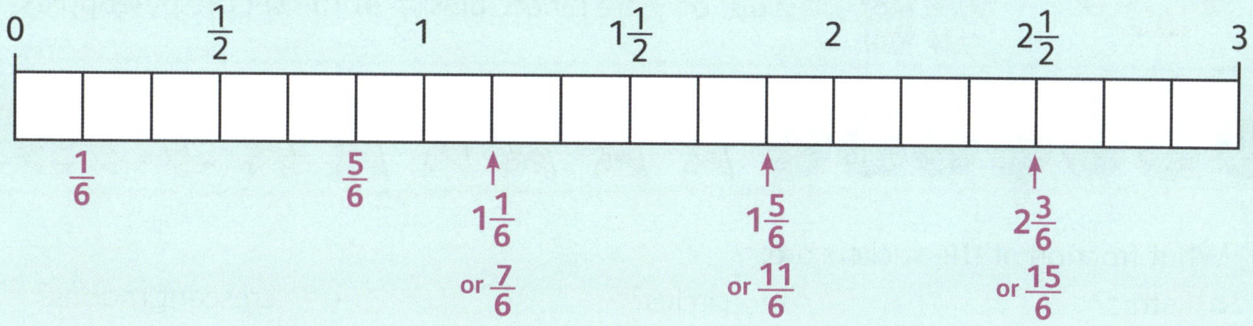

1 Look at the chart of fractions between 0 and 3 above. Identify:

a whole numbers
b proper fractions
c mixed numbers
d improper fractions.

2 Two halves make a whole. How many halves make:

a 4 wholes?
b 5 wholes?
c 10 wholes?

3 Write these improper fractions as mixed numbers.

a $\frac{9}{2}$
b $\frac{7}{2}$
c $\frac{15}{2}$

4 Look at the chart. Write the missing numbers as mixed numbers and as improper fractions.

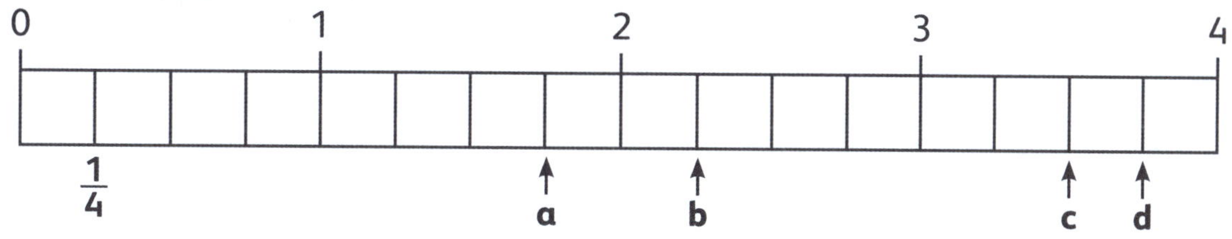

5 You have learned how to write a set. Look at these numbers and decide which of these will be members of the same sets. There will be four sets. Write them in your notebook.

17 $\frac{3}{4}$ 21 $33\frac{1}{3}$ $\frac{12}{7}$ $\frac{24}{5}$ 100 $\frac{9}{10}$

$18\frac{7}{16}$ $3\frac{1}{2}$ $\frac{5}{11}$ $\frac{6}{8}$ $\frac{9}{4}$ $\frac{15}{17}$ 52

6 Find and write the mixed number with its matching improper fraction.

$1\frac{1}{3}$ $1\frac{1}{7}$ $1\frac{1}{5}$ $2\frac{1}{2}$ $3\frac{1}{2}$ $1\frac{1}{4}$ $1\frac{2}{7}$

$\frac{5}{4}$ $\frac{7}{2}$ $\frac{6}{5}$ $\frac{9}{7}$ $\frac{8}{7}$ $\frac{5}{2}$ $\frac{4}{3}$

A fraction of a set

Key maths idea

There are 5 pieces of fruit in this set.
3 out of 5 are bananas. $\frac{3}{5}$ of the set are bananas.
2 out of 5 are pineapples. $\frac{2}{5}$ of the set are pineapples.

1 What fraction of the stickers are:

a stars? **b** circles? **c** crescent moons?

2 What fraction of the circles are black and white?

3 What fraction of the stars are pink?

4 Look at this set of fruit and answer the questions.

What fraction of the fruit are:

a bananas? **b** mangoes? **c** pineapples?

d mangoes and bananas together? **e** bananas and pineapples together?

Adding and subtracting fractions

Key maths idea

When you add or subtract fractions, follow these steps:

Step 1 Make sure the denominators are the same.

Step 2 Work with the numerators to calculate.

Step 3 Simplify the answer if possible.

Example

$\frac{3}{6} - \frac{1}{6} = \square$ The denominators are already the same.

$\frac{3-1}{6} = \frac{2}{6}$

This can be simplified:

$\frac{2}{6} \div \frac{2}{2} = \frac{1}{3}$

$\frac{3}{6} - \frac{1}{6} = \frac{1}{3}$

Remember: to simplify a fraction, you divide the top and the bottom by the same number. Can you remember why this works? Think about dividing a number by 1!

1 Adam and Hakeem each has a pizza that is split into 8 pieces, but they do not eat them all. After dinner, Adam's pizza has $\frac{3}{8}$ left and Hakeem's pizza has $\frac{2}{8}$ left. What fraction of a pizza do they have together?

Hint: you can draw a diagram or fraction chart to help you.

2 Gemma has a slab of chocolate divided into 16 blocks, but she has eaten some already. She has $\frac{11}{16}$ of the slab left and then she eats another 3 blocks.

a How much is left?

b What is the simplified fraction?

3 Calculate. Simplify your answers as much as possible.

a $\frac{2}{7} + \frac{4}{7}$ **b** $\frac{9}{11} - \frac{5}{11}$

c $\frac{5}{15} + \frac{2}{15}$ **d** $\frac{9}{20} + \frac{2}{20}$

e $\frac{9}{10} - \frac{5}{10}$ **f** $\frac{3}{4} - \frac{1}{4}$

g $\frac{4}{12} + \frac{5}{12}$ **h** $\frac{4}{18} + \frac{4}{18}$

4 Yohan has made some mistakes in his working. Explain his mistake. What would the correct answer be?

$$\frac{5}{10} + 1\frac{1}{10} = \frac{6}{20}$$

Talk about

When you are going to add or subtract fractions, you need to make sure they have the same denominators. Look at this problem:

$\frac{4}{5} + \frac{2}{3}$

How can you change the denominators so they are the same?

Finding the lowest common denominator (LCD)

Key maths idea

When the denominators are not the same, you need to use equivalent fractions to make them both into fractions with the same denominator. This is easy when the fractions are from the same fraction family:

$\frac{1}{2} + \frac{1}{4}$

$\frac{1}{2} \times \frac{2}{2} = \frac{2}{4}$

$\frac{1}{4} + \frac{2}{4} = \frac{3}{4}$

Sometimes the fractions are not in the same family. You use multiples of the denominators until you can change them to equivalent fractions with the same denominator. This is called finding a **lowest common denominator (LCD)**.

Example

$\frac{3}{5} + \frac{1}{4}$

$\frac{3}{5}$ and $\frac{1}{4}$ have denominators that are not from the same family.

Think about multiples of 5: 5, 10, 15, 20, 25 …

Think about multiples of 4: 4, 8, 12, 16, 20, 24 …

The first multiple that is common to both denominators is 20.

$5 \times 4 = 20$ So you need to use $\frac{4}{4}$ for the fraction over 5:

$\frac{3}{5} \times \frac{4}{4} = \frac{12}{20}$

$4 \times 5 = 20$ So you need to use $\frac{5}{5}$ for the fraction over 4:

$\frac{1}{4} \times \frac{5}{5} = \frac{5}{20}$

Now you can add: $\frac{12}{20} + \frac{5}{20} = \frac{17}{20}$

Key word

lowest common denominator (LCD)

1 Calculate the answers.

a $\frac{2}{4} + \frac{2}{5}$ **b** $\frac{4}{5} + \frac{1}{2}$ **c** $\frac{1}{2} + \frac{3}{7}$

d $\frac{7}{10} - \frac{1}{2}$ **e** $\frac{5}{4} - \frac{2}{3}$ **f** $\frac{3}{6} + \frac{3}{4}$

g $\frac{3}{4} - \frac{1}{8}$ **h** $\frac{2}{6} - \frac{1}{5}$ **i** $\frac{1}{3} + \frac{1}{10}$

Adding and subtracting mixed numbers

Key maths idea

Example 1

$5\frac{1}{7} + 5\frac{2}{7}$

The denominators are the same, so you can simply add the whole numbers together, and add the fractions together: $5\frac{1}{7} + 5\frac{2}{7} = 10\frac{3}{7}$

Example 2

$4\frac{15}{20} - 3\frac{12}{20} = 1\frac{3}{20}$ The denominators are the same, so you can subtract, working first with the whole numbers then with the numerators.

Example 3

$4\frac{1}{4} - 1\frac{3}{4}$

The denominators are the same, but $\frac{3}{4}$ is greater than $\frac{1}{4}$. How can we subtract? We can convert the numbers to improper fractions: $\frac{17}{4} - \frac{7}{4} = \frac{10}{4}$ $\frac{10}{4} = 2\frac{2}{4}$

Then we can convert the improper fraction back to a mixed number and simplify: $2\frac{2}{4} = 2\frac{1}{2}$ $4\frac{1}{4} - 1\frac{3}{4} = 2\frac{1}{2}$

Example 4

$4\frac{2}{3} + 7\frac{1}{6}$

Express the fractional parts with the same denominator, by using LCD: $\frac{2}{3} \times \frac{2}{2} = 4\frac{4}{6}$ $4\frac{2}{3} + 7\frac{1}{6} = 4\frac{4}{6} + 7\frac{1}{6}$

$4\frac{4}{6} + 7\frac{1}{6} = 11\frac{5}{6}$

Example 5

$6\frac{5}{9} - 3\frac{1}{3}$

$\frac{1}{3} \times \frac{3}{3} = \frac{3}{9}$ Now both fractions have the same denominator. $6\frac{5}{9} - 3\frac{3}{9} = 3\frac{2}{9}$

1 Add the fractions.

a $1\frac{12}{20} + 3\frac{5}{20}$ **b** $12\frac{4}{9} + 3\frac{1}{9}$ **c** $11\frac{1}{5} + 3\frac{1}{5}$

d $2\frac{3}{5} + 4\frac{1}{5}$ **e** $6\frac{1}{3} + 1\frac{1}{6}$ **f** $9\frac{2}{5} + 11\frac{3}{10}$

g $14\frac{1}{4} + 20\frac{1}{8}$ **h** $24\frac{2}{4} + 16\frac{5}{8}$ **i** $52\frac{5}{9} + 22\frac{1}{3}$

2 Subtract the fractions.

a $8\frac{11}{12} - 5\frac{10}{12}$ **b** $4\frac{3}{4} - 3\frac{1}{4}$ **c** $7\frac{5}{9} - \frac{4}{9}$

d $9\frac{7}{8} - 6\frac{5}{8}$ **e** $3\frac{5}{6} - 2\frac{3}{6}$ **f** $4\frac{1}{6} - 3\frac{1}{12}$

g $6\frac{7}{8} - 3\frac{3}{4}$ **h** $19\frac{1}{3} - 11\frac{4}{12}$ **i** $4\frac{7}{12} - \frac{1}{2}$

Problem solving with fractions

1 Jonathan buys 10 ice creams – 6 strawberry and 4 chocolate.

- **a** What fraction of the ice creams are strawberry flavoured? Write this fraction in its simplest form.
- **b** He gives half of the ice creams to his friends and takes the rest home. How many does he give to his friends?

2 Brianna made 5 cups of tea. She used $\frac{3}{4}$ teaspoonful of sugar for three of the cups and $\frac{1}{2}$ teaspoonful for 2 of the cups. How many teaspoonful of sugar did she use in all? Write your answer as a mixed number.

3 Mr Johnson had a meeting at his home and 6 people attended. He put 12 cookies on a plate. Mr Johnson didn't eat any of the cookies so they were shared between the other people.

- **a** How many cookies does each person get?
- **b** What fraction of the cookies does each one get?
- **c** Two people do not eat any of their cookies. How many are not eaten?
- **d** What fraction of the cookies are not eaten?

Maths detective: Sharing fractions

How could you split a pizza among 5 people so that each person gets a different fraction, with none left over?

Maths detective: Tangrams

You will need:

- a square piece of card
- a ruler
- a pencil.

A tangram is an ancient Chinese maths game. It has seven pieces that can fit together to make a square. The shapes are triangles, a square and a **parallelogram**.
You can get tangram sets with pieces made from wood or plastic, or you can make your own from card.

Case 1

Make your own tangram set.

Use these hints to make your own tangram set:

- With a pencil, sketch lines that divide the square into two identical right-angled triangles. You do this by joining the opposite corners.
- Repeat this with the other two corners. You have now divided your square into quarters.
- Use your ruler to sketch lines that divide your square into four quarters, each in the shape of a square.
- To make the smaller shapes, you will need to mark the half-way mark along the sides of the square card. Use these to help you draw the sketch lines for the remaining shapes.

Key word
parallelogram

Case 2

Each of the two larger triangles in the tangram make up one quarter of the square.

1 If you put the two large triangles together to make one bigger triangle, what fraction does each triangle make?

2 Express each of the other shapes as a fraction of the whole square.

3 Express each of the smaller triangles as a fraction of one of the larger triangles.

4 A tangram puzzle gives you a picture to make with the tangram pieces. You have to use all 7 pieces to make the shape.

For example, here is a puzzle of a goose:

The solution looks like this:

Try these:

a rabbit

b cat

Real-world maths: Fractions of shapes

1 A hexagon is a six-sided polygon. Jeffrey made these three different hexagons to show fractions:

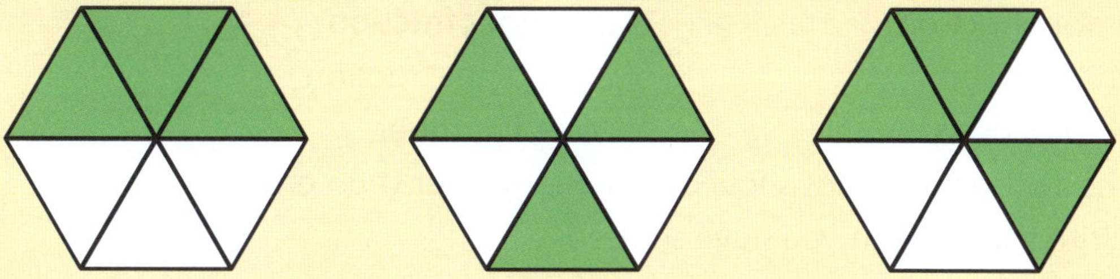

He says: 'Each triangle is $\frac{1}{6}$ of the hexagon. In each hexagon, I have shaded $\frac{1}{2}$ of the hexagon.'

Trace the hexagons. Sketch three different ways you could show $\frac{1}{3}$ of the hexagon.

2 Look at this honeycomb shape.

a How many hexagons make up the shape?

b What fraction is 1 hexagon out of the whole shape?

c What fraction is 1 triangle out of the whole shape?

d How many triangles would make half of the whole shape?

3 Trace the honeycomb shape.

a Use three different colours to shade the honeycomb in a pattern.

b Use fractions to describe the pattern you have made.

4 Begin with a square sheet of paper or card. Fold it in half, then in half again, and twice more. You can fold it into squares or into triangles.

Unfold your sheet and use the fold lines to help you draw lines that divide the paper into different fractions. Use different colours to shade and label these fractions:

a $\frac{1}{8}$ b $\frac{3}{16}$ c $\frac{1}{4}$ d $\frac{1}{16}$

You will need:
- a square sheet of paper or card
- coloured markers.

5 Discuss with a partner how you could fold a piece of paper to show thirds, sixths and twelfths. Try it out with a sheet of paper. Share your results with the class.

What I have learned

1 a Fold a piece of paper into 8 equal parts.

b In each part, write one of these words and an example:

numerator	denominator
proper fraction	improper fraction
unit fraction	equivalent fraction
like fractions	mixed number.

c Shade the parts that show a number greater than one.

d Write the fraction you have shaded.

2 Look at this fraction chart. The first square is shaded; it shows $\frac{1}{12}$ of the whole strip. What fraction do we make if we shade up to each of the arrows? Match the following fractions to the correct letters on the chart.

$\frac{3}{4}$ $\frac{1}{3}$ $\frac{1}{4}$ $\frac{11}{12}$

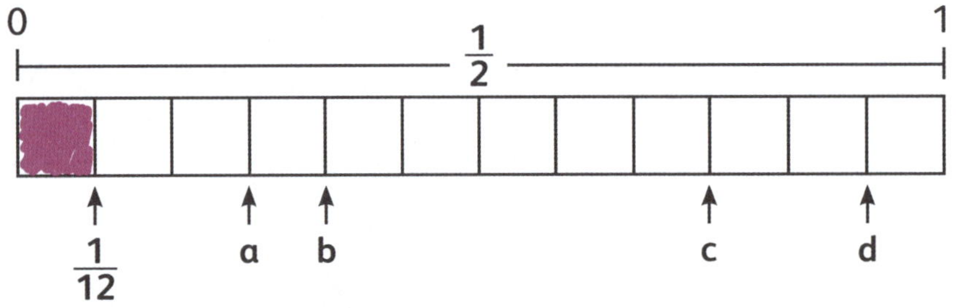

Practice questions

1 Which of the following fractions are not whole numbers?

a $\frac{1}{2}$ b $\frac{2}{1}$ c $\frac{4}{4}$ d $4\frac{1}{2}$ e $\frac{9}{10}$ f $\frac{100}{200}$ g 150

2 Which of the following are not proper fractions?

a $\frac{12}{15}$ b $\frac{12}{11}$ c $\frac{12}{100}$ d $12\frac{1}{2}$ e $\frac{122}{125}$ f $\frac{12}{12}$

3 Explain in your own words each of the following terms and give three examples.

a proper fraction b improper fraction c mixed number

4 Are the fractions in each pair equivalent fractions? Write 'yes' or 'no' for each pair.

a $\frac{12}{24}$ and $\frac{1}{2}$ b $\frac{50}{100}$ and $\frac{1}{5}$ c $\frac{2}{4}$ and $\frac{1}{8}$ d $\frac{10}{15}$ and $\frac{2}{3}$

5 Draw a set that has 3 triangles, 4 squares and 5 rectangles.
What fraction of the set is:

a triangles? b squares? c rectangles?

d triangles and squares together? e rectangles and triangles together?

6 Calculate the answers.

a $\frac{4}{4} - \frac{2}{3}$ b $\frac{6}{12} + \frac{3}{4}$ c $\frac{3}{4} - \frac{1}{8}$

d $12\frac{2}{4} + 17\frac{5}{8}$ e $63\frac{5}{9} + 23\frac{1}{3}$ f $24\frac{11}{12} - 12\frac{10}{12}$

3 Self check

Check you can do this:	If you have difficulty, try this:
1 Look at a fraction or number, and say whether it is a whole number, proper fraction, improper fraction or mixed number. Try it with these fractions: $\frac{1}{2}, 9\frac{3}{4}, \frac{4}{3}, \frac{8}{9}, \frac{4}{4}$	Go to page 27 and refresh your memory about the different kinds of numbers and fractions.
2 Recognise like fractions (fractions with the same denominator). Which of these are like fractions? $\frac{3}{7}, \frac{9}{11}, \frac{7}{3}, \frac{1}{3}, \frac{1}{7}, \frac{5}{7}, \frac{5}{5}$	Go back to page 32 and look at the section on like fractions. Also look at the fraction chart on page 30.
3 Write equivalent fractions. Try to write three fractions that are all equivalent to $\frac{3}{4}$.	Look back at the work from page 31 and also pages 33 and 34. Use fraction charts to help you. Also look at the work on LCD on page 38.
4 Order fractions with different denominators. Try it with this set: $\frac{1}{2}, \frac{9}{10}, \frac{3}{5}, \frac{3}{4}$	Go back to page 32 to refresh your memory about comparing and ordering fractions.
5 Add and subtract proper fractions and mixed numbers. Try these: $4\frac{2}{3} + 1\frac{4}{5}$ $7\frac{5}{6} - \frac{13}{4}$	Go back to page 37 and refresh your memory about calculating with fractions with the same denominators, and look at pages 38 and 39 if you want to revise adding and subtracting fractions with unlike denominators.

Chapter 4 Estimation and calculation

In this chapter you will explore this focus question:

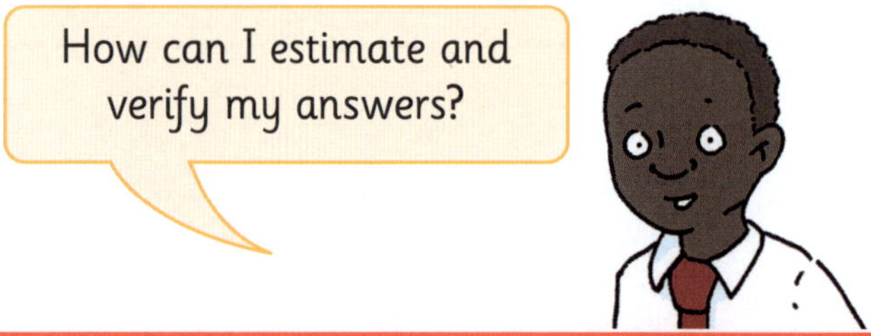

To answer this question, you will:

- make reasonable estimates to help you solve problems
- add and subtract 2-digit numbers mentally
- use the commutative and associative properties.

Starting point

Key words
addition
sum
subtraction
difference
digit

Remember: when you add numbers together, the answer is called the sum. When you subtract one number from another, the answer is called the difference.

1 a $51 + 49 = 100$. Write three more **addition** sentences with pairs of 2-digit numbers, that each make a **sum** of 100.

b What do you notice about the four digits in each addition sentence?

Hint: make them into pairs.

2 a $45 - 35 = 10$. Make three more **subtraction** sentences with pairs of 2-digit numbers, that each have a **difference** of 10.

b What do you notice about the **digits** in the ones places, and the digits in the tens places, in each sentence?

Revisit what you know about estimating and calculating

1 Write a sentence for each word. Give an example using 2-digit numbers. Stick them into your maths dictionary.

add	subtract	estimate
total	sum	difference

2 Round these numbers to the nearest ten.

a 54 **b** 22 **c** 13 **d** 96
e 12 **f** 45 **g** 69 **h** 48

3 Estimate the answers by rounding to the nearest ten. Show your working out.

a 91 + 42 **b** 34 + 39
c 49 + 58 **d** 25 + 19
e 35 + 64 **f** 26 − 17

Remember: when rounding to the nearest 10, look at the ones. If the digit is 4 or less, round down. If it is 5 or greater, round up.

4 Check the actual answers to Exercise 3 using base ten blocks, or sticks that you can make into bundles of ten and loose ones. Use your blocks to model each 2-digit number, then add or subtract.

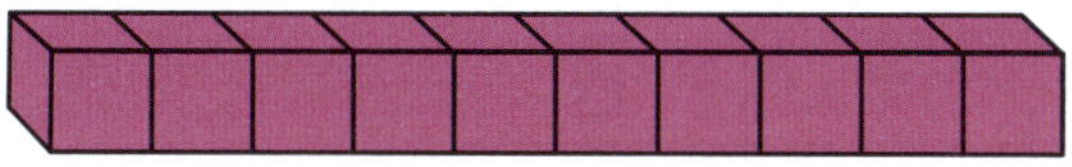

Whenever you read a word problem, think about:

- What information do I have?
- What am I trying to find out?
- How can I use the information to work out the answer?

5 Akiel watches a movie. It is 136 minutes long. 23 minutes are taken up by adverts. How long was the actual movie?

6 Using the calculations below, write a word problem for each one. Then let your partner solve them.

a 154 + 18 **b** 66 + 21 + 110 **c** 418 − 32

Working with base ten blocks and other materials

Key maths idea

You do not need special blocks to model numbers. Here are some different materials you can use.

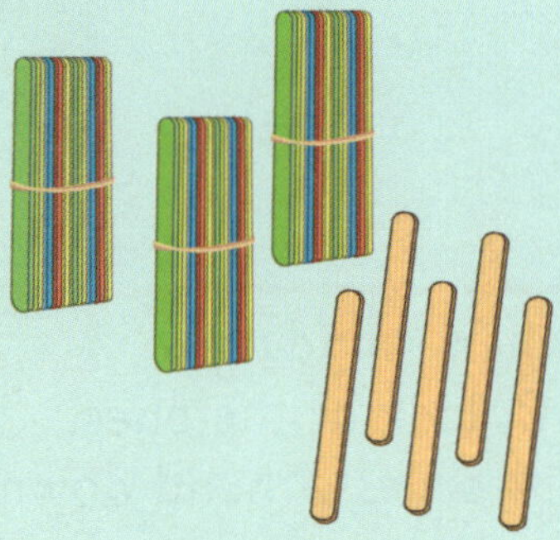

fudge sticks

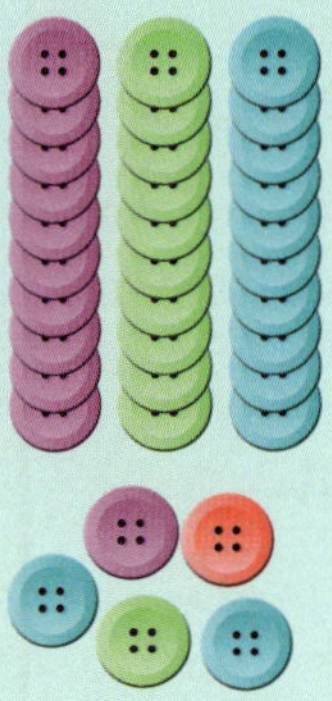

sewing buttons

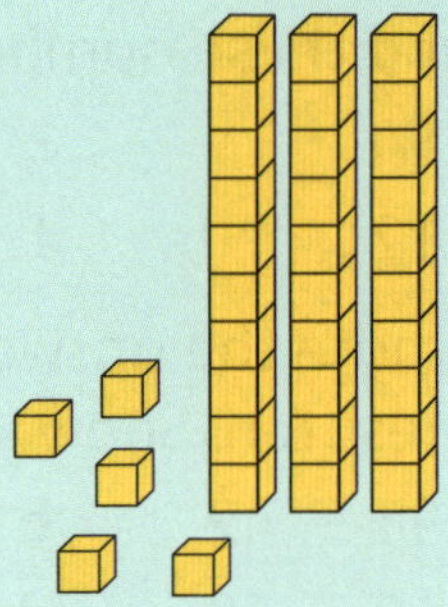

plastic blocks

For bigger numbers, you can work in a different way. Think about how we would represent the number 341. Choose three different colours of blocks. Select one colour to show hundreds, another for tens and another for ones. On a piece of paper, draw the three columns. Then place the blocks in the columns like this:

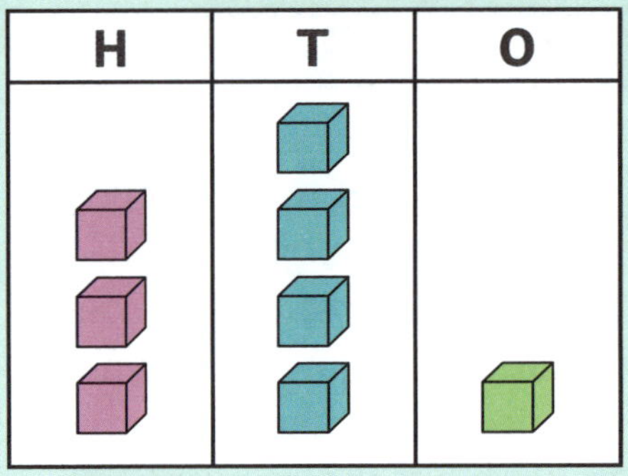

1 These sums are modelled with base ten blocks. Write each addition sum using numbers and work out the total.

a

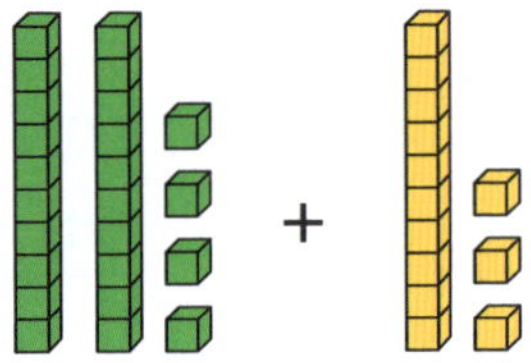

b

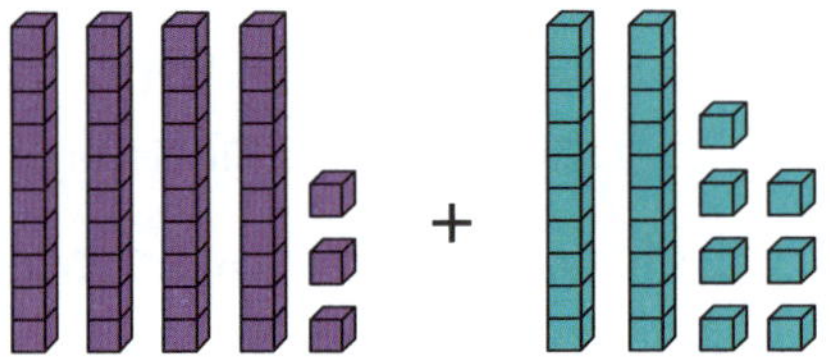

c

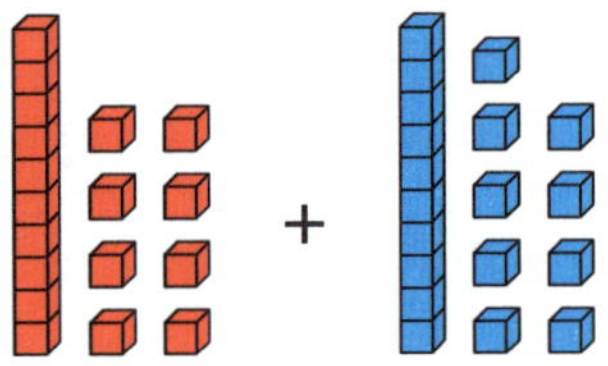

d

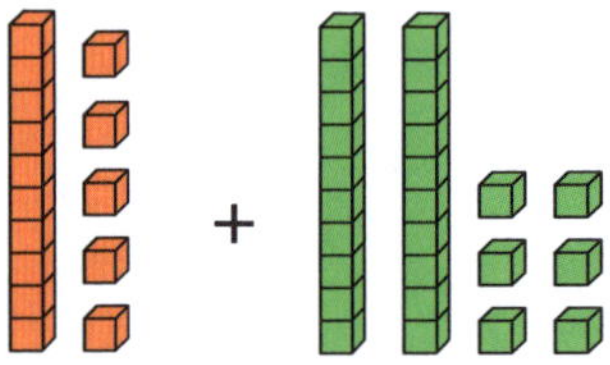

Hint: for some of the sums, you may need to regroup the ones into new tens.

Mental calculation

Key maths idea

It is useful to be able to work out the answers to sums in your head, without using a calculator or working in columns. This is called working **mentally**. Here are three **strategies** you can use to help you work mentally:

- using known facts (such as number facts, doubles, halves or multiples)
- using 'friendly numbers'
- imagining jumps on a **number line**.

Look at these examples.

These students have used number facts about tens to help them solve these addition questions.

$7 + 4 + 3 = \square$ $\qquad$ $12 + 19 + 28 + 11 = \square$

$7 + 4 + 3 = \square$

$10 + 4 = 14$

$12 + 19 + 28 + 11 = \square$

50

$50 + 10 + 10 = 70$

These students decided to break down numbers into parts that easily work together. This is also called using 'friendly numbers'.

$24 + 35 = \square$ $\qquad$ $97 - 16 = \square$

$24 + 35$

$20 \quad 4 \qquad 30 \quad 5$

$24 + 35$

$= 20 + 30 + 4 + 5$

$= 59$

$97 - 16$

$90 \quad 7 \qquad 10 \quad 6$

$97 - 16$

$= (90 - 10) + (7 - 6)$

$= 80 + 1$

$= 81$

Key words

mentally

strategies

number line

These students made some quick number line sketches to help with their mental working.

67 + 16 = ☐

Move forwards in tens. Then add the ones.

56 − 37 = ☐

Move backwards in tens. Then minus the ones.

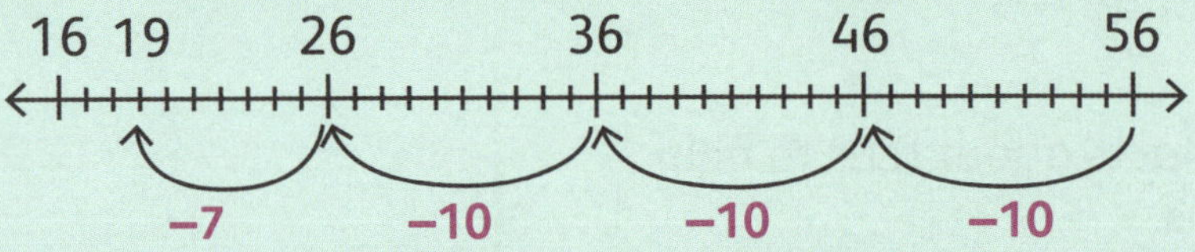

These are only some of the ways to calculate mentally. Choose the ways that work the best for you.

1 Work in pairs to add or subtract numbers mentally.

a Write a 2-digit number on a piece of paper. Swap numbers and then spend half a minute writing as many different ways of making that number as you can. Do this a few times with a new number each time.

For example, 18 could be made like this: 10 + 8, 1 + 17, 20 – 2, 9 + 9, 8 + 10 …

b Write simple addition and subtraction sums with 2-digit numbers.
Take turns to add or subtract mentally.
Talk to each other about the methods you used to get to your answers.

2 Work mentally to find the answers.

a 23 + 19	**b** 50 + 50	**c** 0 + 30
d 90 − 50	**e** 47 + 43	**f** 89 − 34
g 81 + 34	**h** 67 − 36	**i** 76 + 11
j 54 + 9	**k** 66 − 7	**l** 43 + 6
m 60 − 20	**n** 40 + 20	**o** 60 − 30
p 100 − 90	**q** 67 + 22	**r** 53 + 25
s 83 − 18	**t** 25 − 19	**u** 54 + 9

Rounding off

Key maths idea

Round numbers end with one or more zeros. We use **rounding off** to give an **approximate** statement of an amount and to make it easier to calculate.

I have about 500 students in my school.

I've saved about $2 000 of my pocket money.

There were approximately 1 000 people at the stadium.

Words such as 'about', 'around', 'approximately' or 'more or less' may tell us the numbers are rounded off.
When you round off:

- Find the digit in the rounding place. For example, if you are rounding to the nearest thousand, look at the thousands place.
- Look at the digit to the right. If the digit to the right is 0, 1, 2, 3, 4, then leave the digit in the rounding place as it is.
- If the digit to the right is 5, 6, 7, 8, 9, then add 1 to the digit in the rounding place.
- Change all the digits to the right of the rounding place to 0.

Example

Round 4 763 to the nearest hundred.

4763 — Number 7 is in the hundreds place.

4800 — We round up, so we write 8 instead of 7, and 0 in the places to the right.

Key words

round number
rounding off
approximate

1 Round off each amount to the nearest thousand.

a 6729 people **b** 4599 chickens **c** 1134 cars

2 William says he made about 200 blocks of fudge for the school fair. He rounded it off to the nearest hundred. Which of the following could not be the actual number he made?

189 245 272 243 143 200 300

Estimation

Key maths idea

When you **estimate** an answer you make your best guess to get as close to the real answer as you can. Estimation is partly guessing but it mostly needs you to use some methods that you have already learned.

Example 1

There are 158 marbles in this jar. If I pour half into another jar, estimate how many will be in each jar. What is the easiest way to work it out?

If we round off the marbles to 160, it is easy to calculate.

Half of 16 is 8, so half of 160 is 80.

Example 2

What is 1 354 + 5 726?

Step 1: Round and estimate first.
1 000 + 6 000 = 7 000 The estimated sum is 7 000.

Step 2: Add the numbers.
1 354 + 5 726 = 7 080

Step 3: Check.

The answer is close to the estimate, so it seems reasonable.

Key word
estimate

1 Round off, estimate and add. Write your estimated answer and compare it to your calculated answer.

- **a** 143 + 876
- **b** 298 + 345
- **c** 876 + 192
- **d** 9 872 + 4 654
- **e** 99 − 81
- **f** 278 − 234
- **g** 23 + 132 + 97
- **h** 299 + 67 + 12
- **i** 56 − 43

2 A farmer has 88 chickens, 64 goats and 19 pigs on her farm.

- **a** Round the numbers to the nearest 10 and estimate how many animals she has on her farm.
- **b** Calculate the actual number of animals on the farm.
- **c** What was the difference between your estimate and your calculated total?

3 A bus driver travels 33 km to his first stop, 18 km to his next stop and then 24 km to his final stop.

- **a** Round the numbers to the nearest ten and estimate how far he travelled.
- **b** Calculate the actual distance he travelled.
- **c** What was the difference between your estimate and your calculated total?

4 Abisai was given this list to take to the shop.

- **a** Estimate how much money he will need.
- **b** Calculate how much money he will actually spend.
- **c** What is the difference between your estimate and your calculation?
- **d** If Abisai had $2 100 to spend, how much change would he take home?

2 litres of milk	$552
Meat 1 kg	$810
Onions	$120
Bread	$190
Rice (500g)	$ 55.50
Bunch of bananas 1 kg	$ 95.60

Real-world maths: Run a supermarket

Cut out some of the products with their prices, and stick them onto cards. Write the prices on each product card. These will be the 'products' in your store. Take turns being customers and shopkeeper. Each customer chooses three products. The class should work together to work out each customer's total amount.

How does estimating help you at the supermarket?

You will need:

- an advertising leaflet from a supermarket
- scissors
- glue sticks
- card or paper
- marker pens.

The commutative property

Key maths idea

5 + 4 = 4 + 5

When you add, it doesn't matter in which order you add the numbers.

This is called the **commutative property** of addition. It means we can add numbers in any order. Multiplication is also commutative. 2 groups of 3 makes the same product as 3 groups of 2.

$2 \times 3 = 3 \times 2$

Key word

commutative property

1 Without working out the answer, rewrite each sum in a different way that will make the same total.

a 115 + 5 **b** 324 +34 **c** 42 + 145 **d** 250 + 144

2 Without working out the answer, rewrite each multiplication sentence in a different order to make the same product.

a 7×6 **b** 9×8 **c** 10×5 **d** 120×3

3 Use a calculator to check that the answers to Exercises 1 and 2 are correct.

4 Alex and Kyle are collecting cans for the school recycling project. Alex collects 17 bags, each with 12 cans. Kyle collects 12 boxes, each with 17 cans. What was the difference between the numbers of cans they each collected? Work it out mentally!

5 Discuss with a partner why you can't rearrange the order of numbers in a subtraction or division sentence.

The associative property

Key maths idea

A teacher gives her class this shape and asks them to write addition sums to describe how they see the dots.

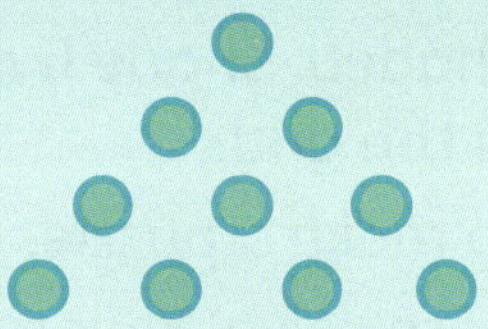

Here are some of their ideas:

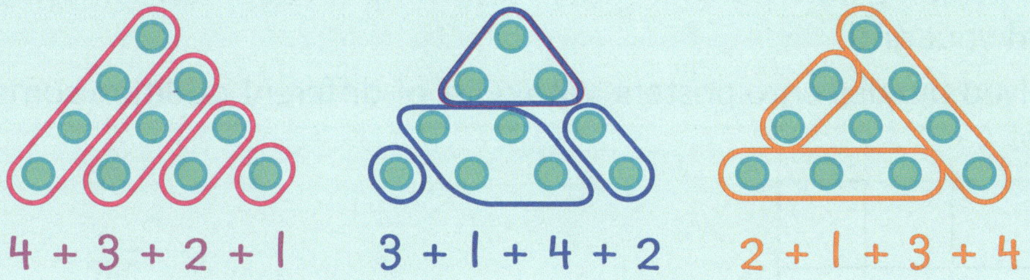

- What do you notice?
- How many other ways can you find to split the dots in this shape?
- Why does it make the same sum, no matter how you split the dots?

The **associative property** says that when we add, it does not matter how we group the numbers.

Next, the students use blocks to show that you can make a product of 12 by multiplying in different ways.

$2 \times 4 = 8$ $\quad 4 \times 3 = 12$

$8 \times 3 = 24$ $\quad 2 \times 12 = 24$

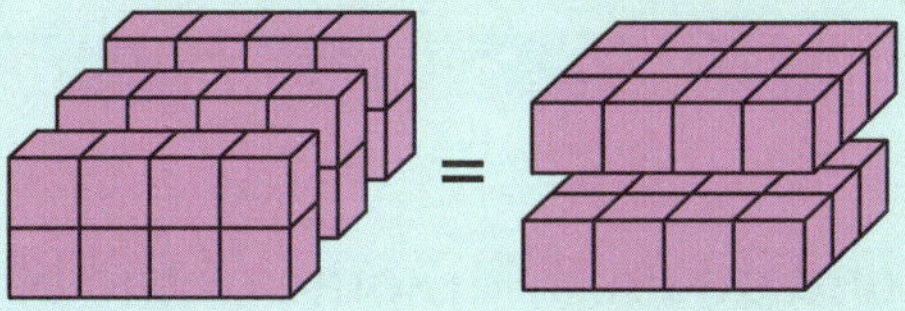

$(2 \times 4) \times 3$ $\qquad 2 \times (4 \times 3)$

Key word

associative property

1 Use flash cards and base ten blocks to show why each statement is true.

a $16 + 14 = 14 + 16$

b $13 + 31 = 31 + 13$

c $3 \times 11 = 11 \times 3$

d $7 + (12 + 10) = (7 + 12) + 10$

e $(4 \times 5) \times 3 = 4 \times (5 \times 3)$

2 Fill in the missing numbers.

a (33 + ____) + 15 = 33 + (21 + 15)

b ____ + (9 + 2) = (17 + 9) + 2

c ____ + (____ + 22) = (99 + 44) + ____ = ____

d 100 + (____ + ____) = (____ + 15) + 20 = ____

e (4 × 5) × 2 = 4 × (____ × ____) = ____

f (____ × 6) × 5 = 3 × (____ × 5) = ____

Maths detective: Grids

Case 1

- Trace or draw this grid onto a sheet of paper.
- One shape has been shaded. Shade 6 more shapes made of blocks. Make sure to use up all the blocks in the grid.
- Cut out your shapes to make a puzzle.
- Give them to a partner to solve. Your partner should also write an addition sum that shows how you shaded your grid.
- As a class, stick the solved puzzles onto posters, with a list of different addition sums that all make a total of 49.

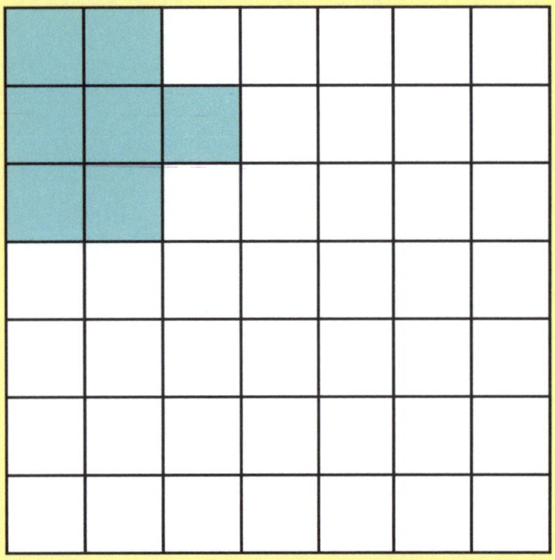

Case 2

Trace or draw the same grid again. Trim some of the blocks so that you can use the grid to show that:

$(5 \times 3 \times 3) = (9 \times 5)$

Is this the same as 3×15? Use your grid to show why.

Case 3

Detective Jones wants to make a grid that can show:

$2 \times (7 + 5) = (2 \times 7) + (2 \times 5)$

- How many blocks does she need for her grid?
- Draw two different possible grids she could use.

What I have learned

1 Dan made mistakes and wrote all the key terms incorrectly for this chapter. Help him to get the terms right. Read the definition and write the correct term.

- **a** Putting two or more numbers together to get the sum. *edition*
- **b** Taking away a smaller number from a larger number. *attraction*
- **c** The result when you minus one amount from another. *defence*
- **d** Making an approximate amount by giving the nearest multiple of ten, 100 or 1 000. *reading off*
- **e** A sensible guess of the answer. *ultimate*
- **f** A way of working in your head. *working minty*

2 Which of these is not equal to 400 + 500 + 700?

a 500 + 700 + 400 **b** 900 + 700 **c** 400 + 500 − 700

3 Which of these is not equal to 10 × (8 × 7)?

a 80 × 7 **b** 70 + 80 **c** (10 × 8) × 7

Practice questions

1 Use the strategies you have learned to work out these number sentences mentally.

a 4 + 7 + 8 + 9 + 1 + 3 + 6 + 2 **b** 50 + 29 + 50
c 85 − 49 **d** 90 − 40
e 48 − 39 **f** 18 + 39 + 41 + 22

2 Round off these numbers to the nearest hundred.

a 348 **b** 289 **c** 801 **d** 442 **e** 760

3 Round off these numbers to the nearest thousand.

a 1 340 **b** 1 673 **c** 4 928 **d** 9 099 **e** 7 429

4 We can make many different addition sums to show the same number. For example: 35 = 10 + 10 + 10 + 5 or 35 = 20 + 15

For each of the following numbers, write five different addition sums that make the same total.

a 47 **b** 58 **c** 84

5 Look at these addition sums represented with base ten blocks. Work out the total for each addition sum.

a
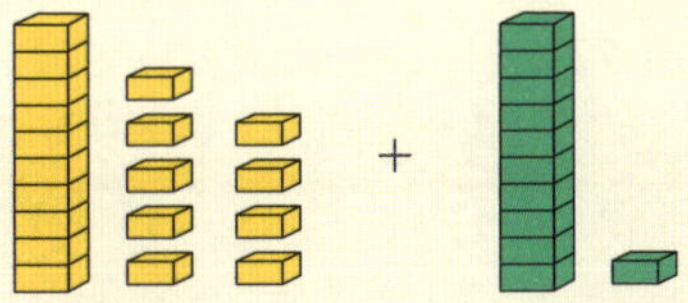

b
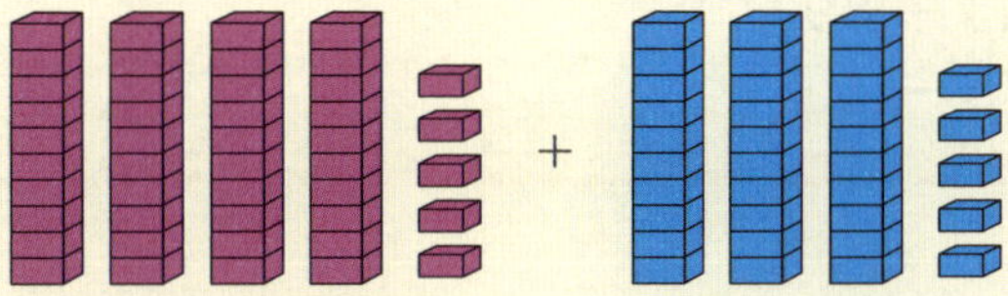

c

d
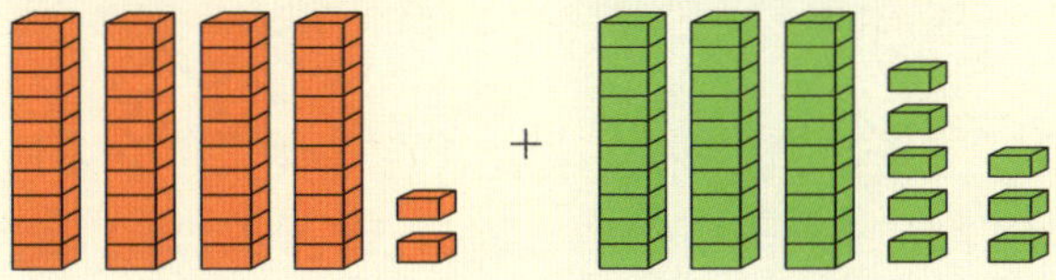

6 Estimate the total for each receipt, then calculate to check if your estimates were reasonable.

a

1 cupcake	$35
1 lollipop	$27
cup of lemonade	$15

b

2 fruit ices	$27 each
1 milkshake	$28

4 Self check

	Check you can do this:	If you have difficulty, try this:
1	Make reasonable estimates. Estimate the answer to these: 457 + 139 2 390 − 478	Go back to the work on pages 51 and 52. Make sure you remember how to round off.
2	Use the commutative and associative property of numbers. Write each of these in a different order that would make the same answer: 15 × 18 5 × 4 × 25 6 700 + 1 386 85 + 190 + 3 655	Revise pages 54 and 55.
3	Add or subtract 2-digit numbers mentally. Try working these out: 43 + 89 75 − 46	Revise the mental strategies on pages 49 and 50.

Unit 2: Measurement

Chapter 5 Measuring length, capacity and time

In this chapter you will explore these focus questions:

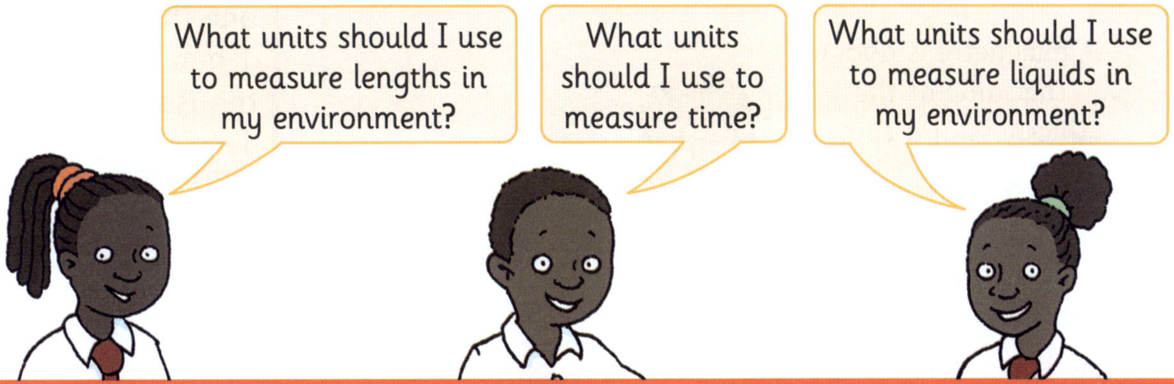

To answer these questions, you will:

- estimate and measure distances and lengths
- estimate and measure capacity and volume
- read and write time, and understand the relationships between units of time
- estimate and measure time.

Starting point

1 What is the same about all three of these measuring instruments? What is different about them?

2 What similarity do you notice between the number lines you have used for counting and calculating, and the markings on the measuring instruments?

3 What would you use each of these instruments for?

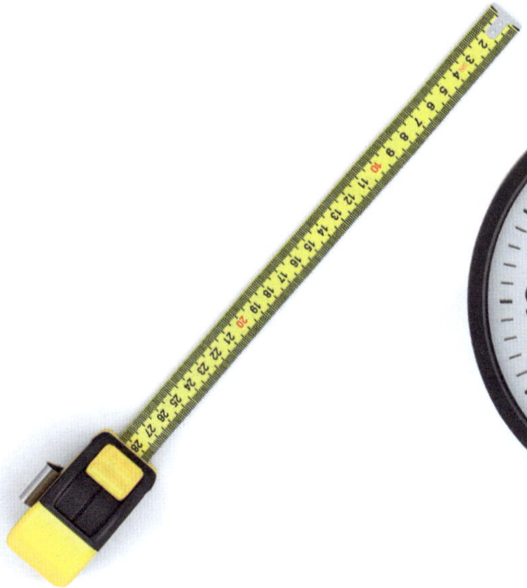

Length

Key maths idea

When we ask how long or wide something is, we are measuring length. The **standard unit** for measuring length is the **metre** (m).

Rulers are marked in **centimetres** (cm) and **millimetres** (mm).

10 mm = 1 cm   100 cm = 1 m

We measure longer distances in **kilometres** (km).

Sometimes we use non-standard units. For example, you might use paces (big steps) to measure the length of your classroom or paper clip lengths to measure a small object.

Remember:
Kilo- means one thousand.
1 000 m = 1 km
Milli- means one thousandth.
1 mm is one thousandth of 1 metre.

Key words
- standard unit
- metre
- centimetre
- millimetre
- kilometre
- kilo-
- milli-

1 Work in groups.

  a Make five paper chains of different lengths.

  b Choose a non-standard unit, such as a paper clip or a strip as of paper. Work out how many units long each paper chain is.

  c Compare your non-standard unit to a ruler. Estimate the actual length of each chain.

  d Measure the chains using the ruler. How close were your estimates?

You will need:
- strips of paper
- glue or tape
- a ruler.

Estimating and measuring lengths

Key maths idea

You have used **estimation** when you worked with addition and subtraction. Estimating lengths is a little different. You estimate length when you guess how long something is. You can do this in different ways:

- guess based on how it looks
- compare it to something you know, such as your hand or a ruler
- measure a portion of the length, then multiply by about how many times that length repeats to make the whole length.

Key word

estimation

You will need:

- string or rope cut to 1 metre long
- a ruler or metre rule
- paper.

1 Work in groups.

a In your classroom, identify three objects that are shorter than a metre, three that are about 1 metre, and three that are longer than a metre.

b Estimate the length of each object.

c Use your 1 metre string or rope to measure the objects.

d Record your findings in a table.

Shorter than 1 metre	About 1 metre	Longer than 1 metre

2 Make a list of five more objects in your classroom you want to measure. Measure them using a tape measure, metre rule, or a string or rope cut to a given length.

Some soft tape measures are also useful for measuring curved things.

3 Use string or a soft tape measure to measure these curved lines.

a
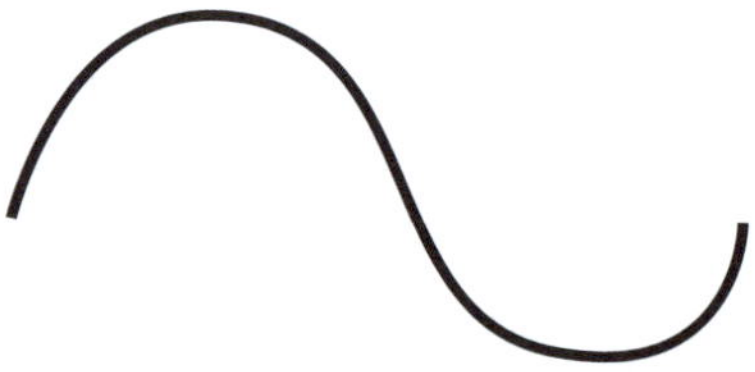

b
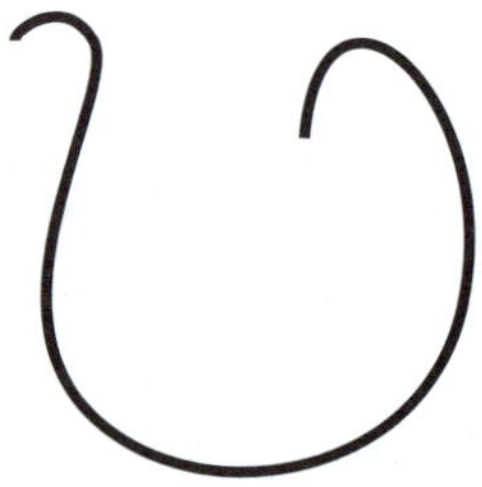

c
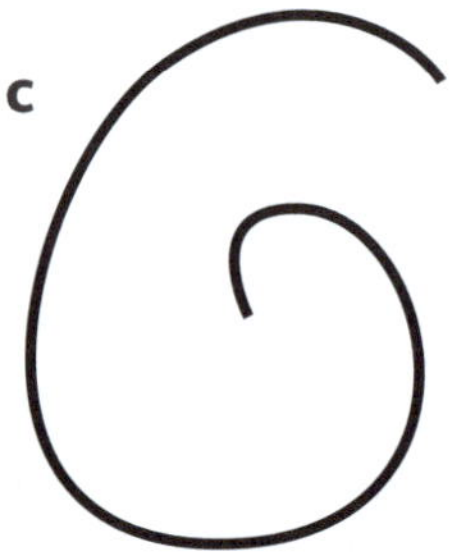

4 Which measuring instrument would you use to measure the following items?

A
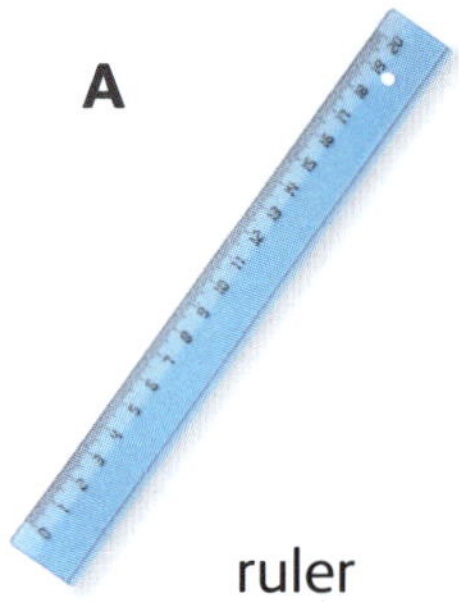
ruler

B

tape measure

C

trundle wheel

D

laser rangefinder

- **a** the length of a fence around a school
- **b** length of fabric to make a dress
- **c** the distance between two towns
- **d** the distance between two countries
- **e** the length of a shoelace
- **f** the length of a sewing needle
- **g** the length of a football field

5 Would you measure these in millimetres, centimetres or metres?

- **a** the length of a book
- **b** the distance to the next classroom
- **c** the length of an ant
- **d** the thickness of a bank card
- **e** the length of your foot
- **f** the height of your chair

More about equivalence

Key maths idea

1 m = 100 cm	To convert m to cm, multiply by 100.
155 cm = 1 m 55 cm	To convert cm to m and cm, write the hundreds as m, and the tens and ones as cm.

1 Convert these measurements to cm.

a 2 m **b** 5 m **c** 4 m 50 cm **d** 3 m 21 cm

2 Convert these measurements to m and cm.

a 257 cm **b** 496 cm **c** 1 000 cm **d** 3 259 cm

3 Write the rule for converting measurements from cm to m and m to cm.

Real-world maths: Throws and jumps

Task 1

A sports event often answers a measurement question about distance or speed, such as:

- running – how quickly can the athletes run a given distance?
- swimming – how quickly can each swimmer swim a given distance, using a specific stroke?

In pairs, write the measurement question that each of these sports events answers:

1 long jump
2 high jump
3 discus

Task 2

For this activity, you will need a paper plane or a frisbee. Go outside. Mark a starting point where everyone will stand. Take turns to throw the object. Measure each throw in metres and centimetres to see how far away it lands from the starting point.

Task 3

In groups, create a mock Olympic long jump event. Think about:

- What will you need to measure?
- What measuring instruments will you use?
- What will you do to make sure the event is fair?

Hold your event. Each student should have a turn to perform their jump.
The others measure and record the jumps, using metres and centimetres in decimal form.

Task 4

Identify buildings and landmarks you see in your community, such as your church, well-known buildings, trees or statues. Estimate the height of these objects so that you can describe them without using measuring instruments. Compare your descriptions with those of your classmates'. Whose do you find most accurate?

Distances on a map

Key maths idea

When we read distances on a map, we use a **scale**. The scale on this map tells us that 1 cm on the map represents 15 km in real life.

Key word
scale

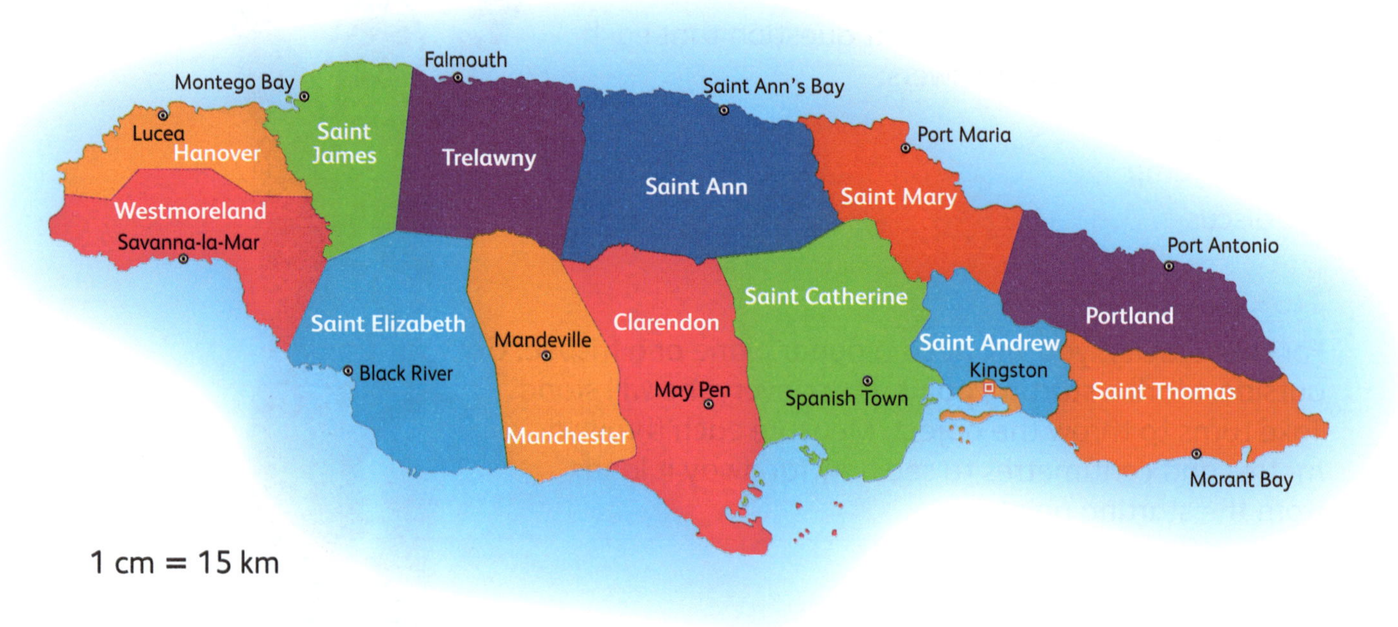

1 Estimate first and then measure the straight line distances between these places on the map.

a Kingston to Falmouth
b Black River to Port Antonio
c Lucea to Port Maria
d Morant Bay to Mandeville
e Spanish Town to Savanna-la-Mar

Do you know the saying 'as the crow flies'? If you say a place is a certain distance away 'as the crow flies' you mean that it is that distance away measured in a straight line.

Capacity and volume

Key maths idea

The **capacity** of the jug is 1 000 ml. That is how much it can hold.

The **volume** of water in the jug is 600 ml. That is the amount of liquid in the jug.

We measure capacity and volumes using **millilitres** (ml) and **litres** (ℓ).

1 000 ml = 1 ℓ

1 000 ℓ = 1 **kilolitre** (1 kl)

You may also see volume and capacity measured in spoons and cups, especially in recipes. Here are some useful amounts to remember:

1 cup = 250 ml 1 teaspoon = 5 ml

1 litre = 4 cups 1 tablespoon = 15 ml

Key words

capacity
volume
millilitre
litre
kilolitre

You will need:

- small medicine syringes, 5 or 10 ml sizes
- medicine measures, 15 ml sizes
- measuring cups and jugs in a variety of small sizes up to 1 ℓ
- a variety of empty cups and jars
- water.

1 Fill a syringe to the 1 ml mark. Check how many times you need to refill it to fill:

a 1 teaspoon

b a 15 ml measuring cup.

2 Line up the cups and jars from smallest to largest. Estimate the capacity of each container and write down your estimates.

3 Fill each container, and then pour the water into measuring cups and jars to check the actual capacity of each container.

4 Record your findings in a table.

Convert between measures of capacity

Key maths idea

Sometimes you need to convert from one unit of measure to another.
For example, a bottle contains 2 litres. How many ml is that?

You know that 1 ℓ = 1 000 ml

$2 \times 1\,000 = 2\,000$

The bottle contains 2 000 ml.

- To convert from litres to millilitres, multiply by 1 000.
- To convert from millilitres to litres, divide by 1 000.

Sometimes you may express amounts of liquid in litres and millilitres. For example, we can write 2 500 ml as 2 ℓ 500 ml.

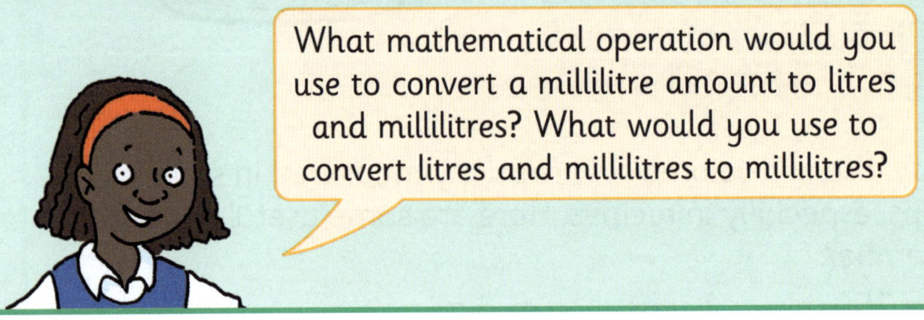

1 Write these amounts in millilitres.

a 1 ℓ **b** 5 ℓ **c** 12 ℓ **d** 9 ℓ

2 Write these amounts in litres.

a 11 000 ml **b** 79 000 ml **c** 2 000 ml **d** 102 000 ml

3 Convert these measurements to litres and millilitres.
For example, 3 686 ml = 3 000 ml + 686 ml = 3 ℓ 686 ml

a 18 030 ml **b** 6 937 ml **c** 12 300 ml **d** 18 450 ml

4 Convert these measurements to millilitres.
For example, 4 ℓ 723 ml = 4 000 ml + 723 ml = 4 723 ml

a 8 ℓ 325 ml **b** 13 ℓ 987 ml **c** 52 ℓ 968ml **d** 437 ℓ 331 ml

5 A decilitre is equal to $\frac{1}{10}$ of a litre.

a How many ml are in a decilitre?

b How many ml are in $3\frac{1}{2}$ decilitres?

c How many decilitres are 800 ml?

Maths detective: Investigate containers

Case 1

Work with a partner. Collect containers of 1 ℓ, 2 ℓ, 5 ℓ and 10 ℓ capacity. You will also need a container or measuring jug with a 500 ml capacity. Count how many 500 ml amounts each of the larger containers can hold.

Case 2

Look at the containers in the picture.

1 What is the capacity in millilitres of container A?

2 What is the capacity of B and C together?

3 How many of D would fill B?

4 How many of C would fill A?

5 Which of the containers could you fill twice to fill A?

6 How much liquid would you need to fill all of the containers to half capacity?

7 How much liquid would you use to fill all of the containers to their full capacity?

A

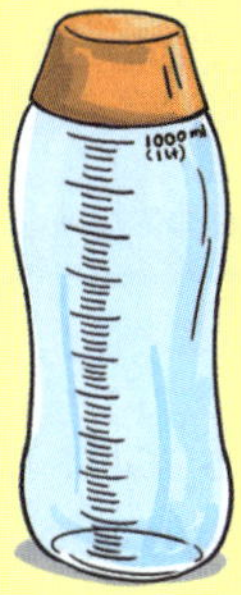

1 ℓ

B

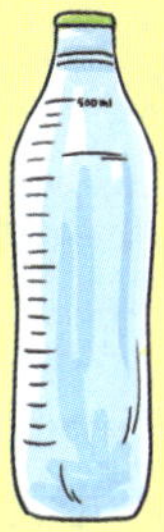

500 ml

C

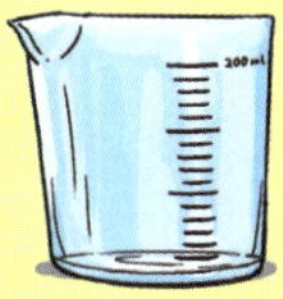

200 ml

D

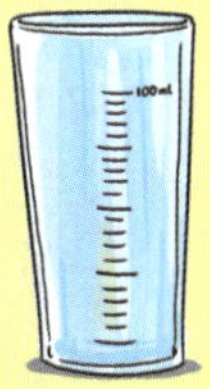

100 ml

Case 3

1 teaspoon = 5 ml

1 cup = 250 ml

1 How many cups make half a litre?

2 How many ml are there in 6 teaspoons?

3 How many litres do 12 cups make?

4 How many ml are there in 3 cups + 7 teaspoons?

Case 4

Your parents invite six guests for dinner. They cook rice and peas. They need $\frac{1}{2}$ cup of uncooked rice per person.

1 How many cups of uncooked rice do they use?

2 To cook the rice they need double the number of cups of water. How many cups of water will they need?

3 What is the correct amount of water in ml?

4 After it is cooked, the rice will double in volume. How many cups of cooked rice will it make?

Case 5

Bell Hill Primary School organised a sports day. Mrs Henry said she would mix big jugs of juice for the children. She found two recipes for mixing the juice:

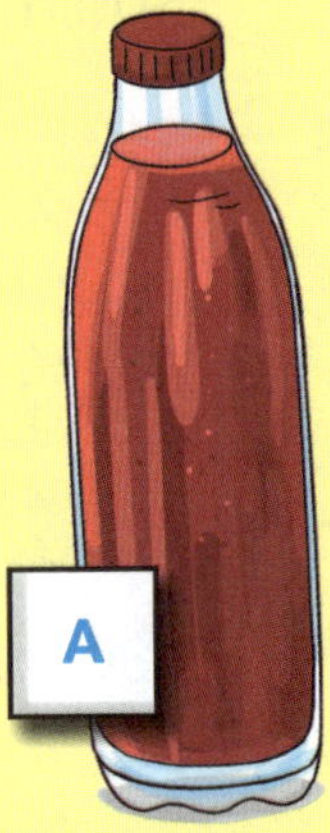

Mix 3 cups of water with 2 cups of concentrate.

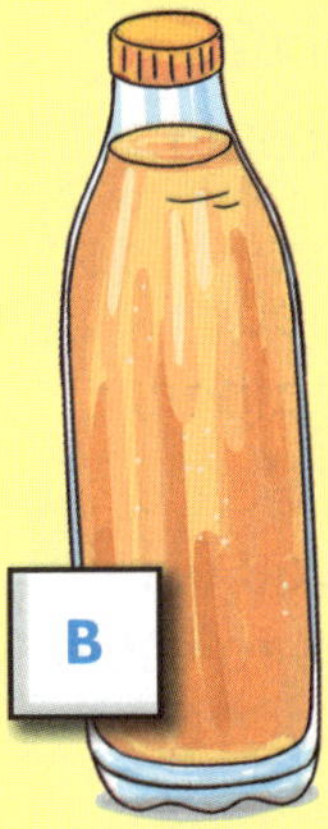

Mix 5 cups of water with 3 cups of concentrate.

1 Complete the following sentences, keeping in mind that 1 cup is 250 ml.

a Recipe A requires _____ ml of concentrate and _____ ml of water.
It makes _____ ml of juice altogether.

b Recipe B requires _____ ml of concentrate and _____ ml of water.
It makes _____ ml of juice altogether.

2 Mrs Henry makes an unequal number of batches of both recipes, for a total of $7\frac{3}{4}$ litres. Work out how many batches she made of recipe A and recipe B.

Case 6

At the end of a busy day a mother runs water into a bath for her two young children to share. The bathtub has a capacity of 60 litres.

1 What is the capacity of half the bathtub?

2 If the tap pours out 5 ℓ every minute, how long will it take to fill the bath to $\frac{3}{4}$ full?

Time

Key maths idea

What similarities and differences do you notice in these watches?

We use watches and clocks to tell the time.

An **analogue** watch shows the time using hands that move around the clock face. The short hand shows the hour of the day. The long hand shows how many minutes past the hour.

A **digital** watch uses numbers to tell the time. The first two numbers give the hour and the numbers after the colon give the minutes past the hour.

These are the units we use to measure time:

60 seconds = 1 minute
60 minutes = 1 hour
24 hours = 1 day
7 days = 1 week
1 month = about 30 days or around 4 weeks
1 year = 365 days (or 366 days in a leap year)

1 Estimate how long each task will take in hours, minutes, seconds:

- **a** write your name
- **b** take a shower
- **c** walk from your classroom to the front of the school
- **d** run around a sports field once
- **e** a bus trip to the next parish.

2 How many minutes are there in:

- **a** half an hour?
- **b** one and $\frac{3}{4}$ hours?
- **c** one quarter hour?

Reading the time

Key maths idea

This analogue clock shows the time as quarter to three. That means it is 15 minutes before 3 o'clock.

A digital clock can use different time formats.

12-hour clock: 2:45 p.m.

24-hour clock: 14:45

When we use the 24-hour clock, we do not need to say a.m. or p.m. 00:00 means midnight. The hours from midnight to noon are 00:00 to 11:59. 12:00 to 23:59 are the hours from 12 noon to midnight.

If you need to convert analogue or 12-hour clocks to 24-hour clocks, add 12 to the hours. For example, 6:25 p.m. becomes 18:25.

12-hour clock	24-hour clock	Analogue
7:00 p.m.	19:00 (say nineteen hundred hours)	seven o'clock
1:30 p.m.	13:30 (say thirteen thirty)	half past one
6:45 p.m.	18:45 (say eighteen forty-five)	quarter to seven

1 Work in pairs. Read the time on these clocks and then write the time in words and in 24-hour clocks.

a 

b 

c 

d 

e 

f 

g 

h 

2 Write these clock times in 24-hour clocks.

a 11:45 p.m. **b** 4:30 p.m. **c** 6:30 a.m.

d noon **e** midnight **f** 20 to 6 in the afternoon

How long does it take?

Key maths idea

If you know the starting time and ending time of an activity or event, you can work out the **duration** – how long it takes. Counting in time **intervals** can help you work out how long something takes. For example:

My maths class starts at 11:40 and ends at 12:10. What is the duration of the lesson?

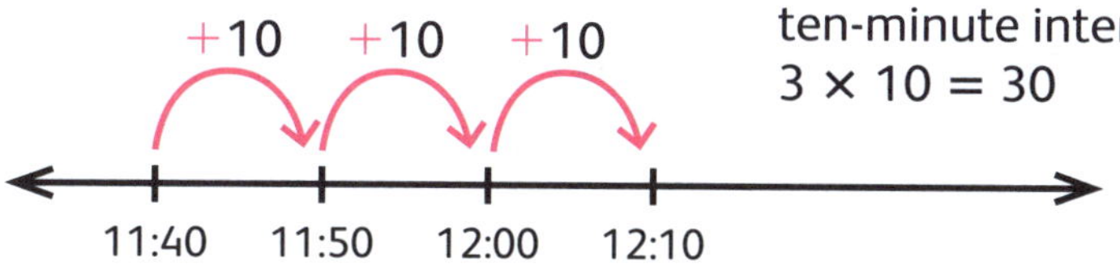

I count on in ten-minute intervals.
$3 \times 10 = 30$

The lesson takes 30 minutes or half an hour.

For bigger time intervals you may count in hours and half hours, days or even months or years.

Key words

duration
interval

1 Draw two sets. Label one of the sets <1 hour and label the other set >1 hour. Choose which of these activities will be elements of each set. Write them in.

cut an apple	eat breakfast	watch a film
a birthday party	travel to a different island	
break time at school	cook an egg	

2 About how long does each activity take? Give an approximate duration in hours or minutes.

a

sleep at night

b

game of cricket

c

bus ride to school

d

beach outing

3 Here are the start and finish times of some different activities. Work out how long each one takes.

	Start time	Finish time
a	4:00 a.m.	7.30 a.m.
b	3:00 a.m.	5:30 p.m.
c	6:19 p.m.	8:19 p.m.
d	10:05 a.m.	11:25 p.m.
e	12:00 p.m.	2:00 a.m.
f	7:30 a.m.	10:15 a.m.

Real-world maths: Solving time problems

1 Jaden's school starts at 7:45 a.m. and ends at 1:00 p.m.

- a How much time does he spend at school every day?
- b How many hours does he spend at school every week?
- c Jaden's sister Amelia goes to a different school. Her school day starts at 7:30 a.m. each day and ends at 3:45 p.m.
- d How many hours does Amelia spend at school each day?
- e How many hours does she spend at school each week?
- f How much more time does Amelia spend at school every day than Jaden?

2 Mrs Campbell travels by bus from home to her job in the city. It takes her 45 minutes to get ready for work. The bus leaves at 7:30 a.m. It takes her 10 minutes to walk to the bus stop. At what time must she get up in the morning?

3 The International Space Station takes 644 minutes to orbit Earth 7 times.

- a About how many minutes does it take to complete one orbit?
- b Now write your answer in hours and minutes.

4 Work in small groups. You will need a clock, a watch or the timer on a mobile phone to measure the time with. In your group choose five things you can do in your space that you would like to time. For example:

- How long can each person skip without stopping?
- How long can each person hold a plank position or balance on one leg?

Draw a table with a list of the names. Add the estimated and measured time of your chosen activity.

What I have learned

Copy and complete this mind map about the different kinds of measurement you have learned about. For each kind of measure (length, time, capacity), list the main units we use, write some questions that we can ask, list measuring instruments you have seen or used, and give an example of at least two measurements.

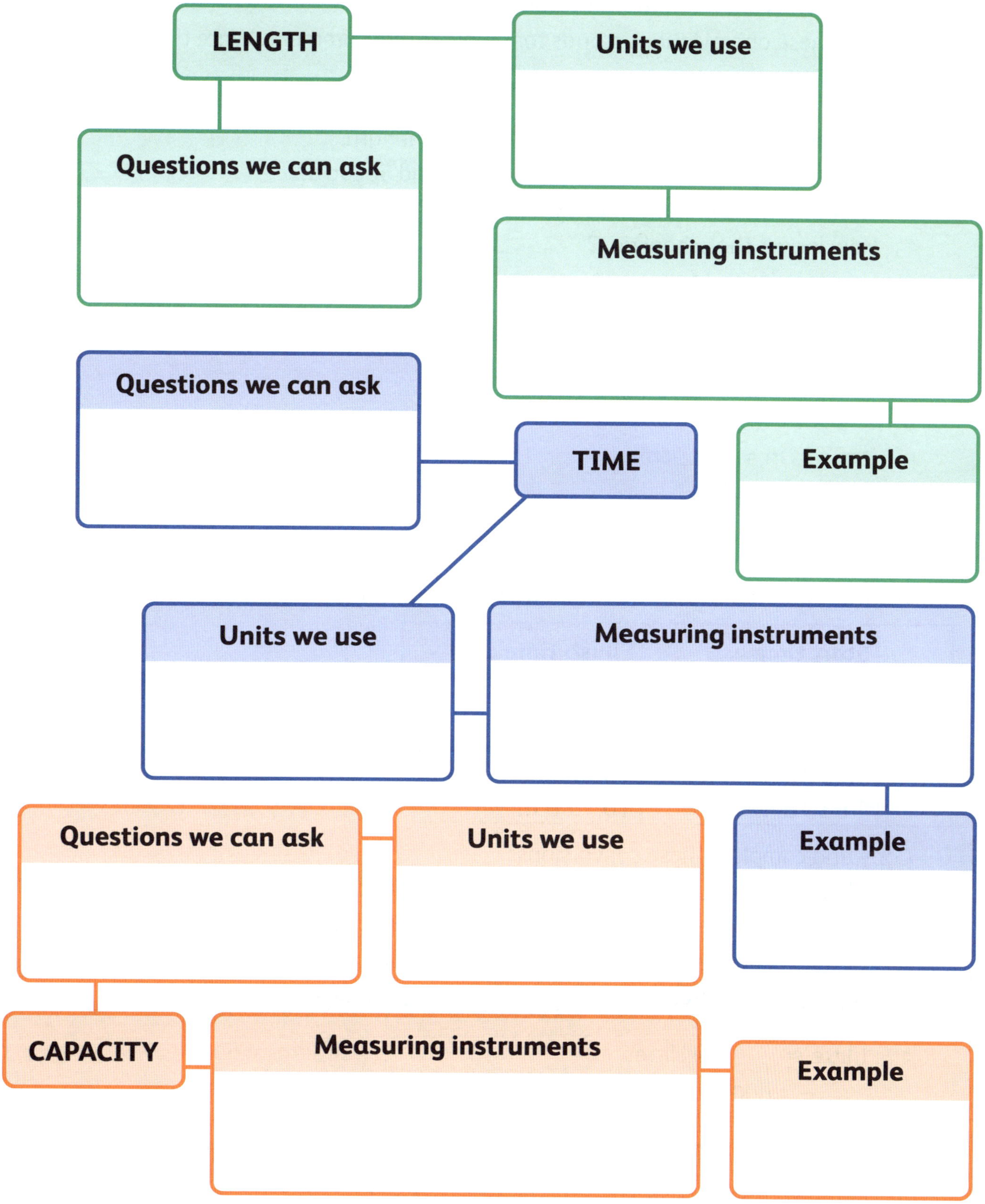

Practice questions

1 Write these units of measurement from the smallest unit to the largest unit.

a metre, kilometre, millimetre, centimetre

b litre, kilolitre, millilitre, centilitre

c minute, second, day, month, hour, year

2 Each of these abbreviations stands for a unit of measurement. Write the unit name.

a mm b ml c min d h e m f ℓ

3 For each of the following questions, say which unit you would use to measure it, and whether the question is about time, capacity or length.

a How far is the distance between Jamaica and St Lucia?

b How much water can you carry in a bucket?

c How long is the school day?

d How long does it take you to run around the school field twice?

e How much medicine is in a medicine bottle?

f How thick is your maths book?

4 Copy and convert these measurements.

a 312 cm = _____ m

b 4 907 cm = _____ m

c 3 m 43 cm = _____ cm

d 12 min 33 s = _____ s

e 7 h = _____ min

f 3 400 ml = _____ ℓ

g 9 ℓ = _____ ml

h 7 m = _____ cm

i 8 116 ml = _____ ℓ _____ ml

j 2 154 cm = _____ m _____ cm

5 Work out how much time has passed from the start times to the finish times.

	Start time	Finish time
a	3:00 a.m.	5.30 a.m.
b	6:00 a.m.	1:30 p.m.
c	12:00 p.m	12:00 a.m.
d	11:30 a.m.	10:15 a.m.

6 Copy and complete these conversions.

a 1 kilometre = _____ metres
1 metre = 10 decimetres
= _____ centimetres
= _____ millimetres

b 1 minute = _____ seconds
_____ hour = 60 minutes
$\frac{1}{2}$ hour = _____ minutes
_____ hour = 15 minutes

c 1 litre = _____ millilitres
_____ = 1 000 ml

7 Seymour has a jug filled with juice. He has 7 glasses and pours 350 ml of juice into each one. After this he still has 200 ml of juice in the jug. What was the capacity of the jug in litres?

8 Lucy is a truck driver. Last week she drove 432 km on Monday, 122 km on Tuesday and 60 km on Thursday. This week she drove 376 km in total. What is the difference between the total distance she drove last week and the total distance she drove this week?

5 Self check

Check you can do this:	**If you have difficulty, try this:**
1 Estimate, measure and record lengths, for example, find the length of a desk in your class.	Go back to pages 60 to 62. Revise the units of length and how to estimate and measure.
2 Measure distances on a map and solve problems using maps.	Revise the work you did on page 64.
3 Estimate and measure capacity or volume using litres and millilitres, for example, say how much a bottle or container can hold.	Revise pages 65 to 68.
4 Read and write time from digital and analogue clocks.	Go over the work on pages 69 to 72.

Chapter 6 More about measurement

In this chapter you will explore these focus questions:

To answer these questions, you will:

- estimate and measure mass in grams and kilograms
- estimate and measure temperature in degrees Celsius
- explore the relationships between units
- convert between units and discuss the meanings of the different prefixes.

Starting point

You have worked in earlier grades with **grams** and **kilograms**. These are the units we use to measure **mass**.

Elephants are the largest land mammals.

Remember that **kilo-** means 1 000 and 1 000 g = 1 kg.

We also have a bigger unit of mass, the **tonne**, which is used for measuring heavy masses. 1 tonne = 1 000 kg.

- If possible, watch a presentation or online video about elephants.
- What mass does a fully grown elephant reach?
- What is the mass of a newborn elephant?
- Find out more facts about elephants, such as the mass of food they eat in a week, a month or a year.

Key words

gram
kilogram
mass
kilo-
tonne

Grams and kilograms

Key maths idea

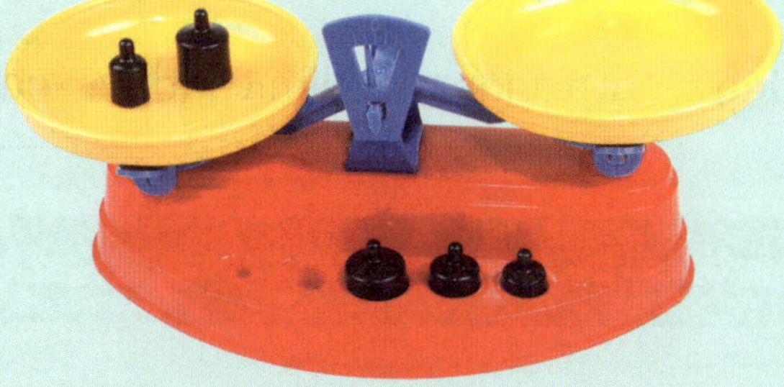

What kind of scales have you seen at home, at school or in stores? How do they work?

You know that 1 000 g = 1 kg.
When items are lighter than 1 kg, we measure them in grams. A kitchen scale can usually measure masses of up to 3 kg. In the activity on this page, you will need to use a kitchen scale or a two-pan balance to measure in grams and kilograms.

You will need:

- a kitchen scale, balance scale or spring scale
- a variety of small objects for weighing
- grocery items, such as tins or dry beans containers.

1 **a** Discuss the objects you have brought for weighing.
  b Guess which will be the heaviest and which will be the lightest.
  c Arrange them in order from lightest to heaviest and estimate the weights.

2 **a** Weigh the object you thought would be heaviest. How accurate was your estimate?
  b Weigh the object you thought would be lightest.
  c Now that you have actual masses for two of the objects, discuss whether you want to change any of your estimates.

3 Weigh the rest of the objects.
Make a table like this to describe the objects you weighed.

Object	Estimate	Actual weight

4 Make amounts that you think weigh exactly one gram, for example a crumpled ball of paper, a few paper clips, some raisins. Use the scale to check their masses.

5 Use the collection of grocery items to try these challenges:

- Use three items to get as close as possible to 1 kg.
- Use as many items as you can to get a total of not more than 1 kg 200 g.
- Make 2 kg using as few items as possible.

In your group, set each other challenges for reaching a particular mass limit with as many or as few items as possible.

Real-world maths: Convert a recipe

Stephen found this old recipe in a family book.

Ingredients

3 ½ oz butter
3 ½ oz dark brown sugar
6 oz golden syrup
2 tablespoons ginger wine
2 eggs, lightly beaten
6 oz self-raising flour
4 teaspoons ground ginger
1 oz fresh ginger, peeled and grated
3 ½ oz crystallised ginger, finely diced

Instructions

Heat oven to gas 3 or 170 °C.
Beat together the butter and sugar with a pinch of salt.
Add the golden syrup, ginger wine and eggs. Beat until smooth.
Sift the flour and ground ginger into the mixture, and fold it in with the fresh ginger and crystalised ginger.
Pour it into a small, oiled loaf tin. Bake for 50 to 60 minutes until a fork or skewer comes out clean.
Allow the cake to cool fully before you slice and serve it.

1 **a** How many of the ingredients are given using mass?
b Which of the ingredients are listed using volume?
c What looks different about these units of mass, compared to the ones we have been using?
d Where have you seen these units before?

2 Stephen asks his mother to help him rewrite the recipe in grams. She explains that 1 oz is approximately 30 grams. Use this to work out the gram amounts for the butter, sugar, syrup and flour.

Converting between units of mass

Key maths idea

Sometimes you need to solve problems that give measurements in different units. You need to **convert** the measurements to the same units in order to calculate them.

Example

A pickup truck can carry a load of 2 tonnes. Elijah has 480 kg of bricks and 600 kg stones on his pickup truck. How many more kg can he add to the load?

Elijah draws this table to show what he needs to work out:

2 tonnes		
480 kg	600 kg	?

But we can't subtract kg from tonnes. So we need to convert the tonnes to kg.

1 tonne = 1 000 kg

So 2 tonnes = 2 × 1 000 kg
= 2 000 kg

Now Elijah can work out the rest of his problem.

480 kg + 600 kg = 1 080 kg — The existing load is 1 080 kg.

2 000 − 1 080 kg = 920 kg — He can fit 920 kg more onto the truck.

When you convert between units of mass, you need to remember:

1 000 g = 1 kg

1 000 kg = 1 tonne

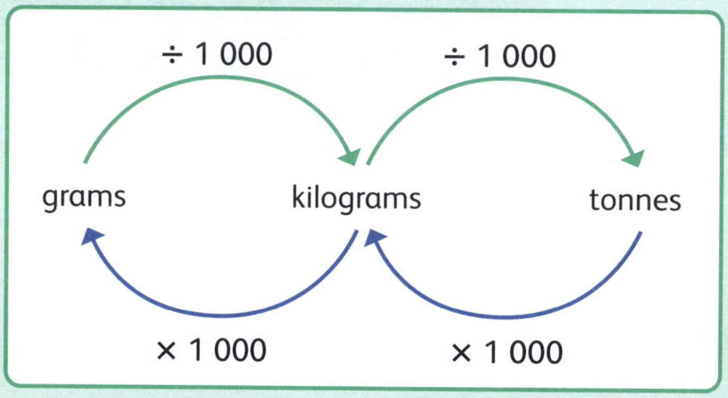

Key words

convert

express

When a question asks you to **express** an amount as a different unit, you need to convert. Sometimes you can do this mentally if you know the relationships between the units.

1 Express the following amounts in grams.
a 1 kg b 10 kg c 100 kg d 1 tonne

2 Express the following amounts in kg.
a 1 tonne b 5 tonnes c 10 tonnes d 100 tonnes

3 Copy and complete with the equivalent mass.
a $\frac{1}{2}$ kg = ____ g b $\frac{1}{4}$ kg = ____ g c $\frac{1}{10}$ kg = ____ g
d $\frac{1}{2}$ tonne = ____ kg e $\frac{1}{4}$ tonne = ____ kg f $\frac{1}{10}$ tonne = ____ g

4 Convert these amounts to kilograms:
a 5 tonnes b 25 tonnes c $2\frac{1}{2}$ tonnes d 100 tonnes
e 1 500 g f 6 000 g g 12 500 g h 14 000 g

5 Convert these amounts to tonnes:
a 7 000 kg b 13 000 kg c 8 500 kg d 9 500 kg

6 Work out the total in grams:
a $\frac{1}{4}$ of a kilogram + $\frac{1}{2}$ a kilogram
b $\frac{7}{10}$ of a kilogram + 450 g
c 150 g + 749 g + $\frac{1}{2}$ kg

7 Solve the problems.
a A farmer produces 875 kg of onions, 940 kg carrots and $\frac{1}{2}$ tonne of potatoes. What is the total mass of the produce?
b A builder orders 2 550 kg cement, 900 kg sand and 790 kg stones. What is the total mass of the delivery?
c A shipping container of fruit contains 22 tonnes of fruit. The container is packed with crates, and each crate weighs 24 kg. What is the maximum number of crates that can fit in the container?

8 Write your own rule for:
a converting from grams to tonnes
b converting from tonnes to grams.

Maths detective: The metric system

Do some detective work to understand how our system of measurement works.

Case 1

1 These units have been mixed up. Sort them into metric and imperial measures. You may need to use the internet, a dictionary or any other source. Draw up a table with two columns to show which system each unit belongs to.

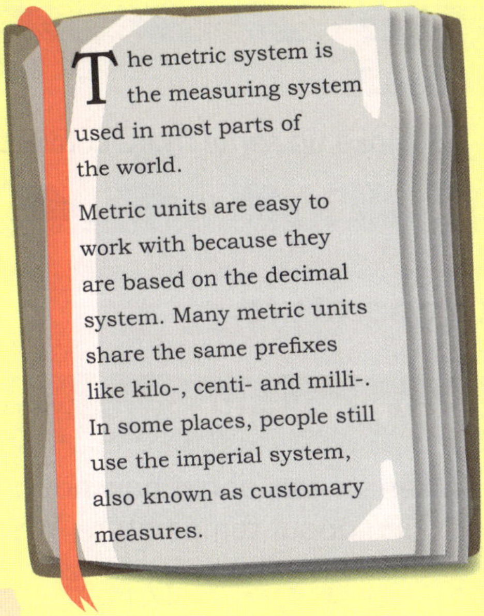

gallon centimetre foot yard
pint kilometre kilolitre ounce
gram millilitre inch litre quart
kilogram pound millimetre mile

2 Use the internet, look at books or ask an adult to help you work these out.

a How many cm are equal to one inch?

b How many feet are equal to one metre?

c How many grams are equal to one ounce?

Case 2

Another page in the book has this table of units from the metric system. Use the clues to solve this puzzle.

- The base unit is the unit all the other units in the row are based on.
- There are three units on each side of the base unit.
- The units in each column have the same prefixes.
- As you move to the right in each row, each unit is worth 10 times more than the one before. So a decalitre (dal) is equal to 10 litres.

1 Copy and complete the table, using these clues.

Length			decimetre (dm)			hectometre (hm)	kilometre (km)
Mass		centigram (cg)			decagram (dag)		kilogram (kg)
Capacity	millilitre (ml)			litre (ℓ)	decalitre (dal)		

2 Work out the answers to these conversions.

a 5 dag = ____ g

b 1 000 ℓ = 1 ____

c 200 ____ = 2 km

d 500 ml = 5 ____

Case 3

1 Make up some more conversions of your own and challenge a partner to solve them.

Temperature

Key maths idea

When we talk about how hot or cold something is, we are talking about temperature. We measure **temperature** in **degrees Celsius (°C)**.

Water boils at 100 °C.

A hot pot of brewed tea is around 60 °C.

Water freezes at 0 °C.

The instrument used to measure temperature is called a **thermometer**. Older thermometers contain a liquid called mercury in a thin tube. As the temperature gets higher (warmer), the mercury moves up the tube. You can read the temperature from the scale alongside the mercury. Modern thermometers give the temperature as a reading on an LED screen.

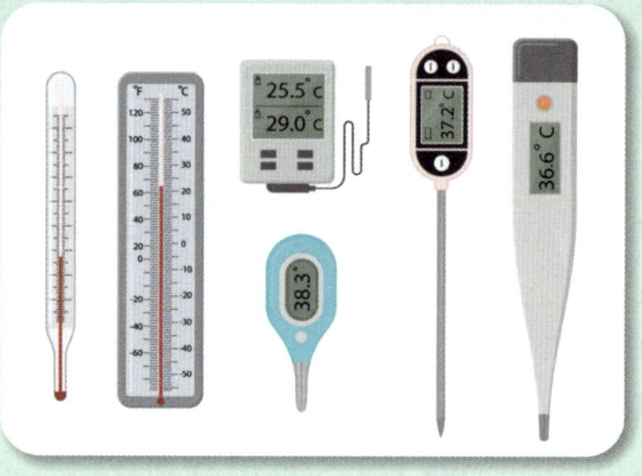

Key words

temperature

degrees Celsius (°C)

thermometer

Work in pairs. Use the information about temperatures above to help you choose the correct answer.

1 The temperature of a healthy human body is about:

a 3 °C b 37 °C c 73 °C.

2 The temperature of a cold milkshake is:

a lower than 10 °C b around 20 °C c around 100 °C.

3 A comfortable room temperature is between:

a 17 and 23 °C b 29 and 39 °C c 0 and 15 °C.

Working with temperatures

Key maths idea

Example

On Monday the temperature was 19 °C. On Tuesday it was 24 °C.
What was the difference in the temperatures?

Look at the different ways these students worked out the answer.

A

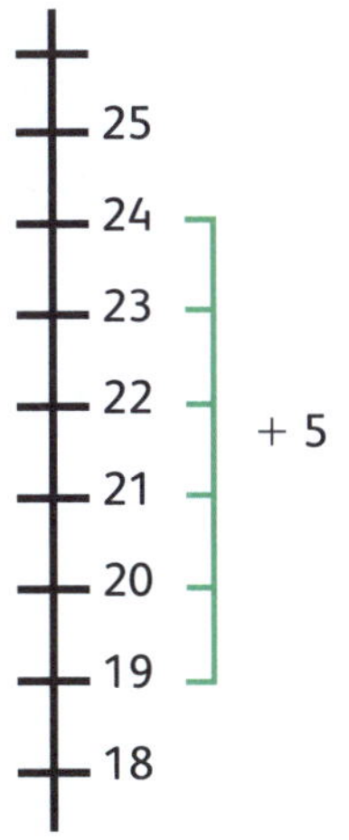

I used a number line.

B

24 − 19 = ?

$$\begin{array}{r} \overset{1}{\cancel{2}}\overset{1}{4} \\ -19 \\ \hline 5 \end{array}$$

I used a subtraction.

C

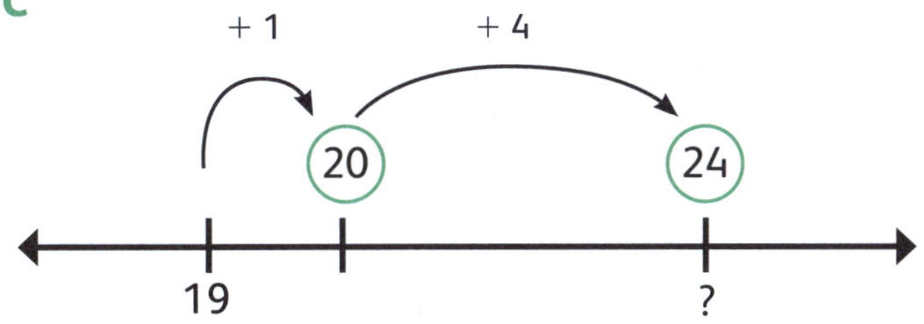

I used 'friendly numbers'.

Remember: the higher the number, the warmer the temperature.
The lower the number, the cooler the temperature.

1. The temperature at 10:00 a.m. is 14 °C. It rises by 10 °C by noon. What is the temperature at noon?
2. A diver measures the sea temperature at the surface, where it is 23 °C. She measures the temperature again several metres underwater, where the temperature is 5 °C cooler. What is the underwater temperature reading?
3. The water in a kettle is 100 °C when it is boiling. Five minutes later, it has cooled by 20 °C. What is the new temperature?
4. The temperature in a greenhouse is 32 °C at 2:00 p.m. It drops by 2 °C each hour after that until 11:00 p.m. What is the temperature at 6:00 p.m.?

5 A weather forecast often gives the maximum and minimum temperatures for the day. The maximum is the highest temperature expected that day. The minimum temperature is the lowest temperature expected that day. The difference between the temperatures is called the range.

SUN	MON	TUE	WED	THU	FRI	SAT
°C	°C	°C	°C	°C	°C	°C
▲25	▲27	▲24	▲21	▲22	▲25	▲23
▼19	▼20	▼16	▼15	▼16	▼20	▼18

a Which day will have the highest possible temperature?

b Which day has the greatest difference between the high and low temperatures?

c Which day has the lowest possible temperature?

d Which days have a range of six degrees?

6 Read the temperatures on these thermometers in °C. Then write them down.

a

b

c

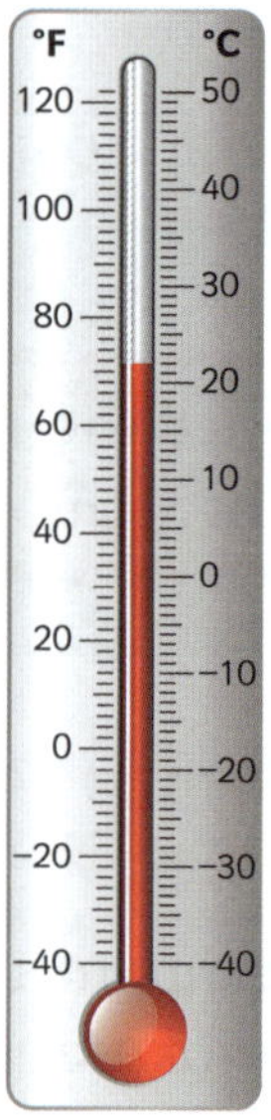

d

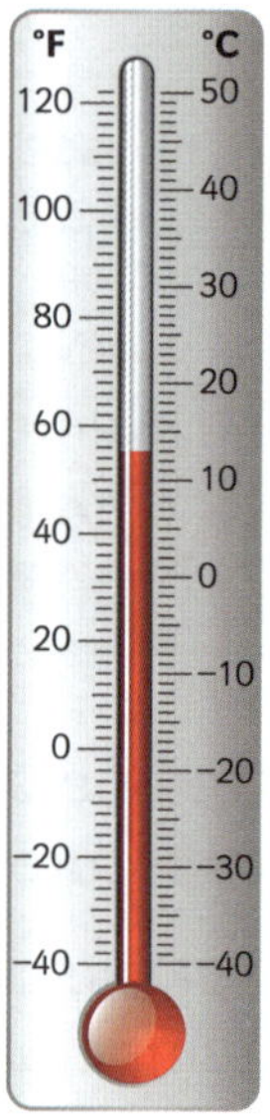

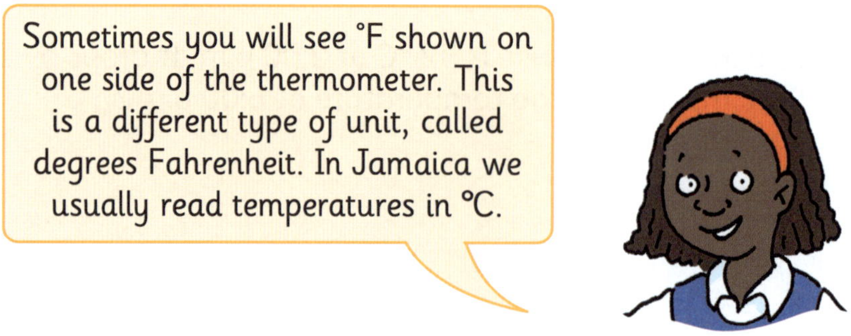

7 Your teacher will show you a real thermometer and will set up some cups with water at different temperatures – cold (with ice), at room temperature, warm and hot. Practise reading the temperature from the different cups.

8 As a class, record the temperature at one place in your school. Do this each day for a week. Write down the temperature each day. Use your findings to draw a bar graph.

Calculating with units

Key maths idea

The **prefix** of each unit can help you know how to convert between units.

Prefix	Means
milli-	one thousandth
centi-	one hundredth
deci-	one tenth
kilo-	one thousand

Key word

prefix

1 Work in a small group. Talk about this picture of the ladder method of converting metric units. Make up a catchy song or poem that can help you remember the information.

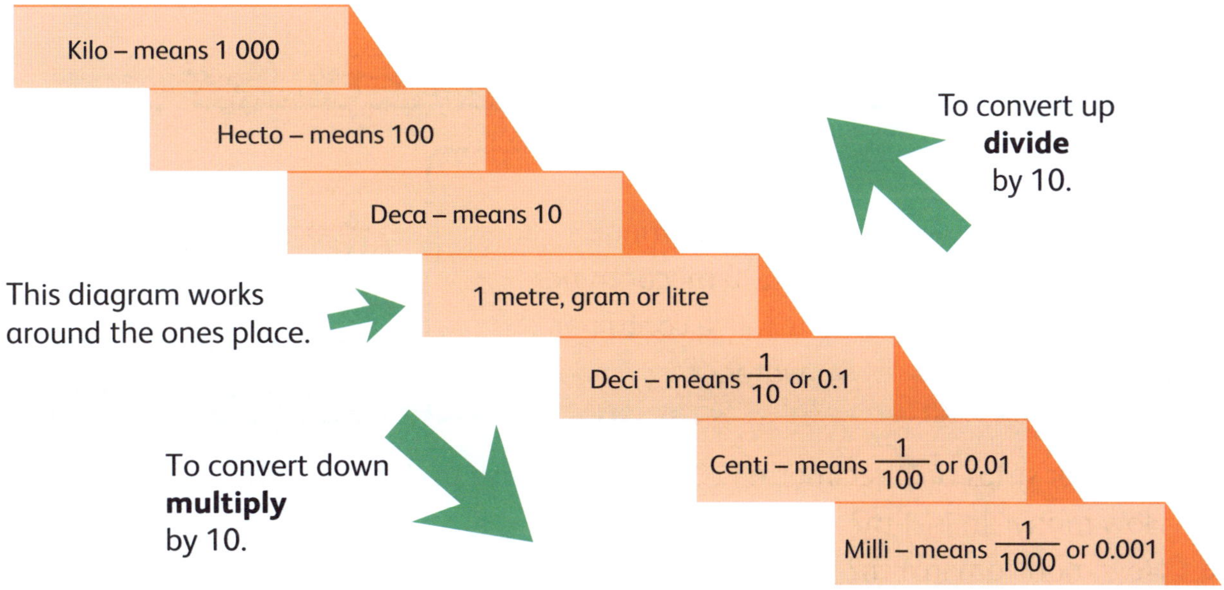

Converting between units

Key maths idea

These diagrams can also help you when you need to convert between units.

- Look at the blue arrows and numbers. When you change from a **bigger** unit to a **smaller** unit, you must multiply.
- Now look at the red arrows and numbers. When you change from a **smaller** unit to a **bigger** unit, you must **divide**.

Example 1

Jack ran 2 389 m and Josie ran 3 km. How much further did Josie run?

3 km × 1 000 = 3 000 m

3 000 − 2 389 = 611 m

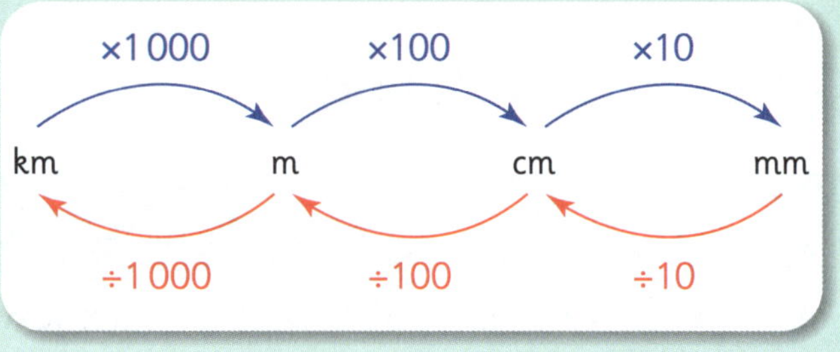

Example 2

Dan has a 5 kg bag of flour. He uses 685 g for making bread. How much is left?

5 kg × 1 000 = 5 000 g

5 000 g − 685 g = 4 315 g

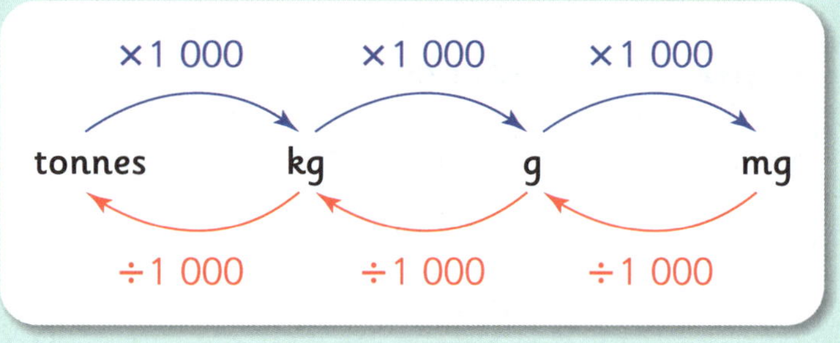

Example 3

Gemma has a bottle that holds 625 ml water. She refills the bottle 3 times. How many litres and millilitres does she use?

625 ml × 3 = 1 875 ml

1 875 ÷ 1 000 = 1 ℓ 875 ml

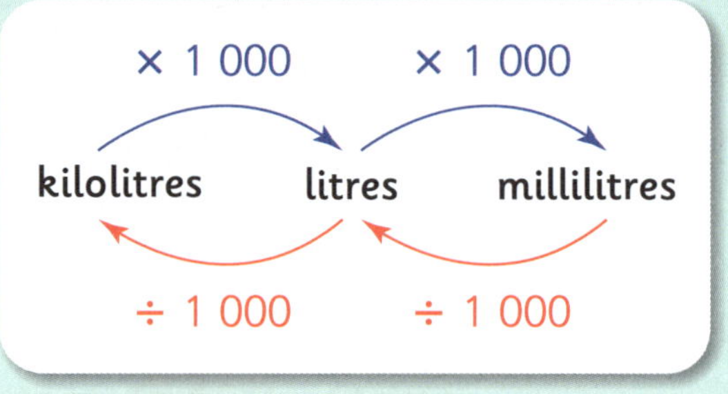

You will need:

- paper strips
- a ruler
- scissors
- tape
- a 10 ml syringe
- a 100 ml container
- a 1 litre container
- water.

When you divide a number by 1 000, the decimal point moves three places to the left.

1 Make paper strips of 1 decimetre (10 cm). Use these to demonstrate:

  a How many cm in 1 m?

  b How many dm in 1 m?

2 Measure out 10 syringes of water. Use these to fill a 100 ml container. Use the 100 ml container to fill the 1-litre container.

- a How many ml in 1 ℓ?
- b How many dl in 1 ℓ?

3 Work in pairs to complete the following tasks.

- a Discuss when you use the units of time, capacity, temperature, length and mass in your daily life.
- b Draw a table like this. Make notes about ways you use these forms of measurement:

Time	**Capacity**	**Temperature**	**Length**	**Mass**

- c Over a period of 7 days, observe the ways in which you use each type of measurement. Add more notes to your table.
- d Present the table to the class.

4 Express 1 kilogram in:

- a grams
- b milligrams
- c decagrams.

5 Copy each measurement and then convert it to find the missing numbers.

- a 3 km = _____ m
- b 2 cm = _____ mm
- c 4 kg = _____ g
- d 8 ℓ = _____ ml
- e 5 m = _____ cm
- f 3 tonnes = _____ kg

6 Convert each measurement to find the missing numbers.

- a 6 000 g = _____ kg
- b 3 000 kg = _____ tonnes
- c 5 000 ml = _____ ℓ
- d 1 323 mm = _____ m
- e 2 604 g = _____ kg
- f 5 431 cm = _____ m

7 Convert each measurement to find the missing numbers.

- a 130 000 m = _____ km
- b 6 km = _____ m
- c 2 100 mm = _____ cm
- d 3 ℓ = _____ ml
- e 2 050 g = _____ kg
- f 10 cm = _____ mm

Real-world maths: Calculations with units

Use what you have learned about units of measurement to solve these problems.

1 The Williams family is driving home after their holiday. They drive 126 km in the morning and 209 km in the afternoon. They left at 10:25 a.m. and arrived home at 3:00 p.m. after a quick stop along the way for lunch.

   a How many kilometres did they travel altogether?

   b Now write your answer in metres.

   c How many hours passed from the time they left to the time they got home?

2 Two builders are working on a building site. One moves a pile of bricks with a mass of 43.7 kg and the other moves a pile of bricks with a mass of 46.2 kg.

   a What is the total mass of the bricks they moved?

   b They are going to need triple the amount of bricks. What will the total mass of bricks be?

   c Convert 43.7 kg and 46.2 kg into grams.

3 An environmental disaster has happened. A ship is leaking 25 ℓ of oil into the sea every 10 km of its trip.

   a If the ship travels 3 000 km, how many litres of oil is spilled into the sea during this time?

   b Work out how many kilolitres this would be.

4 A picture is 290 mm wide and 21.5 cm long. The frame is 0.29 m wide and 23 cm long. The wood part of the frame is 25 mm wide.

   a Convert all the measurements to the same unit.

   b How many centimetres wide is the frame?

   c How many centimetres wide is the picture?

   d How much longer is the frame than the picture?

   e Will the picture fit into the frame? Explain your answer.

What I have learned

Copy and complete this mind map about the different kinds of measuring you have learned about. For each kind of measure (mass and temperature), list the main units we use, give some questions that we can ask, list measuring instruments you have seen or used, and give an example of at least two measurements.

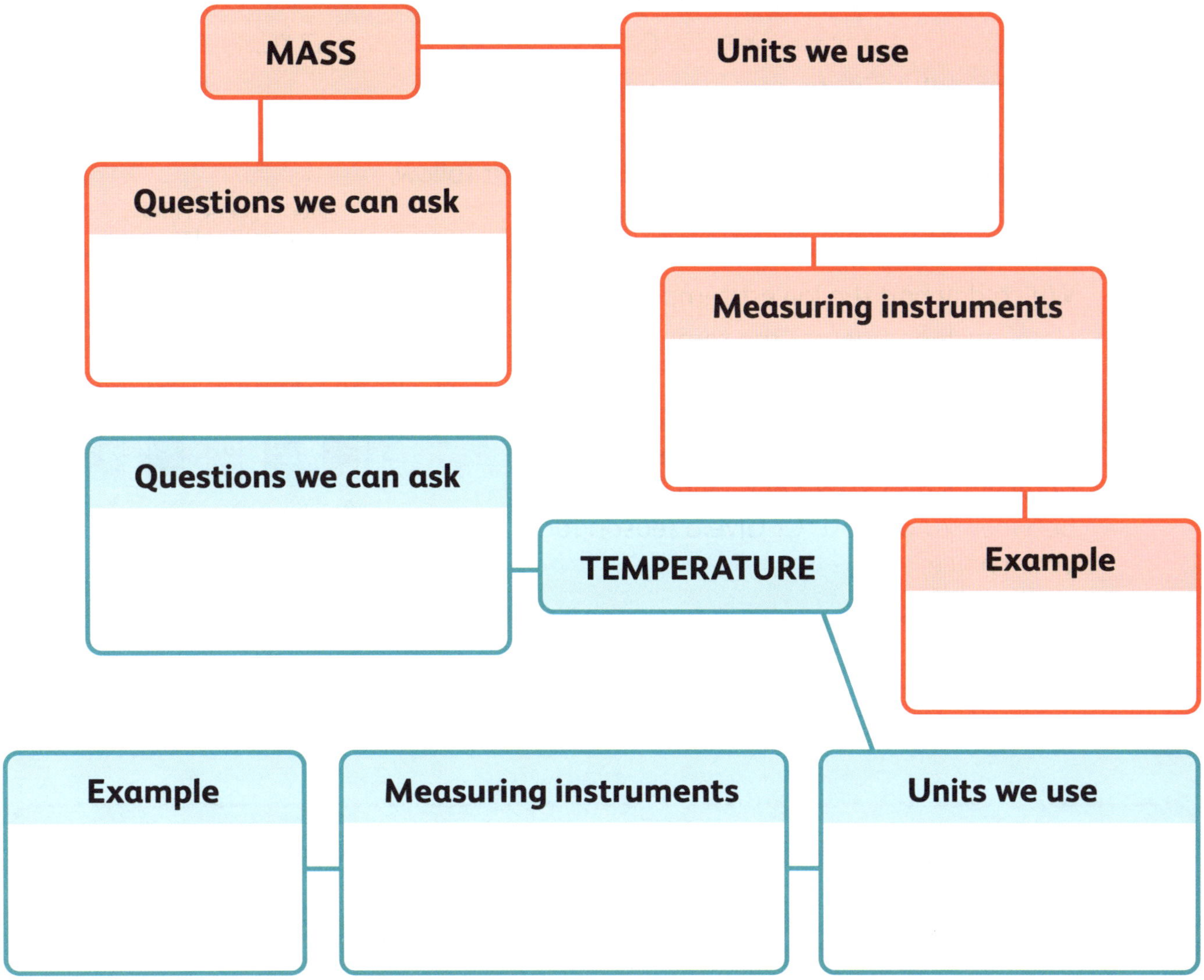

Practice questions

1 Write the unit of mass that is equal to:

a $\frac{1}{1\,000}$ of 1 kg b 1 000 kg c 1 000 mg

2 Write the best approximate mass for a pineapple.

a about 10 g b about 1 kg c about 5 kg

3 Write the best approximate mass of an orange.

a about 100 g b about $\frac{1}{2}$ kg c about $\frac{1}{4}$ tonne

4 Write the best approximate mass of a watermelon.
- a about 30 g
- b about $\frac{1}{3}$ tonne
- c about 3 kg

5 The temperature of a healthy human body is about:
- a 37 °C
- b 50 °C
- c 72 °C.

6 The temperature of boiling water is about:
- a 100 °C
- b 79 °C
- c 43 °C.

7 Room temperature in summer is around:
- a 11 °C
- b 44 °C
- c 24 °C.

8 Look at this bar graph. Answer the questions that follow.
- a Which days had temperatures higher than 25 °C?
- b Which day had the lowest temperature?
- c How much higher was the temperature on Tuesday than on Monday?
- d Which do you think is the most likely prediction for Saturday's temperature: between 5 and 10 °C; between 0 and 10 °C; between 20 and 30 °C? Give a reason for your choice.

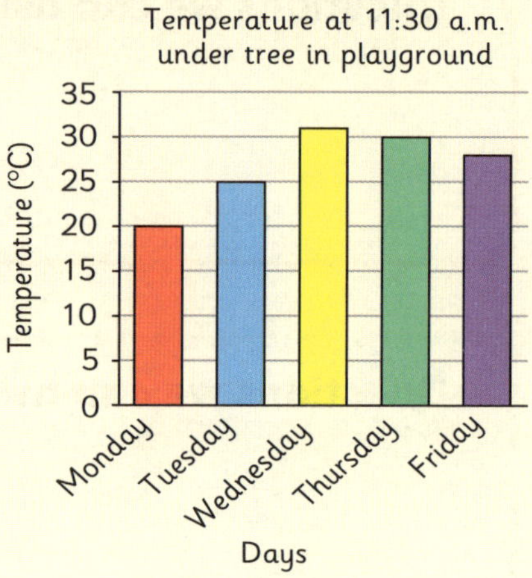

6 Self check

Check you can do this: 	If you have difficulty, try this: 
1 Estimate, measure and record mass using grams or kilograms, or kilograms and grams, and read a scale.	Go back to pages 77 and 78. Revise the work you did on estimating and measuring mass.
2 Understand that 1 tonne = 1 000 kg.	Revise the work you did on pages 79 and 80, converting between kg and tonnes.
3 Estimate and measure temperature in degrees Celsius and solve problems involving temperature.	Revise the work on pages 82 to 84, about temperature.

Unit 3: Geometry

Chapter 7 Lines and angles

In this chapter you will explore these focus questions:

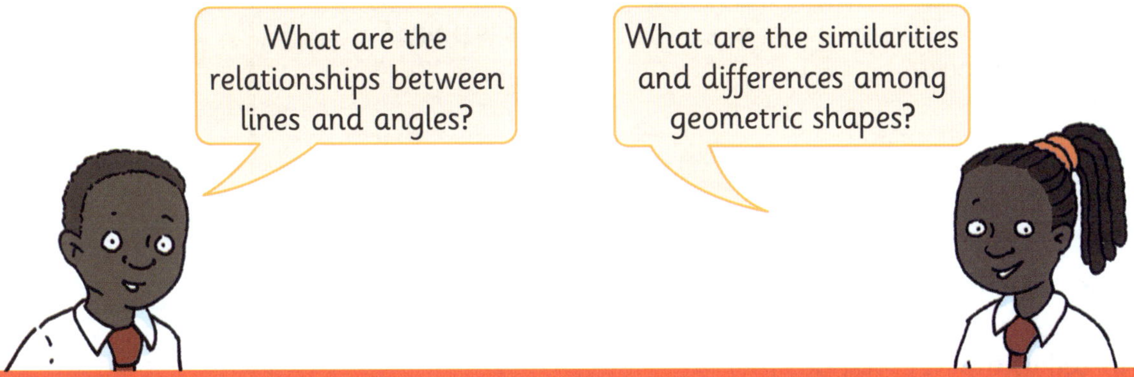

To answer these questions, you will:

- explore and name points, lines, rays and angles
- work with right angles
- identify different types of lines, including parallel and perpendicular lines
- investigate polygons including triangles and quadrilaterals
- identify congruent polygons.

Starting point

Discuss what you can see in the pictures with your class.

- Identify shapes that you can see in the pictures.
- What kinds of lines and angles do you notice?
- Which objects have curved lines? Which have straight lines?
- Identify points where two lines meet.

Revisit points, lines, paths and corners

Key maths idea

Geometry is the part of maths that deals with shapes and space. In earlier grades, you learned about points, lines, paths and corners.

A **point** is a position in space.

This is point A.

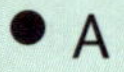

This is point B.

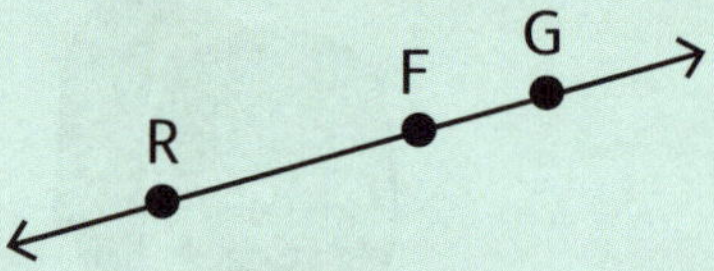

This is a straight line. It passes through points R, F and G. The arrows tell us that it continues in both directions.

This is **line segment** CD. C and D are the endpoints.

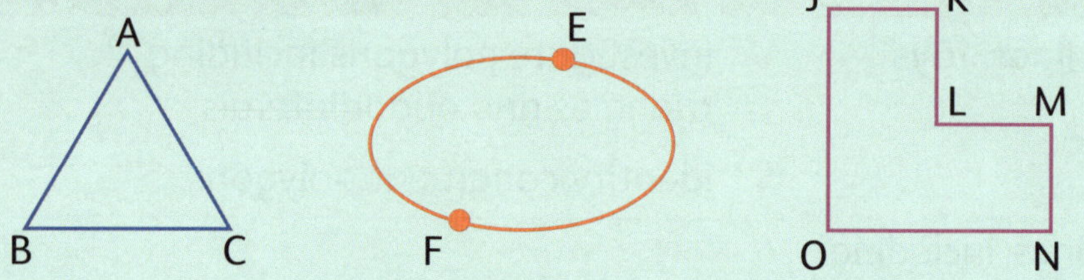

These are **closed paths**.

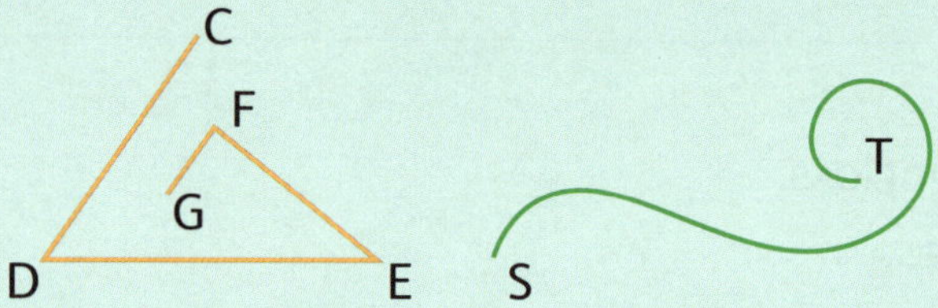

These are **open paths**.

Sometimes lines join at a **square corner**.

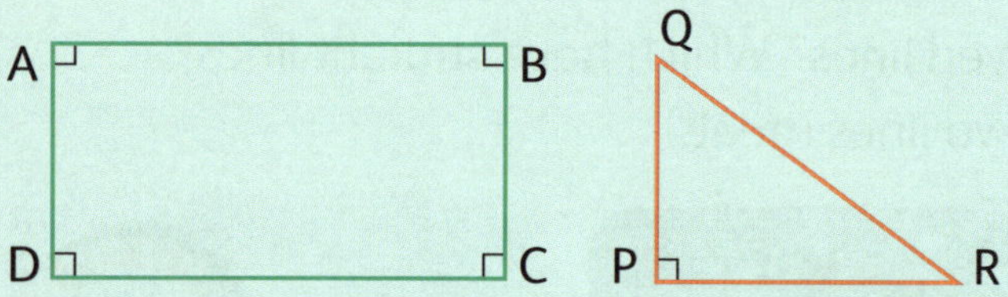

Key words

point
line segment
closed path
open path
square corner

1 Look at the diagram on the right and write down the name for each line segment. How many line segments are there?

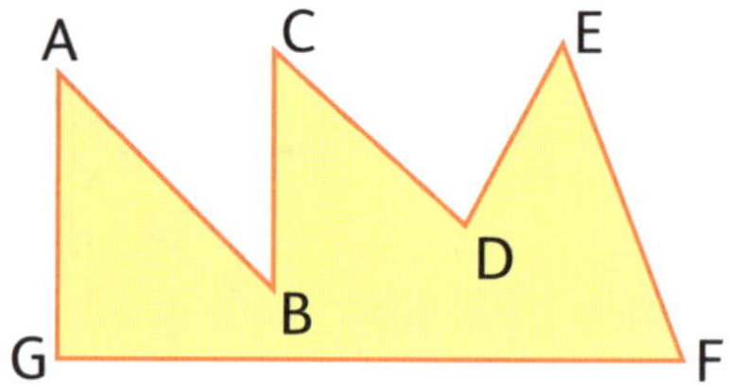

2 You will need some seeds, glue and paper. Glue seeds to make illustrations of the following terms:

a point **b** line segment **c** closed path
d open path **e** square corner **f** curved line.

3 If you have a computer, use drawing tools to illustrate the different kinds of lines you know about.

Rays and angles

Key maths idea

A **ray** is a line that starts from one **endpoint** and continues in one direction. **Intersecting** lines are lines that cross each other (intersect) at a point called the **point of intersection**. This point forms the **vertex** of the angle. When lines intersect or meet, they form an **angle**.

We name angles using points on the 'arms' of the angle (these can be line segments, rays or lines), and the vertex. In <ABC,^B is the vertex.

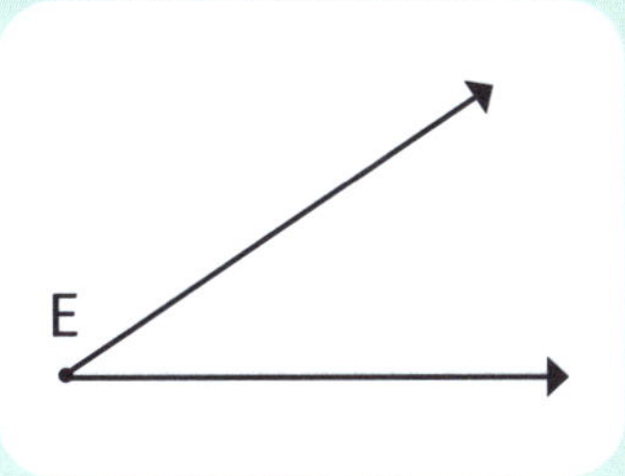

These rays meet at point E.

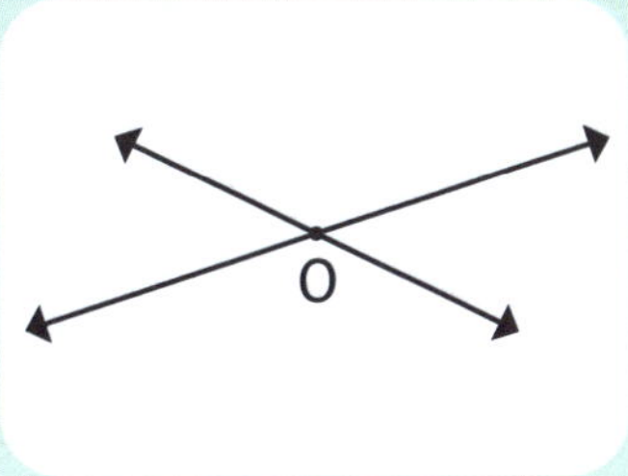

These lines intersect at point O.

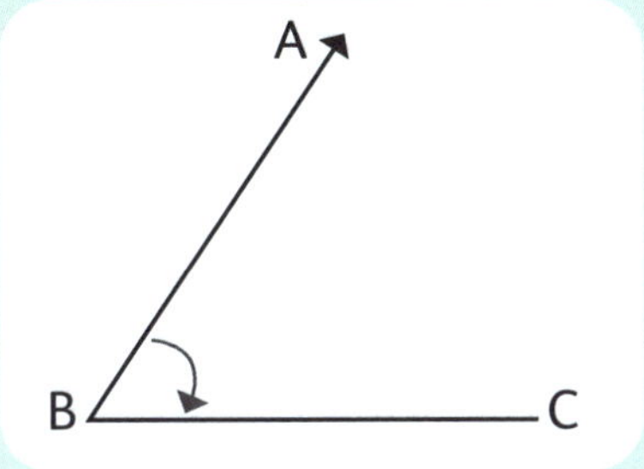

<ABC or AB̂C

Key words

ray
endpoint
intersecting
point of intersection
vertex
angle

Real-world maths: Make geostrips and investigate angles

1 Fasten strips of card, or wooden or plastic strips, to make geostrips like the ones in the picture.

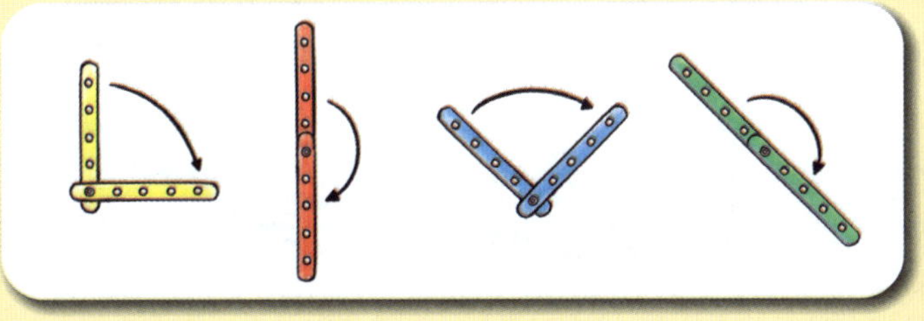

You will need:
- strips of card
- split pins or paper fasteners
- pairs of scissors.

With a partner, look around at objects in your classroom that form angles. Discuss:
- Does the length of the arms change the size of the angle? Why or why not?
- What changes the opening of the angle?

2 Complete the following sentences with the correct term.

ray	line	segment

a KL is a ____ because ____ .

b QR is a ____ because ____ .

c ST is a ____ because ____ .

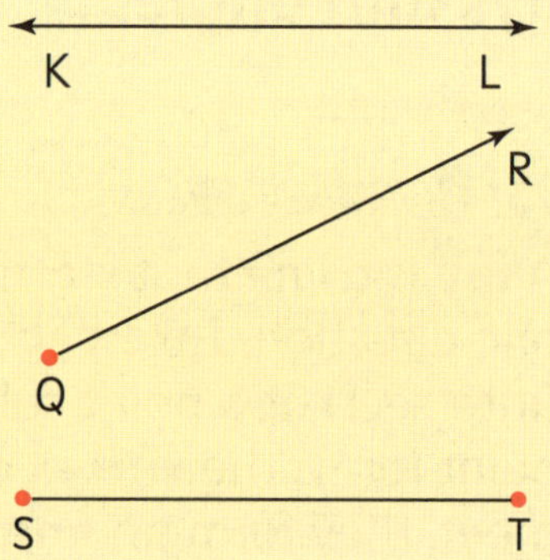

Measuring angles

Key maths idea

We measure angles in **degrees**. We write degrees using the ° symbol.

Remember, an angle is a measure of **turn**.

A **full turn** is 360°. A **half turn** is 180°. It makes a straight line.

A **quarter turn** is 90°. It is also called a **right angle**.

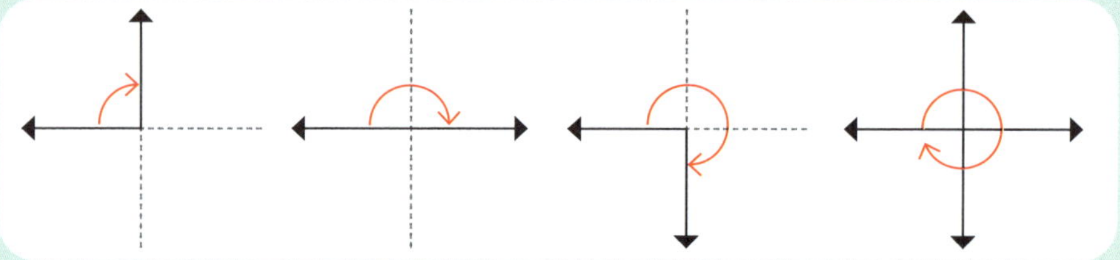

$\frac{1}{4}$ turn	$\frac{1}{2}$ turn	$\frac{3}{4}$ turn	full turn
90° right angle	180° straight line	270°	360°

Key words

degrees
turn
full turn
half turn
quarter turn
right angle

1 Use glue and seeds again. Make examples of:

a a right angle **b** a straight line **c** a $\frac{3}{4}$ turn.

2 Look around your classroom. Some objects create angles when they open or close. Work with a partner to demonstrate angles that are less than, equal to and greater than a right angle. Use the following objects to draw sketches of the angles:

a a pair of scissors **b** a door **c** a book **d** a pair of pencils

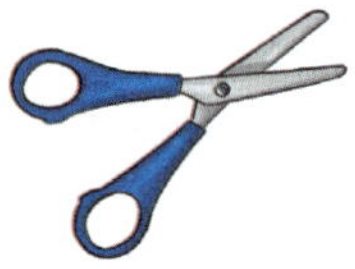

3 In pairs, find five different ways to make right angles using your bodies. Do the same with angles smaller than and greater than a right angle.

4 Make up a song about angles, with actions that show the different kinds of angles. Use the words ray, vertex, degree, turn and right angle.

Maths detective: Investigate right angles around you

Case 1

You will need several sheets of paper.

- Fold your paper in half, then over again, then again. Unfold your paper. Mark all the places where the folds make right angles.
- Take a new piece of paper. Is it possible to fold it over three times without making any right angles?
- Examine the angles you have made. What do you notice about the sizes of the angles?

Case 2

Go on a field trip to an area outside your school. Identify examples of:

- line segments
- objects that are at right angles to the ground
- objects that are at right angles to each other
- lines that form angles greater than a right angle
- lines that form angles smaller than a right angle.

Draw a sketch showing examples of what you observed.

Case 3

1 Look at these pictures of everyday items.

  a Identify the parts that are at right angles to the ground or to another surface. Explain why these right angles are important for the way the object lives or how it functions.

  b Identify other angles (smaller or greater than 90 degrees) and explain the importance of these angles.

2 Draw pictures of some objects from your environment at home or at school. Stick sticks, string or straws to show the angles on your drawing.

Geometry

Describing lines

Key maths idea

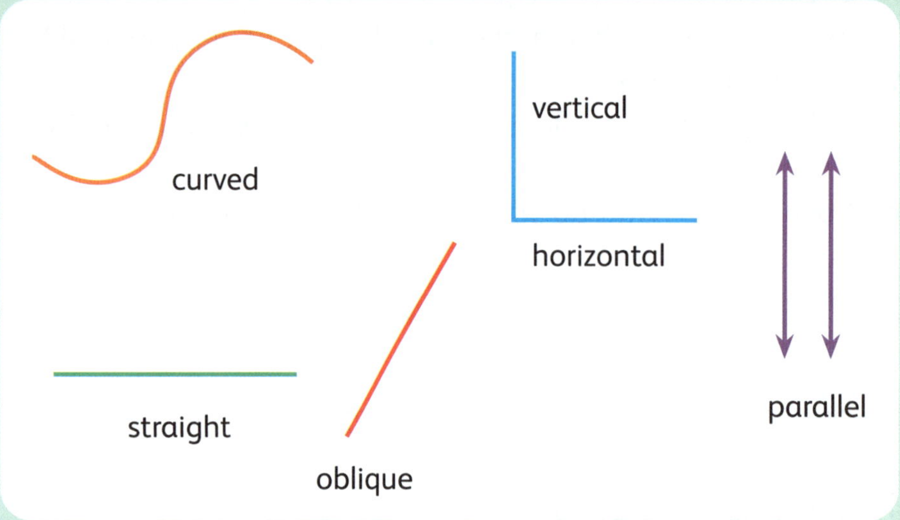

Horizontal lines run from one side to the other or from left to right.
Vertical lines run up and down or from top to bottom.
Perpendicular lines meet at a right angle (90 degrees).
Parallel lines stay the same distance apart and never cross.
Oblique lines run at a slant or an angle.

Key words

horizontal
vertical
perpendicular
parallel
oblique

1 Look at these capital letters.

a Write the letters that have curved lines only.
b Write the letters that have straight lines only.
c Write the letters that have curved lines and straight lines.
d Try to write your name using only straight lines and then only curved lines.

2 Use the words you have learned to describe what kinds of lines you can see in each picture.

a

b

c

d

e

3 Take a walk around your school grounds. Draw pictures of examples of lines, line segments and angles that you can see around you.

4 Use seeds and glue. Make examples of the kinds of lines you have learned about on this page. Remember to name your lines using capital letters. If you have the correct software for your computer, draw different types of lines.

5 Use a straw. Cut it into 9 segments. Design a closed path that has 3 vertical lines and 2 horizontal lines. Stick it onto paper or card.

6 Use a ruler to draw the following. Make sure you label them correctly:

- **a** ray EF and ray ED, meeting at an angle less than 90 degrees
- **b** horizontal line AB, perpendicular to line CD
- **c** point H on line FG
- **d** an angle made up of line segments AB and BC
- **e** vertical line CD
- **f** parallel lines JK and LM.

Exploring polygons

Key maths idea

The word **polygon** comes from the Greek words 'poly' meaning many and 'gon' meaning angle. A polygon is a closed path made from line segments that meet at **vertices** (corners). The line segments that form the outline of the polygon are called the **sides**. Most polygons are named according to their number of vertices.

A **regular** polygon has all its sides equal in length and all its inside angles equal in size.

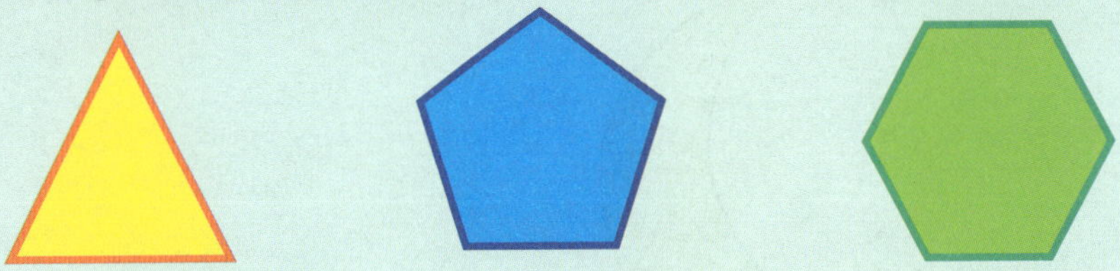

triangle **pentagon** **hexagon**

Key words

polygon
vertices
sides
regular
triangle
pentagon
hexagon
heptagon
octagon
nonagon
decagon

1 Put these shapes into two sets. Label the sets 'regular polygons' and 'non-polygons'.

2 If you have internet or a media player at school, watch a presentation about polygons. Read each of the statements about polygons and explain the mistake.

a All triangles are polygons.

b A polygon may have all sides different lengths and unequal angles.

c Not all squares are polygons.

d A circle is a type of polygon.

3 Look at these shapes with more than six sides:

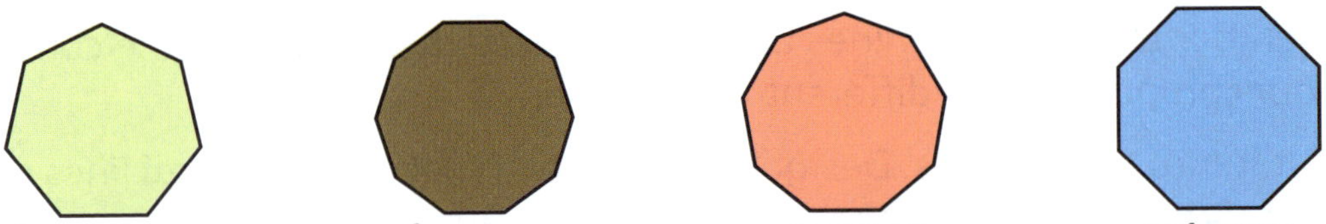

heptagon 7 sides   **decagon** 10 sides   **nonagon** 9 sides   **octagon** 8 sides

Now look at these irregular shapes. Use the terms above to help you identify them.

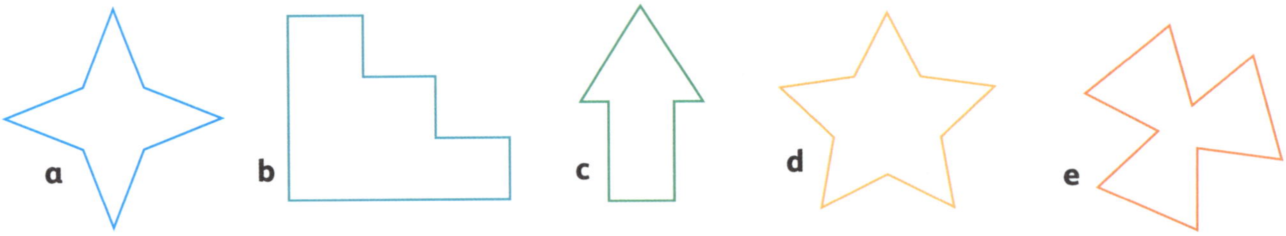

Quadrilaterals

Key maths idea

A **quadrilateral** has four sides and four inside angles. These shapes are all types of quadrilaterals.

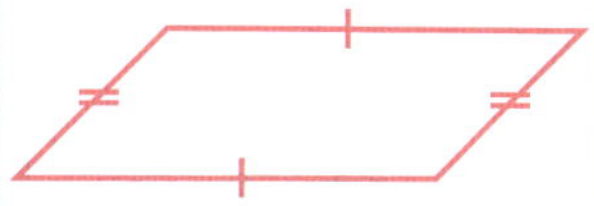

A **parallelogram** has two pairs of opposite sides that are parallel and equal in length.

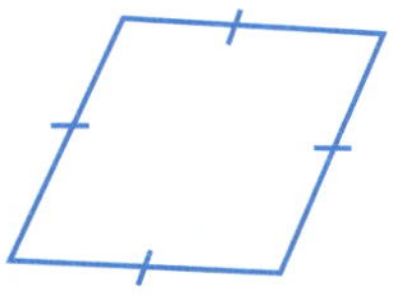

A **rhombus** has four equal sides.

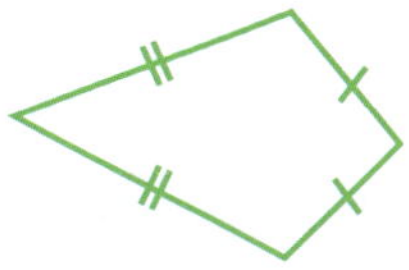

A **kite** has two pairs of equal sides and the equal sides are next to each other.

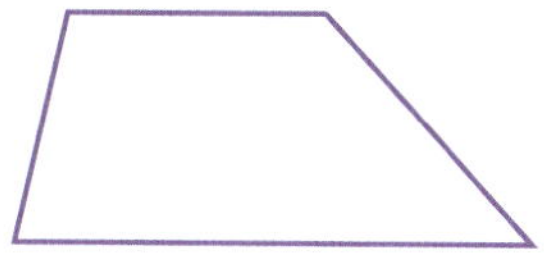

A **trapezium** has one pair of parallel sides.

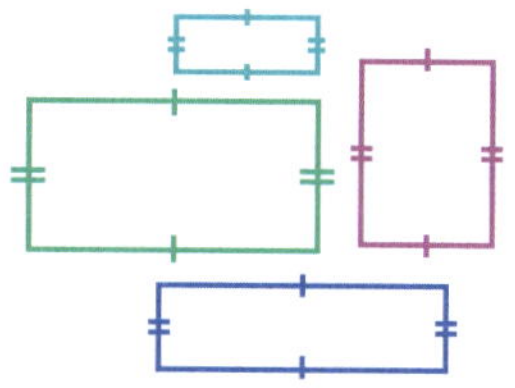

Rectangles have four right angles, and two pairs of sides that are equal in length.

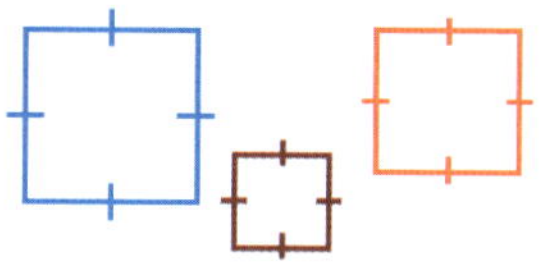

Squares have all their sides equal in length, and all four angles are right angles.

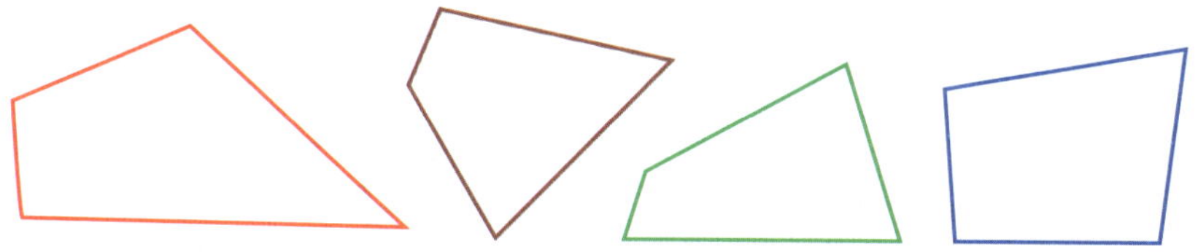

These **irregular** quadrilaterals have no parallel lines, right angles or equal sides.

Key words

quadrilateral
parallelogram
rhombus
kite
trapezium
rectangle
square
irregular

Geometry

1 Work in pairs. Identify each type of quadrilateral. Give reasons for your choice.

a

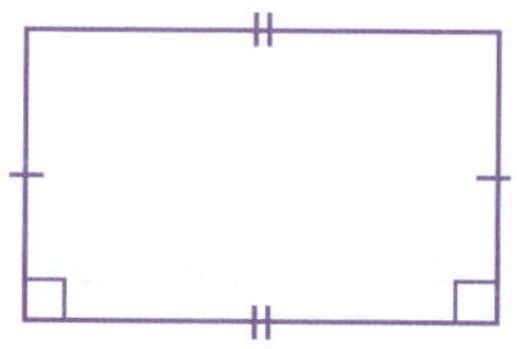

b

c

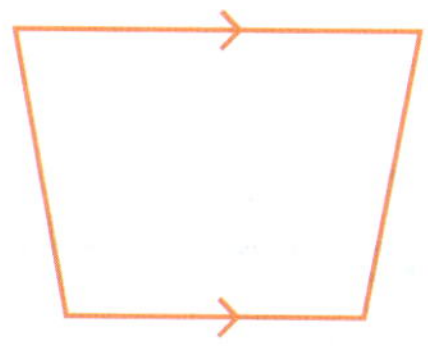

d

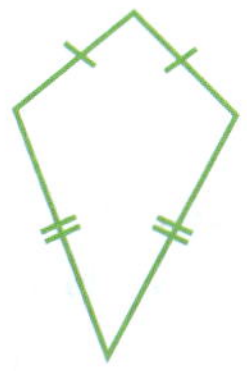

e

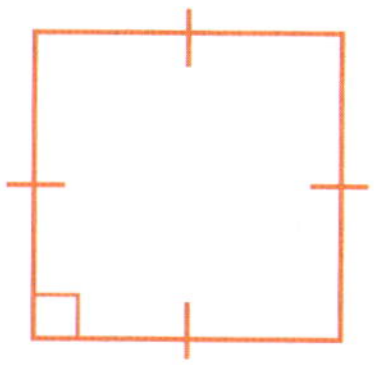

f

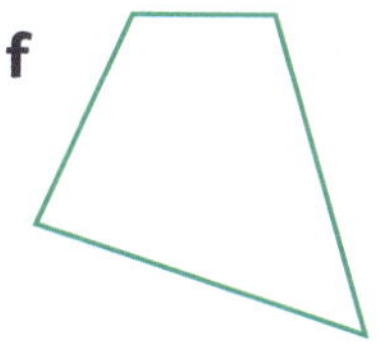

g

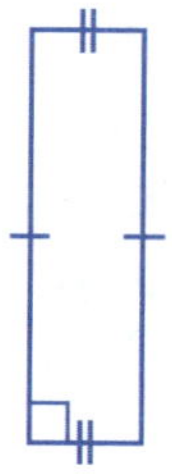

h

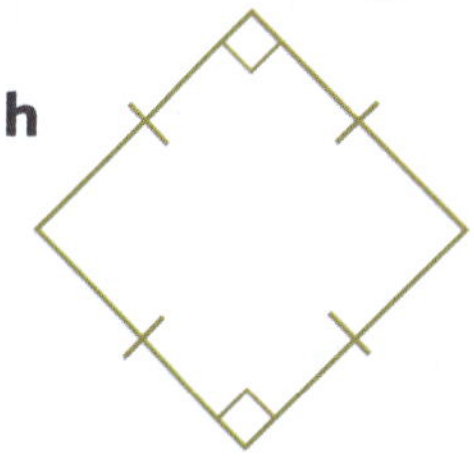

2 Elli has made some mistakes about quadrilaterals. Correct her statements.

a b

Triangles

Key maths idea

All triangles have three sides and three angles. Here are some different triangles.

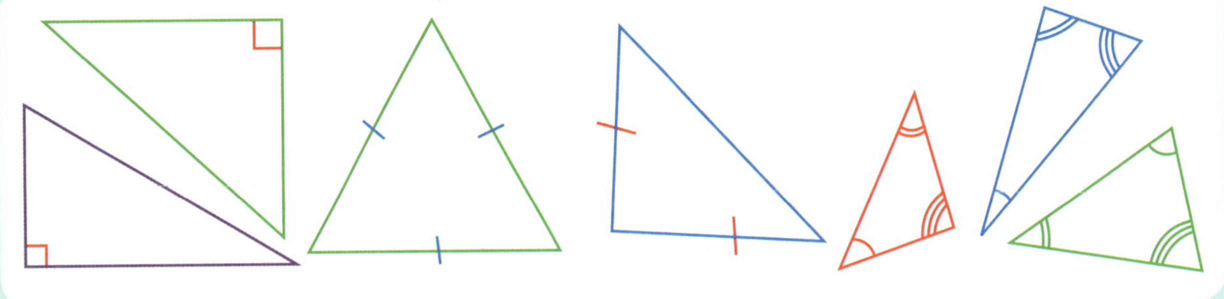

1 Talk about the triangles you can see above.

- **a** Can a triangle have more than one right angle? Try to draw it. What happens?
- **b** What do the small lines on the sides of the triangles indicate?
- **c** Which triangle above is a regular polygon? How do you know?
- **d** Why can't a triangle with a right angle be a regular polygon?

2 **a** Use a ruler and a pencil. Draw and cut out examples of the quadrilaterals you learned about on page 99 – a rhombus, rectangle, square, trapezium and parallelogram, as well as at least two different irregular quadrilaterals.

- **b** Use your ruler to help you draw a straight line from one corner to the opposite corner of each quadrilateral. Cut them in half to make triangles.
- **c** Describe what you notice about the triangles you have made.

You will need:

- a ruler
- a pencil
- a pair of scissors
- card or paper
- glue.

3 Use folding or drawing to explore how you can make triangles and quadrilaterals from other shapes. If you have pattern blocks like the ones in the picture, you can use these to build up different shapes.

- **a** What kinds of triangles can you make by folding a sheet of paper once, or drawing a single line across a quadrilateral?
- **b** What kind of quadrilaterals can you make with a single fold or line across a rectangle?

c Use four triangles to make a single, bigger triangle. What properties do the four triangles need?

d Use four triangles to create a quadrilateral. How can you rearrange them to make a bigger triangle?

e Try to use triangles to create some other polygons you have learned about (such as octagons, decagons and nonagons).

These are pattern blocks.

Can you see how this student has used the polygons to make a bigger polygon?

4 Use triangles and quadrilaterals to create a picture.

5 Measure the sides of these triangles and look at the angles to sort them into three or more groups. Compare your triangle groups with another group's. Explain to each other why you decided to group the triangles in this way.

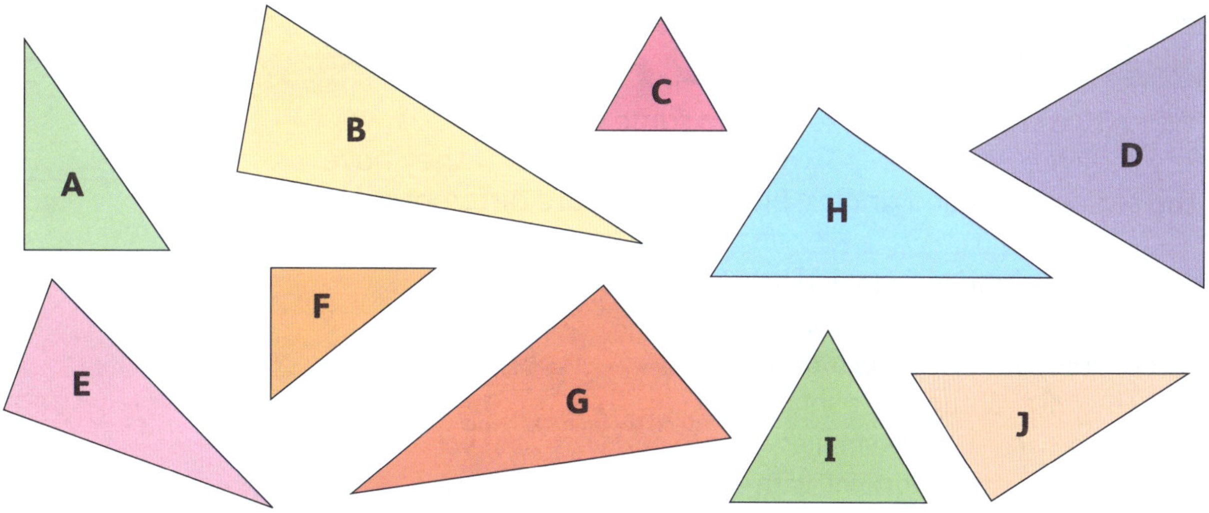

6 Have some fun finding triangles inside a hexagon. Use a hexagon pattern block to complete the tasks below.

a Trace around the hexagon shape until you have six separate hexagon drawings.

b Use a ruler to divide each of the hexagons into triangular parts. Each hexagon should be broken up in different ways but must have at least one triangle shape inside.

Here are two possible ways:

7 Name and draw the quadrilaterals.

a I have at least one pair of parallel sides.

b I have a pair of opposite sides that are the same length.

c I have two pairs of parallel sides.

d I am a parallelogram with a right angle.

e I am a parallelogram with all sides equal.

8 You used tangram pieces in Chapter 3 when you learned about fractions. This student has used all seven pieces to make a parallelogram. Can you find some other ways of making quadrilaterals and triangles using all seven pieces?

Congruent shapes

Key maths idea

When two shapes have exactly the same shape and size, we say they are **congruent**. Two shapes are **similar** if they have the same shape and angles, but their sizes are different.

Look at these hexagons. The four hexagons in the middle are congruent with each other. They are all the same shape and size. The five smaller hexagons are congruent with each other. But the smaller hexagons are similar to the larger hexagons. Their shape is the same, but their size is different.

Key words

congruent
similar

1 Look at each set of shapes. Say whether they are similar, congruent or neither.

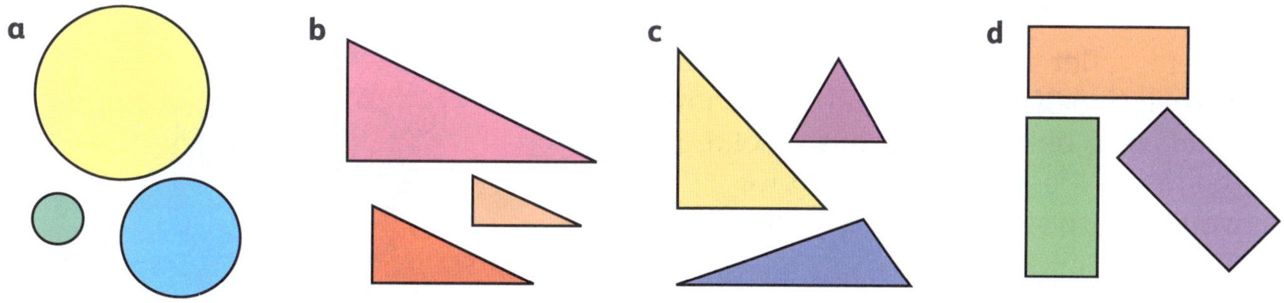

2 Go back to your tangram pieces. Which pieces are congruent? Which are similar?

3 Work in groups. Choose a polygon shape – triangle, rectangle or hexagon. Draw and cut out one set of congruent shapes, and one set of similar shapes. Exchange with another group. Check that each set is really congruent or similar.

4 True or false? Discuss with a partner. Give reasons for your answer.

- a All circles are similar.
- b All circles are congruent.
- c If two squares both have sides 5 cm long, they are congruent.
- d All right angled triangles are similar.

5 Look at the packaging boxes in the picture. Which boxes look similar? Which look congruent? Discuss with a partner.

Real-world maths: Making boxes

A net is a flat shape that we can fold to make into a 3D solid. In maths, we show a net using only the faces of the solid. But in real life, we add tabs that we would use to stick the box together.

1 Joanne is working at a box factory. Her first assignment is to design nets to make cube-shaped boxes. She comes up with one design. Her manager is pleased, but tells her that there are 11 possible nets.

Work in groups. Sketch as many other nets as you can. Can you get the other ten?

2 For each net, identify how many polygons it uses, which are congruent and which are not congruent, and describe the polygons.

What I have learned

Some key words are missing from this paragraph. Work out the missing words. Write them on a large sheet of paper and draw a diagram or write an example that explains each word.

A **r**_____ has one endpoint and an arrow on the other end indicating that it continues in that direction. When two lines **i**_____ or meet, they form a corner called a **v**_____ . The measure of turn between the lines is called an **a**_____ . It is measured in degrees. A **r**_____ measures 90 degrees. Two of these make a **s**_____ and four make a full **t**_____ .

P_____ lines stay the same distance apart and never cross, but **p**_____ lines meet at 90 degrees. We also learned about **v**_____ lines, **h**_____ lines and **o**_____ lines.

A **p**_____ has sides that are made up of line **s**_____ . A four-sided shape is called a **q**_____ . We learned about some different kinds: a **p**_____ , a **r**_____ , a **k**_____ and a **t**_____ . **S**_____ and **r**_____ have sides that meet at 90 degrees.

Shapes are **c**_____ if their shape and size is identical. If their shape is the same but their size is different, they are **s**_____ shapes.

Practice questions

1 Write a sentence explaining the meaning of each term:

a point b line c line segment d ray

2 Look at the street map of Anytown, Jamaica. Some of the streets that run parallel or perpendicular to each other or intersect each other are described next to the map.

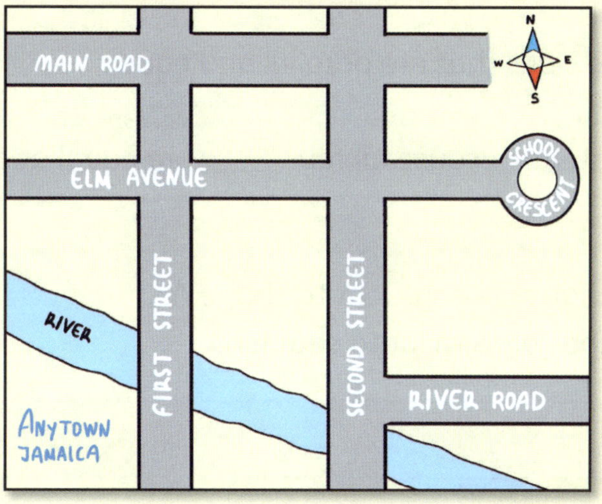

Main Road and Elm Avenue are parallel to each other.
Second Street intersects Main Road and Elm Avenue.
River Road is perpendicular to Second Street.

a Plan and then draw your own street map of an imaginary place. Include at least 2 parallel streets, 4 perpendicular streets and 3 streets that intersect others.

b Name the streets.

c Colour parallel streets purple.

d Colour perpendicular streets red.

e Colour streets that intersect yellow.

3 Describe or sketch something from your environment that has:

a right angles

b angles greater than a right angle

c angles smaller than a right angle.

4 Sort these shapes and draw them in two different sets. Name one set 'polygons' and the other set 'non-polygons'.

5 Choose the correct words to complete the sentences.

square	four	squares	parallelogram	triangles
sides	sides	regular	rectangles	angles

a All shapes with three sides are called _____ .

b _____ and _____ are quadrilaterals.

c A _____ has two pairs of opposite sides that are parallel and equal.

d A rhombus has _____ equal _____ .

e A _____ is the only _____ quadrilateral because all four _____ are equal and all four _____ are equal.

6 Look at this collection of shapes.

a Write the letters of two shapes that are congruent to shape A.

b Write the letters of two shapes that are similar to shape B.

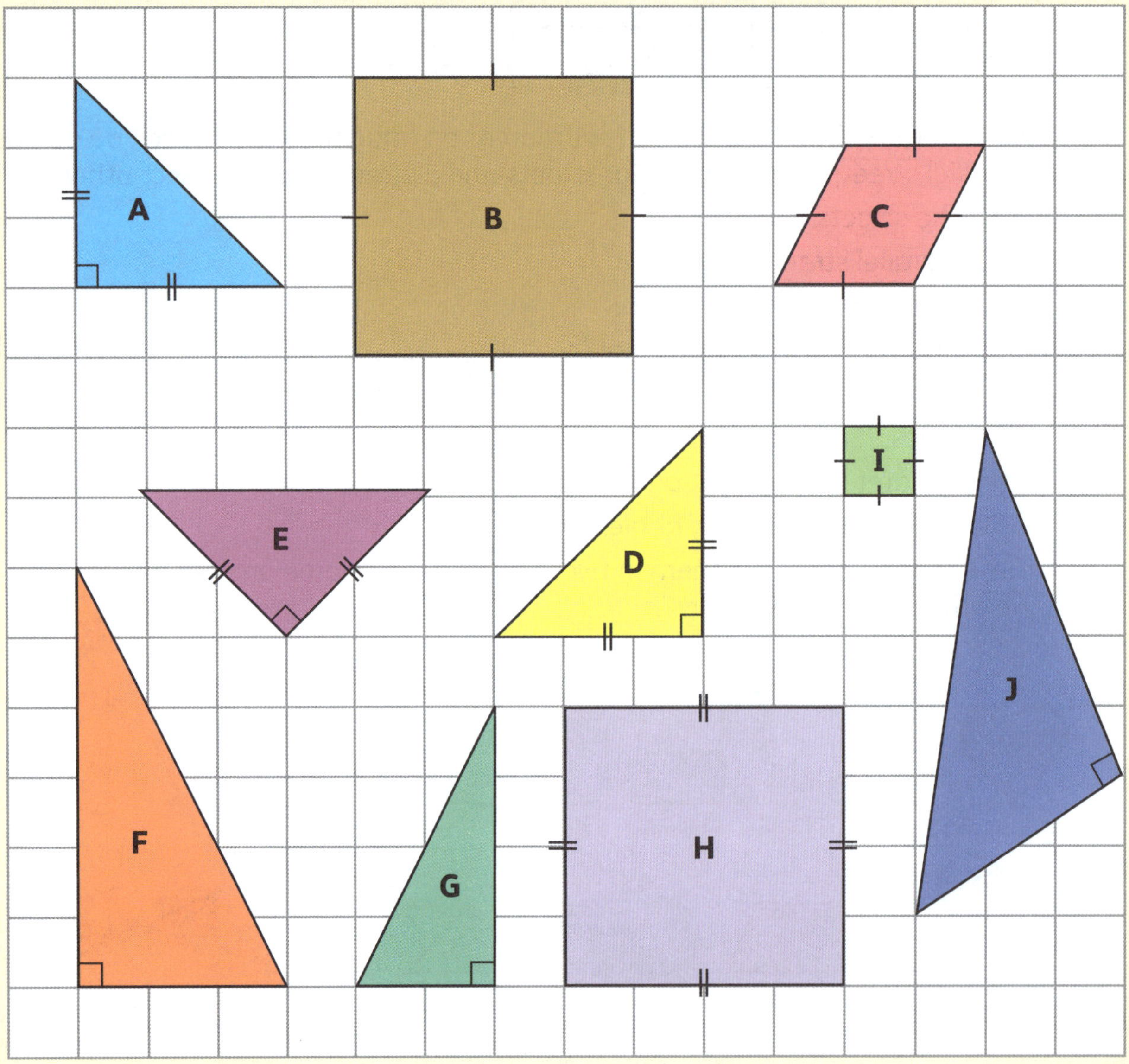

7 Self check

	Check you can do this:	If you have difficulty, try this:
1	Explain what a point, line segment and line are, and do a sketch of each.	Go back to page 92 and revise points, lines, paths and corners.
2	Identify a ray and an angle, and explain how angles are formed and how we name them.	Go back to page 93 and revise rays and angles.
3	Identify right angles and angles greater than or less than a right angle, in drawings and in the environment, and sketch examples.	Revise the work on pages 94 and 95.
4	Identify lines as curved, straight, oblique, horizontal, vertical, parallel and perpendicular and sketch examples.	Revise the work on pages 96 and 97.
5	Identify which shapes are polygons and which are not.	Revise page 98.
6	Describe different kinds of quadrilaterals and triangles, and sketch reasonably accurate examples.	Revise pages 99 to 102.
7	Define congruent shapes and identify pairs of shapes that are congruent.	Revise pages 102 and 103.

Unit 4: Statistics and probability

Chapter 8 Statistics

In this chapter you will explore these focus questions:

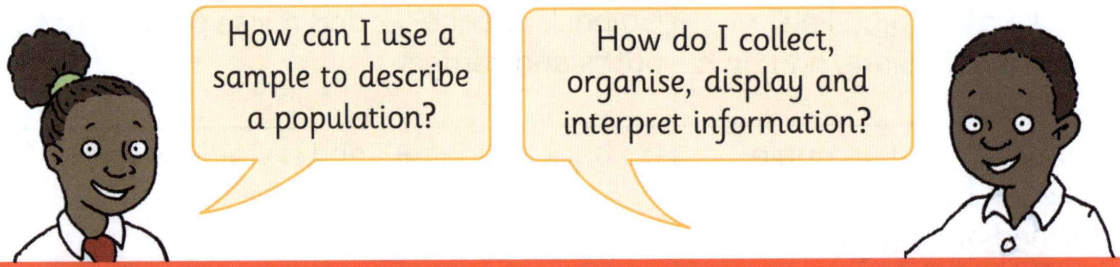

To answer these questions, you will:

- collect samples, and explain the difference between a sample and a population
- conduct surveys to collect data
- classify and sort data.

Starting point

Data is information. You already know some ways to **collect** and **record** data, for example:

- You can take measurements and write them down.
- You can ask questions and record the answers.
- You can use charts and tables to record the data.

1 Measure the heights of nine students from your class in cm. Record them in a table like the one below.

Name	Height in cm

Key words
data
collect
record

2 Write three questions you can ask using the data from your table.

3 Do you remember the kinds of graphs you worked with before? Which type of graph would work best to show this information – a pictograph, block diagram or bar graph?

Graphs and tables

Key maths idea

A **graph** is a way of organising data in a picture so that it is easy to understand. A graph should always have a heading. If the bar has **axes**, these should be labelled. Some graphs, such as pictographs, also have a **key**, to tell us what the symbols on the graph represent.

Example

Some students kept a record of how many hours they exercised over a week. They used **tally marks** (or tallies) to record their hours. Then they organised the information in a **bar graph**.

Name	Hours
Sean	//
Amari	~~////~~ ///
Shawna	~~////~~ ~~////~~ //
Romano	~~////~~ /
Devon	////

Key words

graph
axes
key
tally marks
bar graph

1 Use the bar graph to answer these questions.

a Who exercised for the greatest number of hours?

b How many hours did Sean and Devon do altogether?

c How much longer did Amari exercise than Romano?

2 Here is another bar graph.

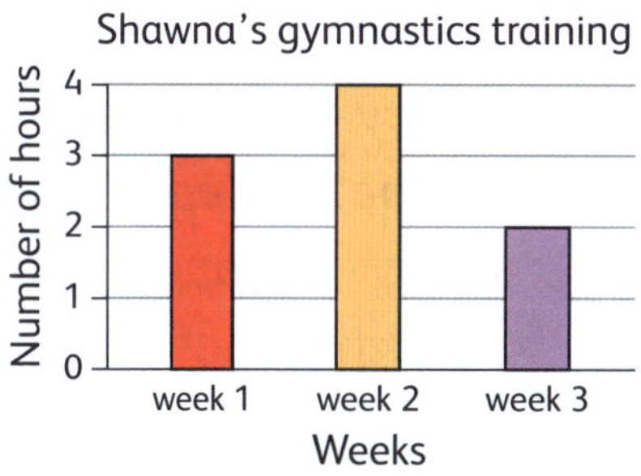

a In which week did Shawna train 2 hours less than the week before?

b How much time did Shawna spend training altogether?

c In Week 4 she trains half as long as in Week 2. What is the total for the month?

3 Read the information on the graph and answer the questions.

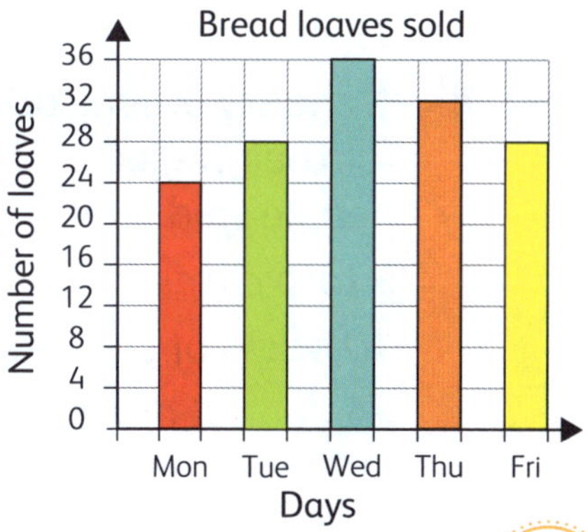

a How many loaves were sold on Monday?

b Which days did the bakery sell the same number of loaves? How many did they sell?

c How many loaves were sold on Monday and Tuesday altogether?

d On which day was the highest number of loaves sold?

e On which days did the bakery sell fewer than 30 loaves?

Statistics and probability

4 Fredrick recorded the weather conditions every day for one month. He recorded his data on a page from a calendar.

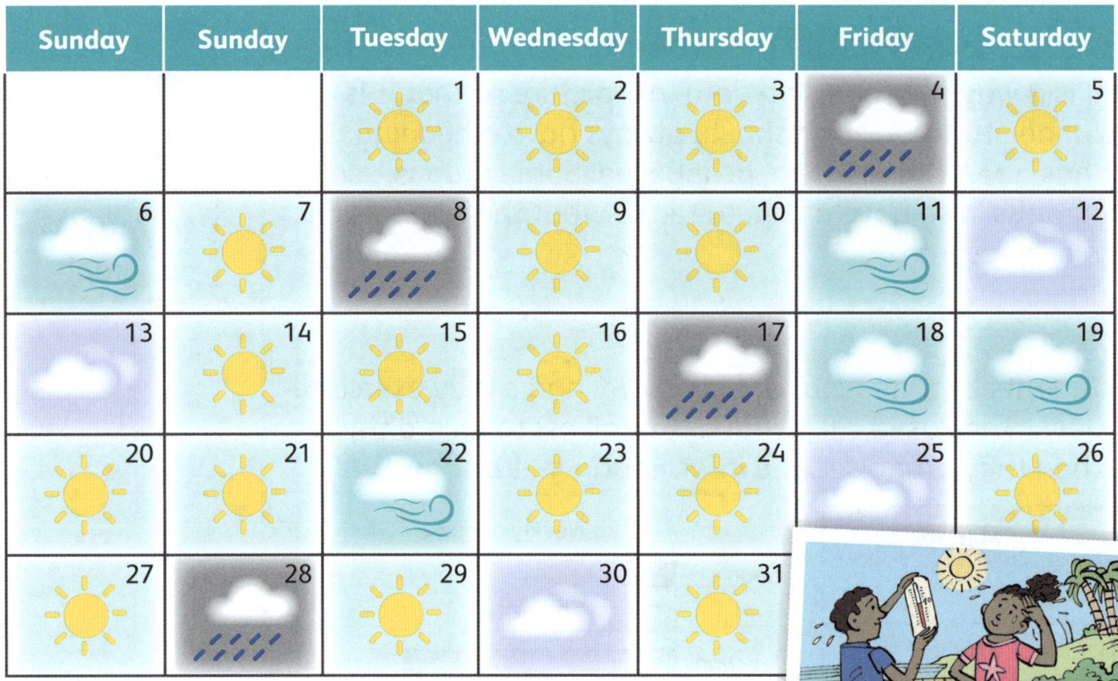

- **a** Draw a table to better organise the data.
- **b** Draw a bar graph that shows the data.
- **c** Write a heading for the bar graph.
- **d** Can you guess in which month Fredrick recorded his data? Give a reason for your answer.

5 Read the information that was collected in these surveys. Choose one of the surveys. Draw a bar graph and a pictograph to show the information.

- **a** A shopkeeper wants to place an order of fruit for his shop. Help him decide how much to order by putting the information about fruit sales onto the two graphs. Each fruit you draw equals 10 fruit sold.

  May – 30 apples, 60 bananas, 10 pineapples

  June – 40 apples, 80 bananas, 20 pineapples

  July – 20 apples, 30 bananas, 20 pineapples

- **b** A school has a recycling project and they have done a survey to see what items are in their recycling bin in one week.

  12 glass bottles *15 tin cans*

  10 newspapers *3 cardboard boxes*

- **c** A charity organisation is having a bake sale to raise money for their projects. This is the data they collected about what people said they would bake for the sale.

  9 toto cakes *22 iced donuts*

  10 packs of cupcakes *4 coconut cakes*

  25 peanut drops

> Remember: a pictograph is a graph that uses pictures or symbols to show the value of the collected information.

Populations and samples

Key maths idea

When you collect data, the **population** is the group of people or objects you choose to study. Usually, you do not collect data about the whole group. You use a **sample**. A sample is a portion of the whole population. The **sample size** is important. If it is too small, it does not give us information that reflects the whole population. A good sample size is around $\frac{1}{10}$ of the population.

Example

A researcher decides to find out the favourite after-school activities at a school. The school has 200 children – 100 boys and 100 girls. The researcher asked these 20 children their favourite activities.

Key words
- population
- sample
- sample size

1 In the example above, what is:

a the school population? **b** the sample size?

2 a What do you notice about the number of girls and boys in the sample?

b How do you think this will influence the findings of the survey?

c What would you do differently to choose a sample?

3 a What is the population of your school?

b What would be a good sample size to use if you wanted to collect data about children at your school?

Biased samples

Key maths idea

Sampling is the way a researcher chooses the sample group. A sample must be **representative** of the whole group. This means that it reflects all the parts of the population. Researchers must take care to prevent bias from affecting the results. **Biased** information is skewed or distorted. This can happen in many ways, such as choosing an inappropriate sample, or presenting questions in a way that influences the participant's answers.

Games of chance, such as bingo and the lottery, are based on random sampling methods.

Sample type	Example
Random sample – each member of the population has an equal chance of being selected	Picking names out of a hat
Purposeful sample – the sample is based on the idea of the study	Counting how many students signed up for each school sports club
Cluster sample – choosing a specific group for the study	Choosing three classrooms and interviewing all the students in those classes
Systematic sample – a sample selected using a system	Asking every tenth student that walks through the the school gates

Key words

sampling
representative
biased

1 Consider the different sample types in the table above.

a Which sample type did the researcher on page 111 use?

b Which sample type would have provided less biased results?

2 Read each scenario. Decide whether the sample size is appropriate. If not, suggest a more appropriate size.

a A scientist is collecting data about a bee population in a certain area. The population is 500 and the researcher identifies 50 bees to observe.

b A tourism office collects data about tourists visiting Kingston over January. The total number of tourists is 30 000. The sample size is 300.

c A shop owner collects data about its customers. The population size is about 2 000. The sample size is 20.

3 Your teacher will help you find out how many students there are in your school, and the number of students in each grade. Imagine you want to find out how many students at your school like to play football.

a What size sample will you choose? Explain your answer.

b How will you make sure that the sample is fair and representative?

c Explain how many students you will interview from each grade.

4 Copy this table and complete the following tasks.

Which school sport would you like to play?

Sport	Number of students	Tallies
Netball		
Football		
Cricket		
Swimming		
Athletics		

a Ask 20 students to choose one of these as their first choice of sport to play.

b Record their choice by making a tally mark in the matching space.

c Work out the total numbers and then arrange the sport activities from the most popular to the least popular.

d Explain why you think this data could be important in a school.

Statistics and probability

Methods of collecting information

Key maths idea

Researchers collect data in many ways and for different reasons. Here are some examples of the way people can collect data:

By letter

By phone

In-person **interviews**

Email **questionnaires**

Apps and online platforms

Observation

Focus groups

Key words
- interview
- questionnaire
- observation

1 Discuss the seven methods of data collection described above.

- **a** Which would be quickest?
- **b** Which would take the longest?
- **c** What are the advantages and disadvantages of each method?
- **d** How would you make sure that the information given is truthful and accurate?
- **e** How could each method introduce bias into the sample?

2 For each situation, suggest which of these methods of data collection would work best. Give a reason for each answer.

observation and measuring	interviews	questionnaires

- **a** How plants grow under different light conditions.
- **b** Students' attitudes about national lockdowns during the COVID-19 pandemic.
- **c** Number of children at school who are interested in joining a choir.
- **d** Opinions of students and staff about making school days longer.
- **e** Contact details of families at a school.

Maths detective: Observation skills

Observation does not only mean looking! When you are doing your detective work in maths and science, you need to use all five of your senses – looking, hearing, touch, smell and sometimes even taste (when it is safe to do so).

Case 1

1 Look at each picture.

a Describe what each person is measuring or observing.

b Discuss with a partner what they could be trying to find out.

Case 2

In science this year, you are learning about what living things need to survive. You know that they need water, nutrients and air. But you want to find out what happens if a plant has too much water.

You find a book of science experiments, but the book is damaged and all you can see is this picture.

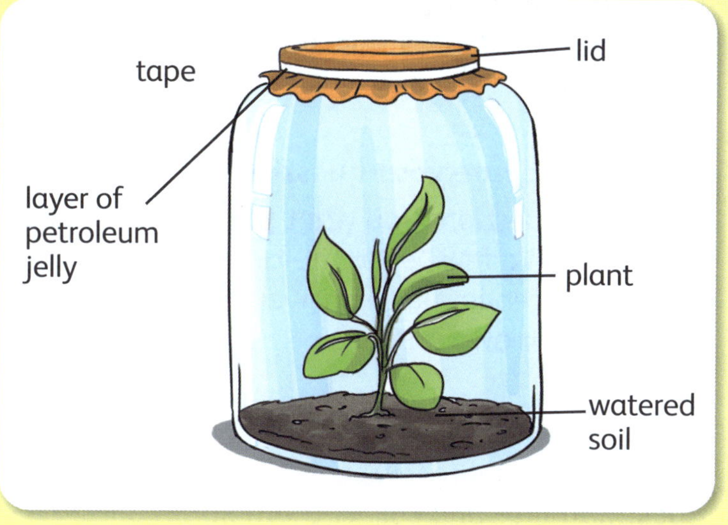

You can see that the experiment requires two jars, each with a plant inside and the lid sealed with petroleum jelly, but you can't make out any of the writing, or the second picture.

Work with a partner to solve the puzzle:

1 a Why do you need to set up two jars, rather than just one, for the experiment? What should be the same and what should be different about the jars?

b What do you think the petroleum jelly is for?

c What do you need to observe during the experiment?

d How long do you think the experiment should last – 1 hour, 1 day or 7 days? Why?

e How will you record your observations?

2 If possible, set up your experiment. Record the observations. Write a short report about what you found out.

Statistics and probability

Case 3

Choose one of these questions or make up some questions of your own that you can answer using observation skills. Collect data to answer your question.

- How much weight are students carrying on their backs in their school bags?
- How many pieces of furniture are there in the classroom?
- How many cars park in the school car park each day?
- What is the total distance a student walks during the school day, going into and out of the classroom and around the school?

Steps for collecting data

Key maths idea

Here are five steps to follow when you collect data using questionnaires, surveys or interviews. Start by thinking about what you want to find out and how you will find it out.

Step 1 Prepare your questions

Write the questions you want to ask. Make them clear and easy to understand. Test them out on a few people before you decide on the final wording. Decide whether you want the interviewee to answer yes/no, to check boxes next to one or more options, or to write their own answers.

Step 2 Collect the data

Conduct your interviews or give out the questionnaires and collect them again once they are complete.

Step 3 Tally the results

If your information involves amounts that you need to add up, you may need to use tally charts.

Step 4 Organise the data

Use tables or graphs such as bar graphs or line graphs.

Step 5 Interpret and describe

Explain how the information you found answers your questions.

Real-world maths: Collecting data

1 The parent committee sent out this questionnaire.

School tuck shop survey

Please draw a circle around your answers.

Class: 3 4 5 6

1 How often do you buy something from the tuck shop?

never seldom once a week more than once a week every day

2 Do you bring lunch food from home?

never seldom once a week more than once a week every day

3 What foods would you like us to add to our menu?

fries hamburgers healthy salad fruit salad whole fruits

- **a** What do you think this questionnaire is trying to find out?
- **b** Why do you think the committee wants this data?
- **c** Suggest two more questions that you could add to this questionnaire.

2 Design a questionnaire to find out about the students in your class:

- **a** the parish in which each student was born and which parish their parents came from
- **b** which infant or pre-school they attended
- **c** the birth dates of students in your class.

Collect the information and use it to make a display about your class.

3 Find out which mode of transport each student in your class uses. Collect the information using a tally chart. Then draw a graph to present the data. You can also make small models of each type of transport for your display.

4 Choose one of the following questions. You will design a questionnaire or interview to gather more data about the question.

- What time in the morning do students in our class get up and what do they spend the most time doing each morning?
- How many students in our class have food allergies or dietary restrictions?
- What are the favourite subjects in our class?
- How many of the students in our class keep pets and how many pets do they have?

What I have learned

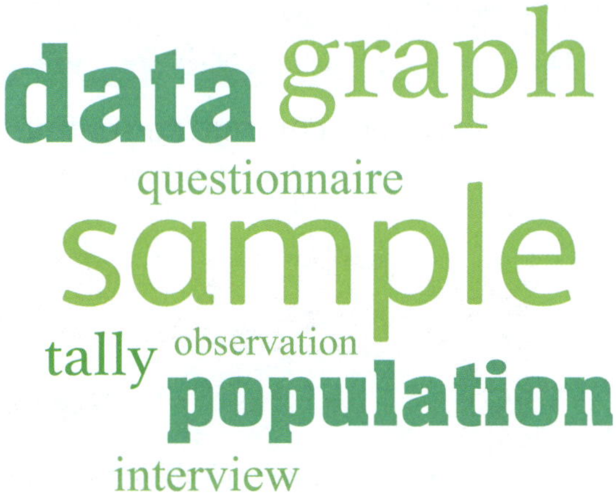

A word cloud is a graphic made from a collection of words. The size of the word indicates how often it is used in the collection. The bigger the word appears in the word cloud, the more often it is used.

1. Three of the terms appear more than 30 times in this chapter. Which three terms are they?
2. Two of the terms appear fewer than 10 times in the chapter. Which terms are they?
3. Write each term in your notebook. Write a few sentences explaining the meaning of the term and giving an example. If you need to refresh your memory, go back through the chapter and find the words where they appear.

Practice questions

1. Explain in your own words what you know about these three methods of data collection and how they work. Write at least two sentences for each one.
   a observation   b interview   c questionnaire
2. Copy this table and write a sentence to explain what each word means.

Word	Definition
a Sample	
b Population	
c Tally	
d Tally chart	
e Bar graph	

3. Irie wanted to find out what type of music a large group of people liked to listen to. Her questionnaire has the question, 'Do you like rap music or reggae music?' written on it. How can you help Irie improve her question?
4. Look at these words.

questionnaire	observation	interview	organise	interpret

   a Draw up a tally chart and use it to count how many times the vowels (a,e,i,o,u) appear in the words.
   b Give a total for each vowel.
   c Which vowel appears the most?
   d Which vowel appears the least?

5 Look at this graph.

a Explain in your own words what the data on this graph shows.

b Which fruit is least popular?

c Which fruit is most popular?

d Which three fruits should the school tuck shop offer if it wants to stock the students' favourite fruits? Give a reason for your answer.

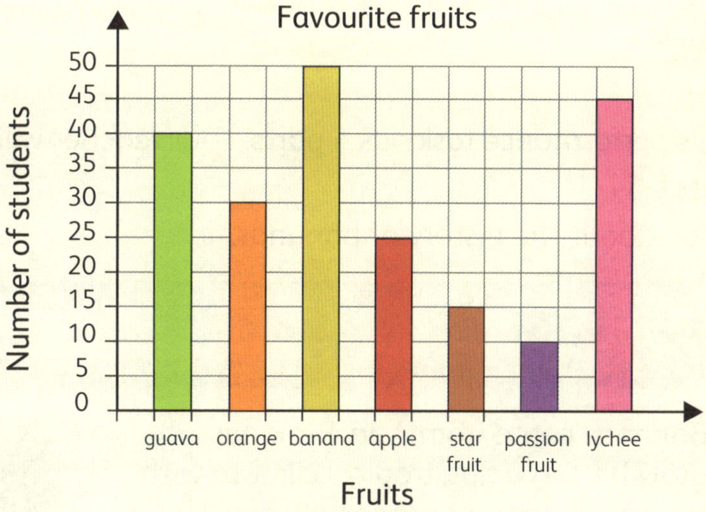

8 Self check

Check you can do this:	If you have difficulty, try this:
1 Interpret data from a table or graph.	Go back to pages 108 to 110, and remind yourself how to look at the information on the bar graphs and tally charts.
2 Explain what populations and samples are, and identify an appropriate sample size.	Go back to pages 111 to 113.
3 Identify different sampling methods and explain why they are fair or not.	Revise the table on page 112.
4 Identify different methods of collecting information.	Revise page 114.
5 Collect data using observation.	Revise the work from page 115.
6 Collect data using questionnaires and interviews.	Revise the work from page 116.

PEP task: All about bananas

This performance task has 4 parts. Your teacher will tell you when to do each part.

Part 1

Read about the history of bananas.

Bananas come from Asia. Bananas grow best in a hot, tropical climate with temperatures that range between 10 °C and 30 °C. Some botanists believe that the Spanish brought bananas to the Caribbean in the 1500s, but others believe that there were bananas in the area before that.

In the 1900s, advances in shipping and refrigeration made it possible to transport bananas across long distances. This meant that Jamaican farms could export bananas to other countries. However, Jamaican banana growers have faced many setbacks, including hurricanes and diseases that wiped out crops. In 1953, the Banana Board was established to support the Jamaican banana industry.

Many different varieties of bananas are grown in Jamaica. These include traditional varieties, such as Robusta, Williams, Grand Nain, Lacatan and Gros Michel. There are also varieties that scientists have developed on the island, including RG1, Tetraploid 1242, Tetraploid 6812 and Highgate. The Bodles Bananas Breeding Station and Research Centre stores and maintains samples of more than 150 varieties of bananas.

Find information in the article to show that each of these statements is true.

1 Bananas grow well at temperatures of 25 °C.

2 There was no Banana Board in Jamaica in 1900.

3 It was difficult to export bananas to far-away places before 1900.

4 There are at least 150 different types of bananas.

Use information from the graph and table on the next page to answer questions 5 and 6.

5 Is each statement true or false? If it is false, change it to correct it.

- a Jamaica's banana production has mostly stayed the same since the 1970s.
- b In the 1990s, Jamaica produced between 100 000 and 150 000 tonnes of bananas each year.
- c The 1960s were a difficult time for banana production in Jamaica.

6 Choose the correct answer a, b or c.

Based on the graph, we can predict that banana production in 2020 was around:

a 200 000 tonnes **b** 100 000 tonnes **c** 50 000 tonnes.

Part 2

Look at the table about banana production in Jamaica.

1 Use the data in the table. Work out the difference in value (in US$) between production in 2013 and 2014.

2 Did banana production increase or decrease after 2013? How can you tell this from the graph?

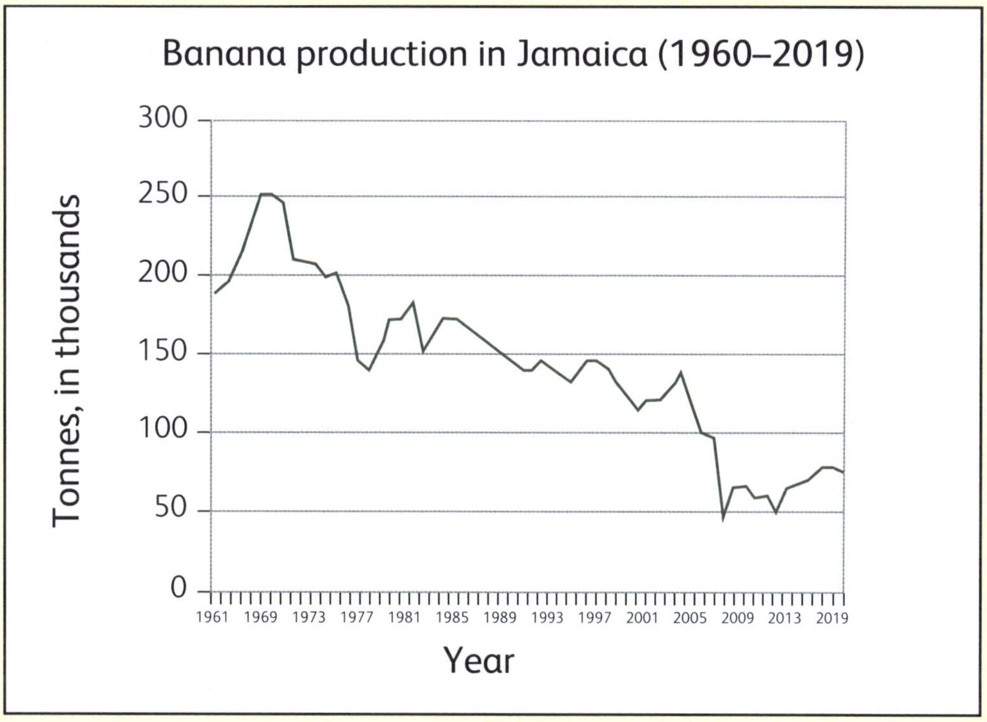

Year and value in US$ (2008–2019)

DATE	VALUE	DATE	VALUE
2019	63,653	2013	37,221
2018	66,381	2012	47,473
2017	64,816	2011	46,660
2016	58,702	2010	53,649
2015	54,810	2009	52,765
2014	51,581	2008	36,200

(Source: https://knoema.com/atlas/Jamaica/topics/Agriculture/Crops-Production-Quantity-tonnes/Bananas-production)

3 Compare the different kinds of bananas in the photograph on page 120.

a What is similar about all these bananas?

b How are they different?

Part 3

Look at this information about the lengths of different types of bananas and answer questions 1 to 3.

The length of some different kinds of bananas are given in the box.

Lady finger $7\frac{1}{2}$ cm	Plantain 30 cm
Gros Michel $22\frac{1}{2}$ cm	Burro banana 15 cm
Manzano 10 cm	Rajapuri 15 cm

Janice drew this bar graph to compare the lengths of different types of bananas. She left off some information.

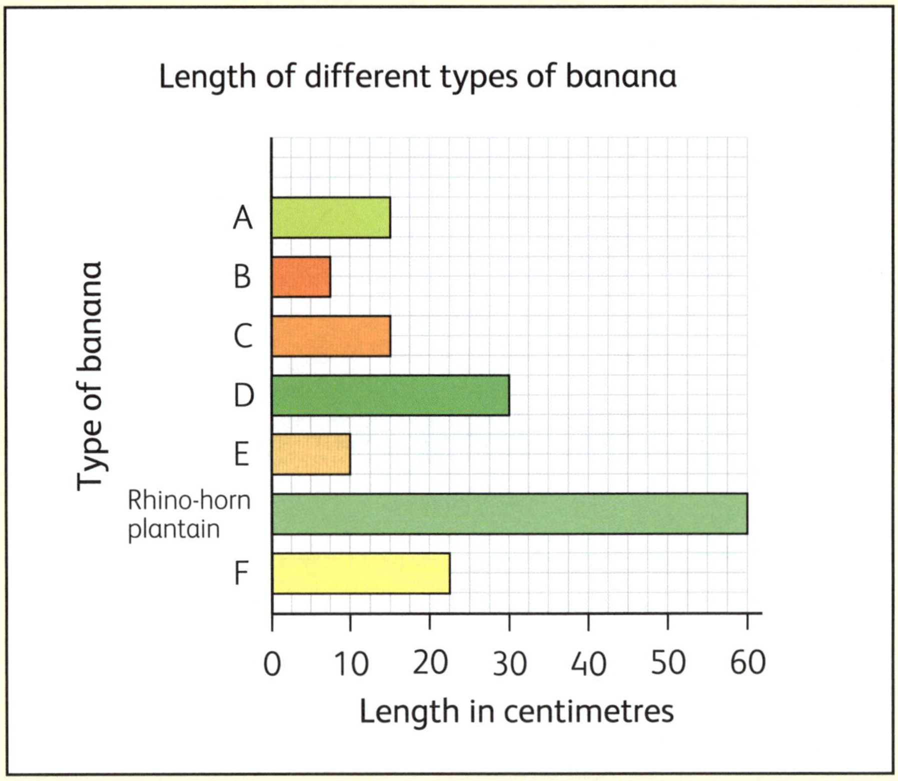

1 Which banana type is shown by bars A to F on this graph?

2 What is the length of a Rhino-horn plantain?

3 Which types of bananas are less than half the length of a plantain?

Part 4

Read the information about bananas and plantains on the diagram and in the table.

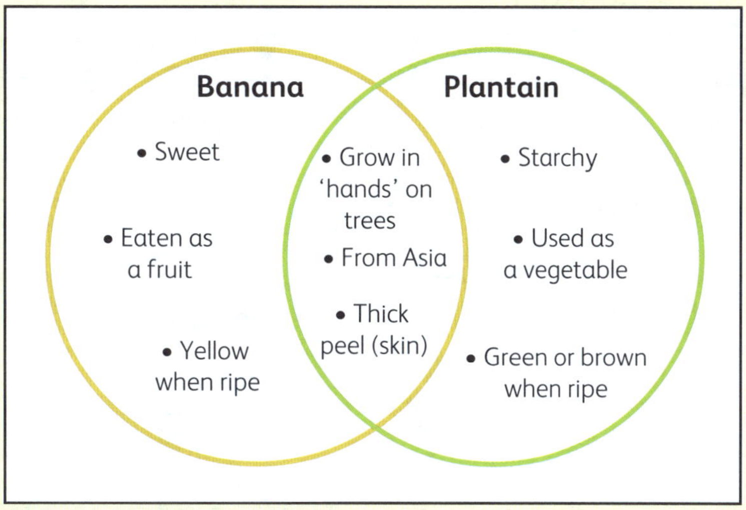

Nutrition facts		
(100 gram raw)	**Banana**	**Plantain**
Calories	89	116
Carbohydrates – sugar	23 grams 12 grams	32 grams 14 grams
Fibre	3 grams	2 grams
Protein	1 gram	2 grams
Fat	0 grams	0 grams
Vitamin C	9 milligrams	11 milligrams
Magnesium	27 milligrams	32 milligrams
Potassium	358 milligrams	465 milligrams

Use the diagram to answer these questions.

1 Which fruit is sweet?

2 Which fruit is used as a vegetable?

3 How do you know that both bananas and plantains come from Asia and that they have a thick peel?

Use the table to answer these questions.

4 Which fruit contains the most carbohydrate?

5 How much more Vitamin C is there in a plantain?

6 If you added 100 g raw banana to 100 g raw plantain how much sugar would there be?

7 How much less potassium is there in the banana?

TERM 2

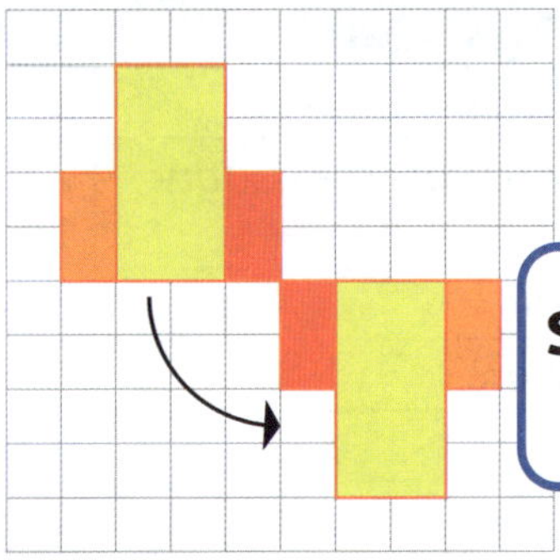

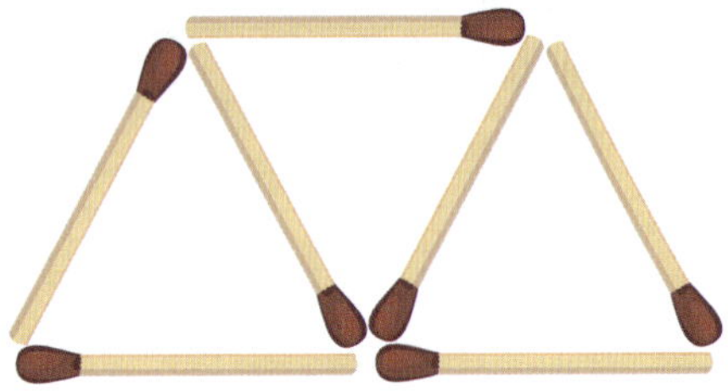

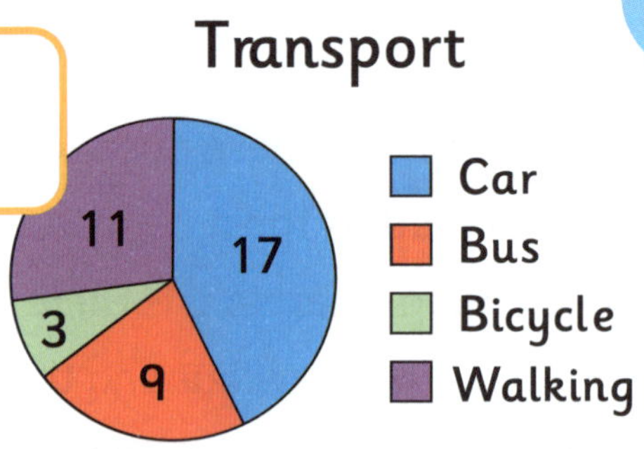

Unit 1: Number

Chapter 9 Multiplication and division

In this chapter you will explore this focus question:

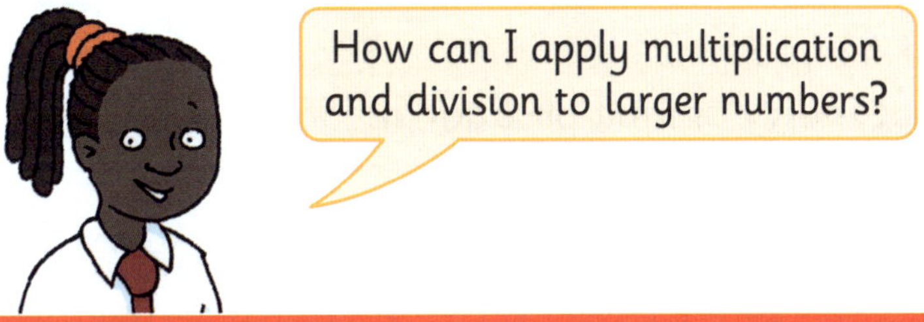

To answer this question, you will:

- recall multiplication tables
- use multiplication facts and a variety of strategies
- multiply numbers with up to four digits by 1-digit or 2-digit numbers
- multiply by multiples of ten
- estimate and calculate to solve two-step problems.

Starting point

1 a How many eggs are in the box? How can you use **multiplication** to work it out?

b How would you work out the number of eggs in two boxes?

c The way the eggs are arranged in rows and columns in the box is called an **array**. How can arrays make working out the number of eggs through multiplying easier?

2 Use groups of counters or small objects in the following quantities. Can you find two or more different ways to arrange them in arrays? Use multiplication sentences to describe the arrays you make.

a 24 b 36 c 20

3 Michelle says: 'You can't make odd numbers into arrays.' Make some arrays to show why she is mistaken.

Key words
multiplication
array

Revisit what you know about multiplication

Key maths idea

$4 \times 3 = 12$

Say 'four times three equals 12'.

4 and 3 are **factors** of 12.

12 is the **product** of 4 and 3.

The **commutative property** of multiplication means that $4 \times 3 = 3 \times 4$.

The first five **multiples** of 4 are: 4, 8, 12, 16, 20.

Key words

factor
product
commutative property
multiples

1 Look at the set of emoticons.

- **a** How many are in each row and in each column?
- **b** Use multiplication to express the relationship between the rows, the columns and the total.
- **c** Identify the factors and product in your answer to part b.
- **d** Work with a partner. How many different arrays can you make for the same number of emoticons? Use counters or objects to help you find different arrangements.

2 Look at this Venn diagram.

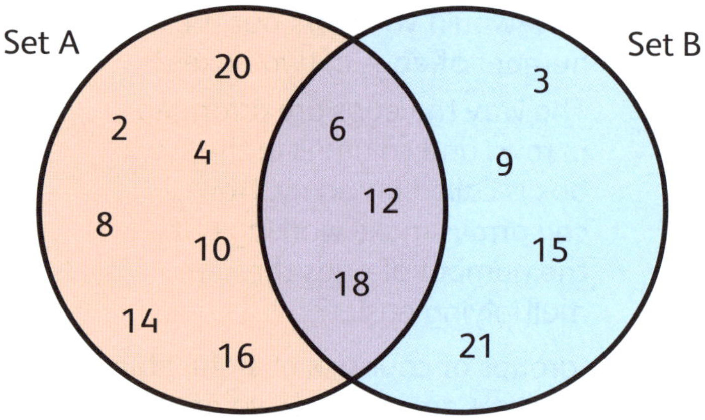

- **a** Write all the elements of set A, in order from smallest to greatest. Use the word 'multiples' to describe the set.
- **b** Write all the elements of set B, in order from smallest to greatest. Use the word 'products' to describe the set.
- **c** Set C is the set of multiples of 5. How many elements of set C can you find in set A or set B?
- **d** Draw your own Venn diagram, showing the intersection of these two sets:
  Set F = {multiples of 5 up to 45}
  Set G = {multiples of 4 up to 40}

How do you remember times tables?

Key maths idea

This grid shows the multiplication table up to 12. The factors are at the top of the columns and at the left side of each row.

It is useful to know the multiplication facts up to 12 by heart, but that does not mean you need to memorise long lists or grids! You can use what you know about number patterns to help you.

	1	2	3	4	5	6	7	8	9	10	11	12
1	1	2	3	4	5	6	7	8	9	10	11	12
2	2	4	6	8	10	12	14	16	18	20	22	3
3	3	6	9	12	15	18	21	24	27	30	33	36
4	4	8	12	16	20	24	28	32	36	40	44	48
5	5	10	15	20	25	30	35	40	45	50	55	60
6	6	12	18	24	30	36	42	48	54	60	66	72
7	7	14	21	28	35	42	49	56	63	70	77	84
8	8	16	24	32	40	48	56	64	72	80	88	96
9	9	18	27	36	45	54	63	72	81	90	99	108
10	10	20	30	40	50	60	70	80	90	100	110	120
11	11	22	33	44	55	66	77	88	99	110	121	132
12	12	24	36	48	60	72	84	96	108	120	132	144

1 Work with a partner. Find the column and the row that shows:

- **a** the 4 times table
- **b** the 12 times table
- **c** the 8 times table
- **d** the 10 times table.

2 What pattern do you notice in the following rows or columns of multiples?

- **a** multiples of 2
- **b** multiples of 5
- **c** multiples of 10
- **d** multiples of 11

3 **a** What is the total number of blocks in the whole grid?

b Patrick says:

Explain what Patrick means.

c Square numbers are numbers that we can arrange in a square array. The first three square numbers are 1, 4, 9. Find square numbers on the grid and describe what patterns you notice about the way they are arranged.

4 Copy and fill in the missing numbers.

a	3 × ____ = 27	and	____ × 3 = 27
b	7 × 6 = ____	and	6 × ____ = ____
c	8 × ____ = 40	and	____ × ____ = 40
d	____ × ____ = 81	and	81 = ____ × ____
e	63 = ____ × 9	and	9 × ____ = 63

Properties of multiplication

Key words

fact family
inverse
associative property
distributive property

Key maths idea

A **fact family** shows all the multiplication and division facts for a set of three numbers. If you know one fact, you can work out all the others, because multiplication and division are **inverse** properties.

Example

$5 \times 9 = 45 \quad 45 \div 9 = 5$

$9 \times 5 = 45 \quad 45 \div 5 = 9$

The **associative property** tells us that when we multiply three or more numbers, the product stays the same no matter how we group the numbers.

$(5 \times 2) \times 3 = 5 \times (2 \times 3)$

Check this with a partner. Is 10×3 the same as 5×6?

Multiplication also has a property called the **distributive property**. Look at this picture:

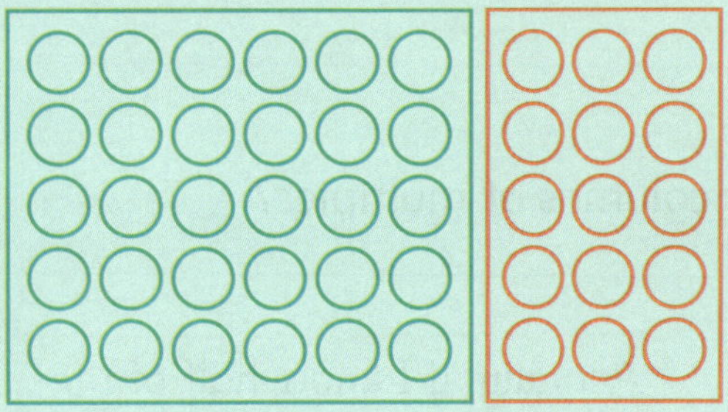

5×9

$6 + 3$

$= (5 \times 6) + (5 \times 3)$

$= 30 + 15$

$= 45$

1 The supermarket is giving out packs of 3 football cards. Kay's mom gets 4 packs on Monday and another 6 packs on Wednesday.

- **a** How many football cards did she get on Monday? Write it as a multiplication sentence first.
- **b** How many cards did she get on Wednesday? Use multiplication.
- **c** How many packs did she get altogether?
- **d** Use the distributive property as shown above to help you work out how many cards she got altogether. Draw a diagram to show your working.

2 Daniel has 8 pairs of socks in his wardrobe and 5 pairs of socks in the laundry basket. Use multiplication sentences to show how many pairs of socks he has:

- **a** in his wardrobe
- **b** in the laundry basket
- **c** altogether.

3 Which of these expressions are not equal to 7×12?

- **a** $(2 \times 12) + (5 \times 12)$
- **b** $(1 \times 11) + (6 \times 11)$
- **c** $(10 \times 2) + (7 \times 10)$

Multiply by 10, 100 and 1 000

Key maths idea

Our number system is a base 10 system. In bigger numbers, each digit in the number is worth ten times the value of the next place to the right.

$125 = 100 + 20 + 5$

$= (1 \times 100) + (2 \times 10) + (5 \times 1)$

When you multiply a number by 10, the number increases by 1 place value:

$3 \times 10 = 30$

When you multiply a number by 100, the number increases by 2 place values:

$3 \times 100 = 300$

When you multiply a number by 1 000, the number increases by 3 place values:

$3 \times 1\,000 = 3\,000$

The number you multiply by is called the **multiplier**.

Key word

multiplier

1 These number machines take input numbers, apply a multiplier and produce an output number. Use what you know about multiplication to work out the missing inputs and outputs.

a

b

c

d

e

f

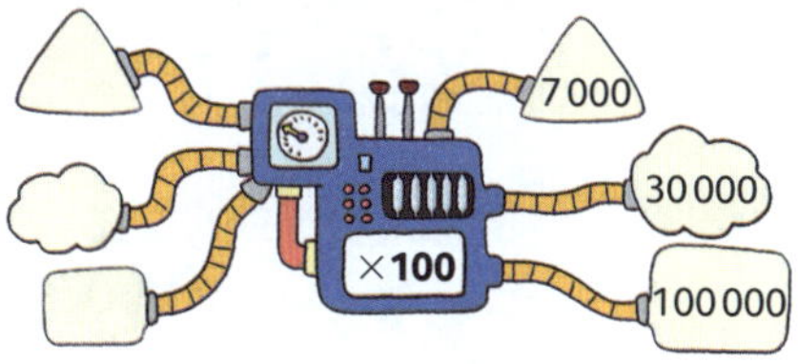

2 Complete this set of multiplication sentences and write each one in words. The first one has been done for you.

a 10 × 10 = 100 Ten groups of ten make one hundred.

b 10 × ☐ = 1 000

c 10 × 1 000 = ☐

d 10 × 10 000 = ☐

e 10 × 100 000 = ☐

Maths detective: Investigate a million

Case 1

How long would it take you to count to a million?
Work with a partner.

You will need:

- a stopwatch
- a calculator.

1 Use a stopwatch to time how long it takes to count to 20. Count, slowly and clearly, to 20. Write down how many seconds it took. Round off to the nearest second.

2 Now work out how long it would take you, if you continued counting at the same pace, to get to:
  a 100 b 1 000 c 1 000 000.

3 Use your calculation to work out how long counting to a million would take, in days (assuming there are 24 hours in a day).

4 Do you think you should try counting to one million? Explain why or why not.

Case 2

This kitchen tap uses 4 litres of water per minute.

1 How long will it take to use 1 million millilitres of water?

2 How long will it take to use 1 million litres?

Case 3

1 Work with a partner. The answers are given. Complete the number sentences.

- 1 million = (300 000 + ☐) × 2
- 1 million = 10 × ($\frac{1}{4}$ of ☐)
- 1 million = ☐ ÷ 10
- 1 million = 250 × ($\frac{1}{2}$ of ☐)

Hint: remember to use inverse operations. Addition and subtraction are inverse operations. Multiplication and division are inverse operations.

2 Write three puzzles of your own that make an answer of one million. Exchange puzzles with another maths detective pair.

Real-world maths: Things you do every day

1 Identify three things you do once a day. For example, you eat breakfast once a day. What else do you do once a day?

2 Identify other things that you do different numbers of time per day. Use your ideas to complete a table like this.

Twice a day:	Three times per day:	Five times a day:
Ten times a day:	Twenty times a day:	More than 20 times a day:

3 Work in groups.

a Guess how many **breaths** you take in one hour. Write down the guess of each person in your group.

b In your group, discuss how you could work out how many breaths you take in one hour and in one day.

c Test out your method and explain it to the class.

d Compare your answers with the rest of the class.

4 Read this article about breath and the body and answer the questions that follow.

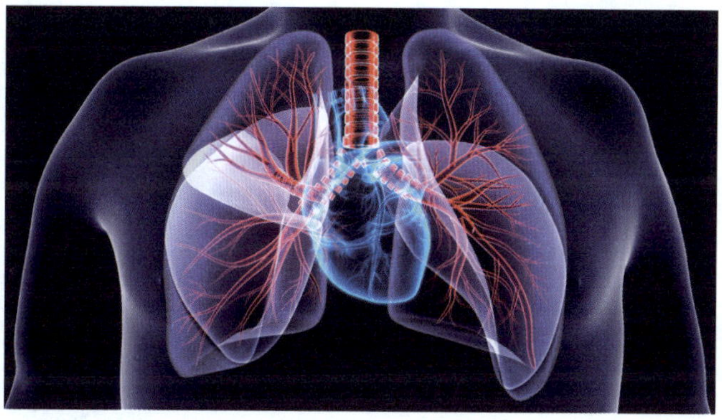

When you **breathe** in, *oxygen* flows into your *lungs*. The lungs transfer oxygen into your blood and release *carbon dioxide*, the waste product that your cells produce. Your heart pumps to push the blood in a cycle around your body. The *arteries* carry *oxygenated* blood from the lungs to the organs. The *veins* carry *deoxygenated* blood from the organs back to the lungs.

When your body is relaxed, you breathe around 15 times per minute, which provides about 12 litres of air. When you do **physical exercise**, your **breathing** speeds up because your muscles need more oxygen. You may breathe between 40 and 60 times per minute, which provides about 100 litres of air.

At the same time, your heart beats faster in order to pump more oxygen to your muscles. When your muscles work harder, they use more oxygen and produce more carbon dioxide.

It is normal to feel 'out of breath' when you exercise hard, but regular exercise makes your muscles stronger and more efficient, improves your *circulation*, and makes your heart stronger. The fitter you get, the less 'out of breath' you will feel when you exercise. Regular exercise can improve your physical and mental **health**. It reduces the risk of conditions such as diabetes, heart disease and depression.

a Draw a diagram or flow chart to show the cycle of what happens to your breathing when you exercise.

b Use the numbers in the article to help you work out about how many ml of air each breath gives your body.

c Think about the different kinds of measurement you learned in Chapters 5 and 6. Give an example of how you might use these five units of measure when you exercise:

litres   metres   kilometres   kilograms   degrees Celsius

Key words
breath
breathe
breathing
physical exercise
health

What kinds of exercise do you enjoy?

Multiplying in different ways

Key maths idea

In a classroom, each chair has a mass of 1 155 g. What is the mass of:

a 3 chairs? **b** 16 chairs?

Latoya used **expanded notation**:

a $3 \times 1\,155$

$= (3 \times 1\,000) + (3 \times 100) + (3 \times 50) + (3 \times 5)$

$(3 \times 50) = 3 \times 5 \times 10 = 15 \times 10$

$(3\,000) + (300) + (150) + (15)$

$= 3\,465$ g

or 3 kg 465 g

$$\begin{array}{r} 3\,000 \\ 300 \\ 150 \\ +\ 15 \\ \hline 3\,465 \end{array}$$

b $16 \times 1\,155$

$= (16 \times 1\,000) + (16 \times 100) + (16 \times 50) + (16 \times 5)$

$(16 \times 50) = 8 \times 2 \times 50 = 8 \times 100$

$(16 \times 5) = 8 \times 2 \times 5 = 8 \times 10$

$(16\,000) + (1\,600) + (800) + (80)$

$= 18\,480$ g

or 18 kg 480 g

$$\begin{array}{r} 16\,000 \\ 1\,600 \\ 800 \\ +\ 80 \\ \hline 18\,480 \end{array}$$

Marcus multiplied using the **column method**:

a

$$\begin{array}{r} {}^{1\,1} \\ 1\,155 \\ \times\ 3 \\ \hline 3\,465 \end{array}$$

b

$$\begin{array}{r} {}^{3\,3} \\ 1\,155 \\ \times\ 16 \\ \hline 6\,930 \\ +\ 11\,550 \\ \hline 18\,480 \end{array}$$

Key words

expanded notation

column method

grid method

1 Work with a partner.

- **a** Discuss what is similar and what is different about Latoya and Marcus's methods.
- **b** Which method looks easier to you?
- **c** Here is another method, the **grid method**. Look carefully at this method. How does it work?

	1 000	100	50	5	
3	3 000	300	150	15	= 3 465

	1 000	100	50	5	11 550
10	10 000	1 000	500	50	+ 6 930
6	6 000	600	300	30	18 480

2 Use any method you prefer to work out these multiplication sentences. First estimate to get an idea of what the answer will be. Then calculate.

a	9 × 265	**b**	435 × 7	**c**	4 × 342	**d**	27 × 13
e	32 × 19	**f**	43 × 30	**g**	1 467 × 13	**h**	987 × 22
i	3 420 × 31	**j**	120 × 20	**k**	5 663 × 5	**l**	2 064 × 16

3 A caterer charges $2 845 per person at an event. How much does he charge for:

a 5 people? **b** 12 people?

4 This is Curtis's homework. Check his answers and correct any that are wrong.

Curtis Hall 17 November

a	150 x 19 = 2 850	b	42 x 3 = 122	c	127 x 9 = 1 140
d	142 x 8 = 1 136	e	25 x 12 = 300	f	402 x 28 = 1 256
g	300 x 15 = 4 300	h	11 x 389 = 4 279	i	864 x 12 = 10 368

Two-step problems

Key maths idea

Laura thinks of a number. She doubles it, and the answer is 100. What number did she think of?

When you double a number, you add it to itself, or you multiply by 2. To get back to the starting number, you would need to work out: what number multiplied by 2 makes 100?

Half of 100 is 50.
Another way to say this, is:

$$100 \div 2 = 50$$

Some problems require you to work in more than one step. Here are some examples of **two-step problems**.

Examples

1 I think of a number. I add 25, then multiply by 6. The answer is 216. What was the number?

$216 \div 6 = 36$ Divide by 6 to undo 'multiply by 6'.
$36 - 25 = 11$ Subtract 25 to get back to the number you started with.

2 Cal buys two gifts. She spends the same amount on each gift. Her change from \$6 000 is \$1 200. How much did each gift cost?

First, we can work out how much Cal spent:

$$\$6\,000 - \$1\,200 = \$4\,800$$

Now we can work out the cost of each gift.

$\frac{1}{2}$ of \$4 800 = \$4 800 ÷ 2 = \$2 400

Key words

inverse operations
two-step problem

When you get a problem like this, you can work backwards, using **inverse operations**.

Think: What am I trying to find out? What information do I already have? How can I use this information to work out the solution?

1 Show your working to solve these two-step problems.

- **a** I am thinking of a number. I double it and add 150, and the answer is 240. What was the number?
- **b** I think of a number. I multiply it by 5 and then subtract 87. I have 163 left. What was the number?
- **c** I think of a number. I multiply it by itself and add 351. The answer is 432. What was the number?

2 Make up your own two-step 'think of a number' problem for a partner to solve.

More two-step problems

Key maths idea

Sometimes, when you have a real-life problem to solve, it requires more than one step to work it out.

Example

An isosceles triangle ABC has sides AB and BC equal in length. The perimeter of the triangle is 40 cm and the base AC is 10 cm. What is the length of AB?

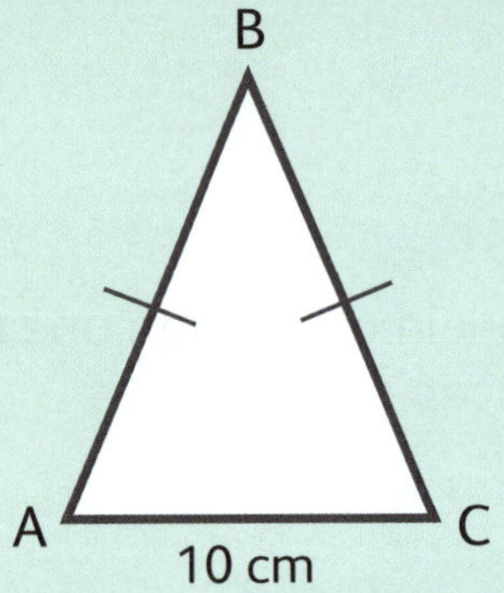

When you look at the problem, think about what you already know and what you need to work out.

The perimeter is the sum of the lengths of all three sides. If we take away the length of the base we get the sum of the remaining two sides:

$40 \text{ cm} - 10 \text{ cm} = 30 \text{ cm}$

We know that the remaining two sides are equal in length. So $2 \times$ length of AB $= 30$ cm.

$30 \text{ cm} \div 2 = 15 \text{ cm}$

So AB $= 15$ cm

You can often make a sketch to help you.

Work with a partner. Solve these problems.

1 Leo is 19 years older than Sofia. Sofia will be 20 in 5 years' time. What is Leo's age now?

2 Melanie has 10 cupcakes now, but when they came out of the oven, she gave one quarter of the batch to her granny and then ate 2 of what was left. How big was the whole batch?

3 I am thinking of two numbers. If I add them together, they make 15, but if I multiply them, they make the same as double 28. What are the two numbers?

4 One side of a rectangle is 18 cm. If the total perimeter of the rectangle is 50 cm, what is the length of each of the other sides?

5 Olivia and Jake share a pack of stickers equally between them. Olivia uses 9 of hers and has 23 left. How many were in the whole pack to start with?

The lattice method

Key maths idea

Teachers in China use a different method for solving multiplication. It is called the **lattice method** because of the pattern that the grid makes when you work in this way.

A lattice is a pattern made of interwoven lines. Usually we see lattice patterns in wooden structures that support plants and also in woven baskets. Here you can use lattices for multiplication!

To use the lattice method, follow these steps.

Example 1

24 × 73

Step 1: Draw the lattice grid

Start with drawing a grid of boxes. We are multiplying a 2-digit number by another 2-digit number, so we need a 2 × 2 grid. Then draw diagonals that go up from the lower-left corner to the top right corner in each box, like this.

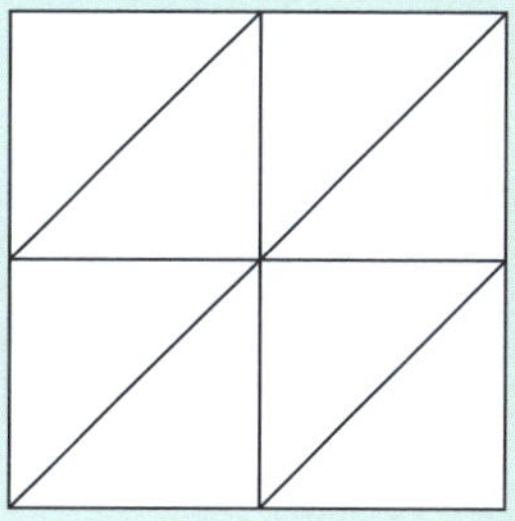

Step 2: Label the lattice

One of the numbers goes along the top of the grid, with each digit above one box. The other number goes down the right hand side, with each digit alongside one box. Notice that the place values run from greatest to smallest, left to right and top to bottom.

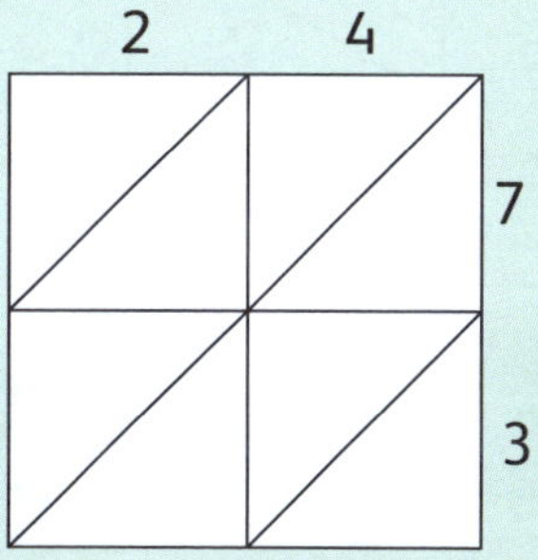

Step 3: Multiply the numbers

For each box, multiply the digit directly above the column by the digit to the right of its row. When you write the product, the ones go below the diagonal line and the tens go above it, like this.

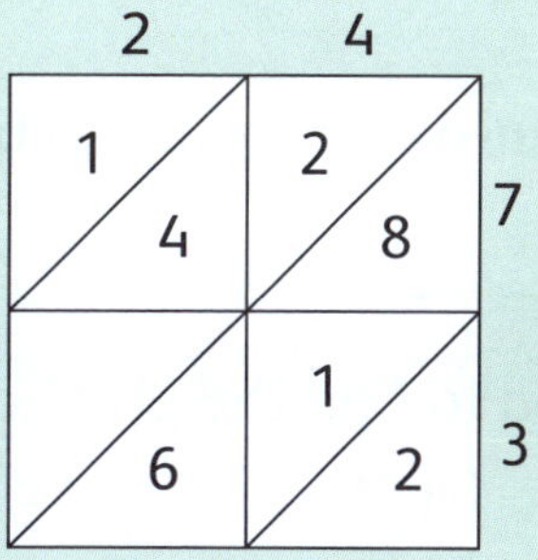

Key word
lattice method

Step 4: Add the products

Now you need to use the diagonal lines. Start in the bottom right corner. The 2 in this corner is on its own, so write the total under that box.

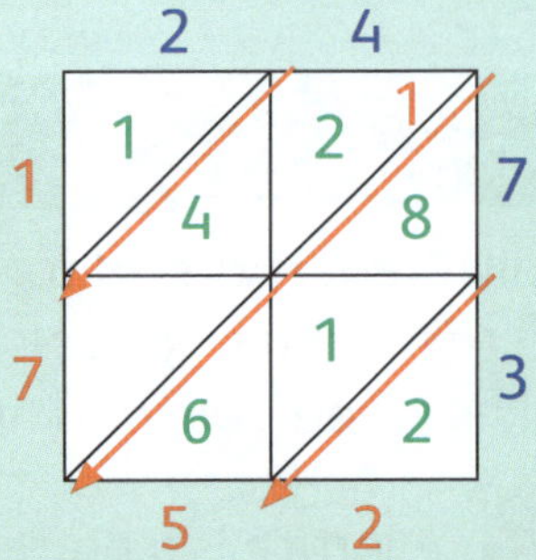

Now move up to the next three triangles. They form a downward diagonal from right to left.

8 + 1 + 6 = 15. Write the 5 under this diagonal as shown and carry the 1 to the next diagonal above it.

1 + 2 + 4 = 7

Write this next to the bottom left box as shown.

Last, move to the top right corner. Again, this number is on its own. Write the 1 next to the top block.

Step 5: Read out the answer

Now you can read out the answer. Start at the top left and finish at the bottom right: 1 752.

So 24 x 73 = 1 752

Example 2

Here is an example with a 3-digit and a 2-digit number:

143 × 82

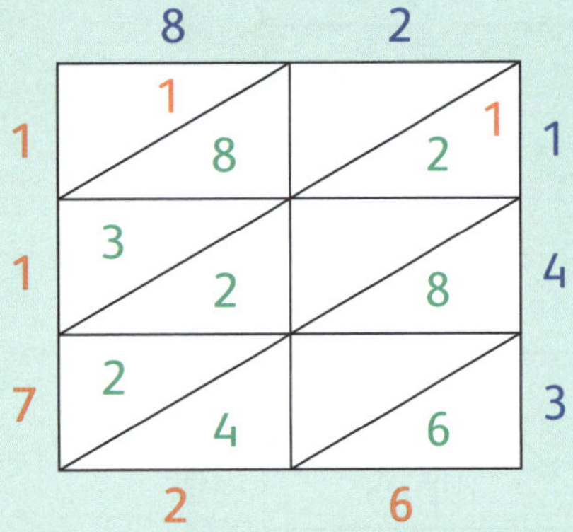

If you calculate with longer numbers, use more columns or rows.

Make sure to read the answer from top to bottom, left to right.

So 143 × 82 = 11 726.

1 Use the lattice method to find the product.

a 34 × 15 **b** 51 × 27 **c** 13 × 32

d 761 × 19 **e** 321 × 22 **f** 31 × 639

2 Malaki sells bags of bananas. Each bag contains 5 bananas.

a How many bananas are there in 7 bags?

b How many bags of 5 can he make if he has 65 bananas?

3 A farmer sells plantain in bunches of 4.

- **a** How many bunches of 4 can she make if she has 52 plantains in one box?
- **b** If she sells 8 bunches of 4, how many has she sold?
- **c** If she sells the 8 bunches for $200 each, how much money does she have?

4 Delroy and some friends are shopping for sunglasses. A pair of sunglasses costs $700.

- **a** Delroy and his friends buy 7 pairs of sunglasses. How much money do they spend altogether?
- **b** Kimberly spends $3 500 on sunglasses. How many pairs of sunglasses did she buy?
- **c** Tamara buys four pairs of sunglasses and pays with three thousand-dollar notes. How much change will she get?

5 Each week, Ronda drives 785 km to deliver vegetables to a market.

- **a** If she does deliveries 34 weeks of the year, what distance does she drive altogether?
- **b** If she shares the driving with her sister, so that each of them do the deliveries half the time, how many weeks would each person do deliveries?
- **c** How many kilometres would each of them drive in a year?

6 There are 300 tickets available for a cricket match. Each ticket costs $200 and 250 tickets are sold.

- **a** How much money was paid for the sold tickets?
- **b** How many tickets were still available to be sold?
- **c** How much money are these tickets worth if they are sold?

What I have learned

Write a sentence and draw a diagram or sketch to explain what each of these terms means.

array	commutative property	distributive property	
associative property	times table	fact family	million

Practice questions

1 Draw two different arrays for each product.

a 18 b 26 c 38 d 45

2 Draw these tables in your notebook and complete them.

a How many legs are there?

Number of birds	1	2	3	5	7	9	10	12	15
Number of legs					14				

b How many paws are there?

Number of cats	1	2	3	5	7	9	10	12	15
Number of paws			12				40		

c How many toes are there?

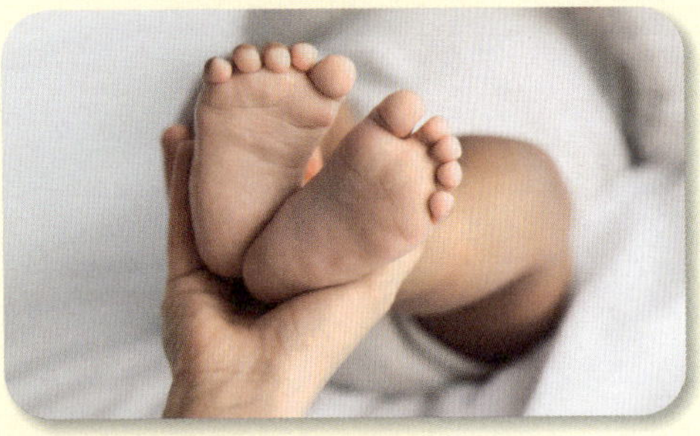

Number of feet	1	2	3	5	7	9	10	12	15
Number of toes	5		15						

3 True or false? Explain how you worked it out. Use the terms commutative, distributive or associative.

a $3 \times (5 + 3) = 3 \times (6 + 2)$ b $7 \times 3 \times 4 = 4 \times 7 \times 3$

c $18 \div 2 = 2 \div 18$ d $3 \times 5 = 5 \div 3$ e $(10 \times 2) + (10 \times 3) = 10 \times 5$

4 Write three different number sentences that are equal to each multiplication sentence. Use brackets, and + and × signs in your number sentences.

a 8 × 7 b 12 × 5 c 9 × 8

5 Each triangle has the product at the top corner and the factors to multiply in the bottom corners. Work out the missing number for each triangle.

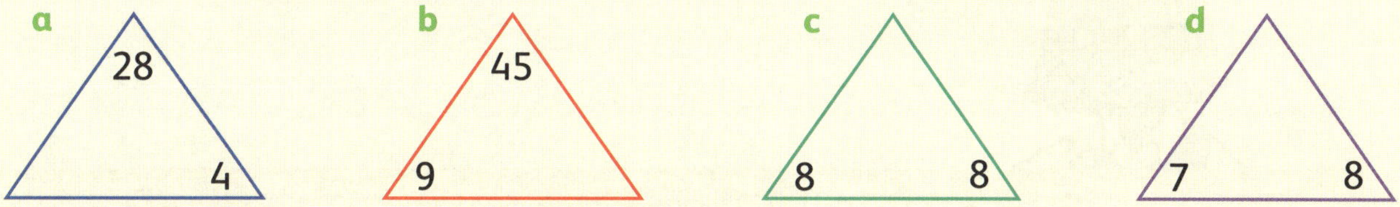

6 Use the inverse property of multiplication and write two multiplication and two division sums for each triangle in Exercise 5.

7 Use any two different multiplication methods to show how to find the product for each of these calculations.

a 63 × 7 b 49 × 19 c 326 × 24 d 9 872 × 33

9 Self check

Check you can do this:	If you have difficulty, try this:
1 Recall and use multiplication tables, for example, recall the products of 4 × 9, 6 × 7, 8 × 9, 4 × 6 and so on.	Go over pages 126 and 127. Remind yourself of some of the patterns you can use for remembering multiplication tables.
2 Explain the associative, distributive and commutative properties of multiplication.	Revise the properties shown on page 128.
3 Multiply any number by 10, 100, 1 000. Use multiplication to express numbers up to millions.	Go back to the work on pages 129 to 130 and revise how we use place value when multiplying by multiples of 10.
4 Use expanded notation, the grid and column method, and the lattice method as strategies for multiplying numbers.	Revise the work on pages 133 to 137.
5 Estimate and calculate to solve two-step problems.	Revise the work on pages 135 to 136.

Chapter 10 Decimals

In this chapter you will explore this focus question:

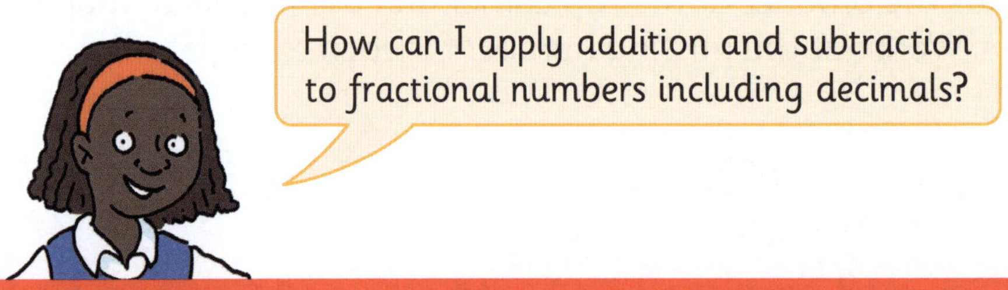

To answer this question, you will:

- revise what you know about different kinds of fractions
- express fractions as decimals and vice versa
- complete sequences of decimal numbers
- convert between different types of fractions (proper, improper, mixed numbers, decimals)
- add and subtract using decimals and other fractions
- solve problems involving decimals and other fractions.

Starting point

Work with a partner. Identify the **fractions** you see in each picture. Use fractions to tell the story about what the children are trying to work out. How do they solve the problem?

Key word
fractions

Revisit what you know about fractions

1 These students have mixed up their definitions of the different types of fractions. See if you can correct each statement. You can go back to Chapter 3 to refresh your memory about the different types of fractions.

Key words

proper fraction	improper fraction
mixed number	like fractions
unit fraction	equivalent fractions

2 Use the fractions from the box on the right.

- **a** List all the proper fractions.
- **b** Write two pairs of equivalent fractions.
- **c** Write two unit fractions.
- **d** Write a pair of like fractions.
- **e** List all the mixed numbers.
- **f** Write two improper fractions.

$\frac{1}{2}$	$4\frac{2}{5}$	$\frac{11}{4}$	$\frac{5}{10}$	$1\frac{1}{2}$
$\frac{9}{10}$	$\frac{1}{3}$	$\frac{2}{5}$	$\frac{9}{10}$	$\frac{11}{10}$

3 Calculate.

- **a** $\frac{3}{7} + \frac{2}{7}$
- **b** $\frac{1}{5} + \frac{1}{5} + \frac{2}{5}$
- **c** $\frac{5}{9} + \frac{2}{9}$
- **d** $\frac{7}{12} + \frac{3}{12}$
- **e** $\frac{2}{10} + \frac{4}{10} - \frac{3}{10}$
- **f** $\frac{6}{8} - \frac{4}{8}$
- **g** $\frac{4}{9} + \frac{4}{9} - \frac{3}{9}$
- **h** $\frac{6}{12} + \frac{6}{12}$
- **i** $\frac{7}{7} - \frac{6}{7}$

4 Work with a partner and subtract the proper fractions from the whole number. Before you begin, discuss with your partner how you can convert the whole numbers to improper fractions with like denominators so that you can do the calculations.

- **a** $7 - \frac{5}{6}$
- **b** $9 - \frac{5}{7}$
- **c** $20 - \frac{10}{11}$
- **d** $8 - \frac{1}{2}$
- **e** $5 - \frac{3}{4}$
- **f** $1 - \frac{9}{10}$

More fraction revision

Key maths idea

Julian used number cards and a toothpick to model improper fractions.

Example 1

He says:

You can think of this in different ways:

- five out of three equal parts
- a set of five divided into sets of three
- five thirds

He uses counters to show the parts:

$\frac{3}{3} = 1 \quad + \quad \frac{2}{3}$

Example 2

- 12 out of 5 equal parts
- 12 divided into groups of 5
- 12 fifths

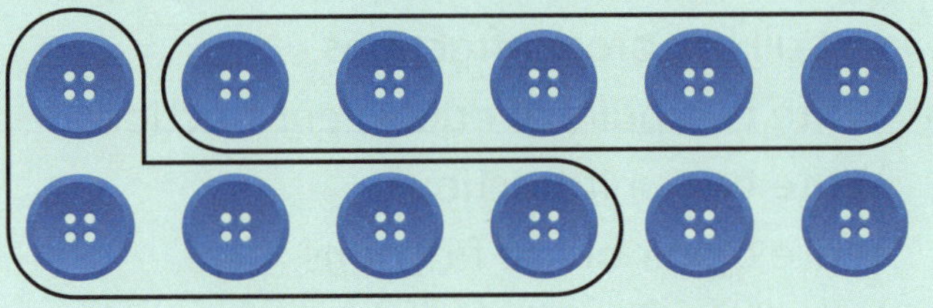

2 groups of 5 and two left over:

$\frac{5}{5} + \frac{5}{5} + \frac{2}{5} = 2\frac{2}{5}$

1 Look at these sets of shapes. Write down the mixed number for each set.

a

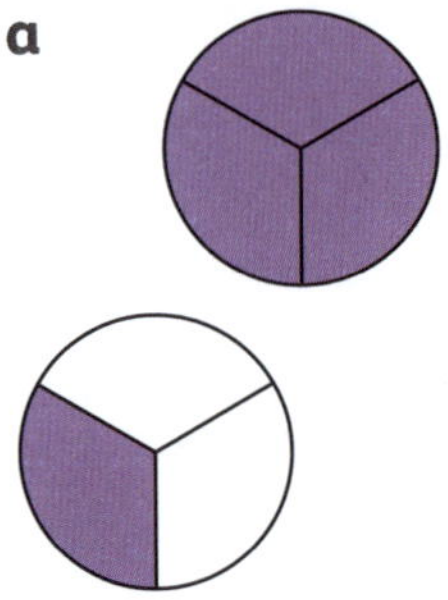

b

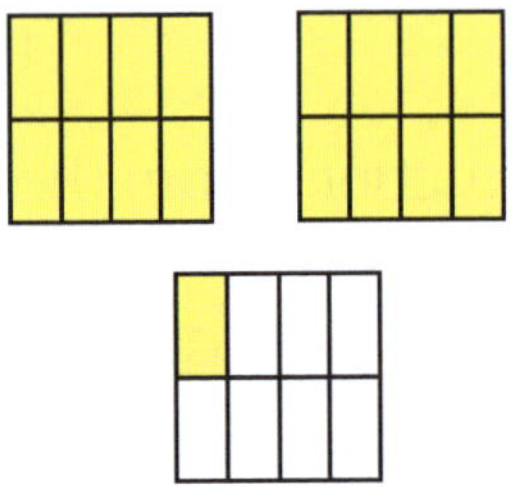

c

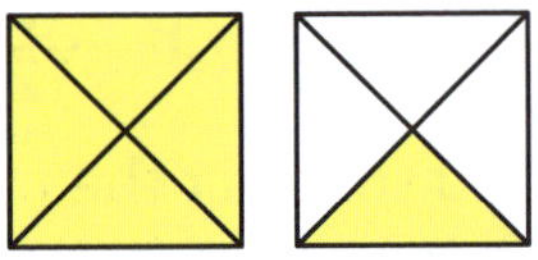

d

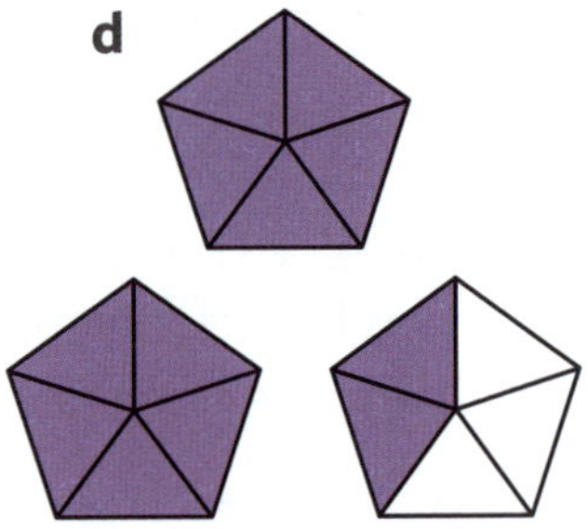

2 Which is longer:

a $1\frac{3}{4}$ years or 18 months?

b 65 minutes or 1 h 15 min?

c 140 s or $1\frac{1}{2}$ minutes?

d 9 days or 220 hours?

3 Convert the improper fractions to mixed numbers.

a $\frac{9}{5}$ **b** $\frac{14}{6}$ **c** $\frac{18}{7}$ **d** $\frac{18}{4}$

e $\frac{21}{5}$ **f** $\frac{51}{8}$ **g** $\frac{123}{50}$ **h** $\frac{64}{10}$

4 Change these mixed numbers to improper fractions.

a $1\frac{2}{5}$ **b** $3\frac{3}{7}$ **c** $8\frac{1}{7}$ **d** $8\frac{1}{4}$

e $10\frac{1}{11}$ **f** $5\frac{1}{8}$ **g** $2\frac{22}{25}$ **h** $6\frac{5}{6}$

5 Work with a partner. Take turns to explain how to convert improper fractions to mixed numbers and mixed numbers to improper fractions. Use words like *numerator*, *denominator*, *whole*, *part*, *improper fraction*, *mixed number*.

Maths detective: Modelling improper fractions and mixed numbers

Case 1

Use your toothpicks, cards and counters to model these improper fractions.

a $\frac{14}{3}$ **b** $\frac{10}{4}$

c $\frac{19}{5}$ **d** $\frac{18}{10}$

You will need:

- small cards with the digits 0 to 9
- some toothpicks
- small objects, such as buttons or counters.

Case 2

Use your counters to work out which of these are not whole numbers.

1 $\frac{15}{10}$ **2** $\frac{12}{3}$ **3** $\frac{18}{6}$ **4** $\frac{25}{10}$ **5** $\frac{30}{6}$

Case 3

1 Look at your answers to Case 2. What rule can you find for working out whether an improper fraction is a whole number?

2 Which is greater: $\frac{14}{5}$ or $\frac{16}{6}$?

3 What is the difference between $4\frac{1}{2}$ and $\frac{17}{4}$?

Tenths

Key maths idea

Decimals or **decimal fractions** are numbers that show fractional parts using place value. We use a **decimal point** after the ones place. Digits that appear to the right of the decimal point are fractions of a whole. The first place to the right of the decimal point has a place value of **tenths**.

$\frac{1}{10}$	$\frac{2}{10}$	$\frac{3}{10}$	$\frac{4}{10}$	$\frac{5}{10}$	$\frac{6}{10}$	$\frac{7}{10}$	$\frac{8}{10}$	$\frac{9}{10}$	$\frac{10}{10}$
0.1	0.2	0.3	0.4	0.5	0.6	0.7	0.8	0.9	1.0

Key words

decimals
decimal fractions
decimal point
tenths

1 In the diagram, three of the ten blocks are shaded. For each shape, write what fraction is shaded. Write it as a proper fraction and as a decimal.

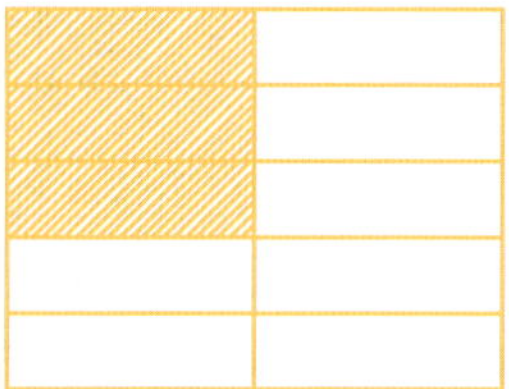

$\frac{3}{10} = 0.3$

a

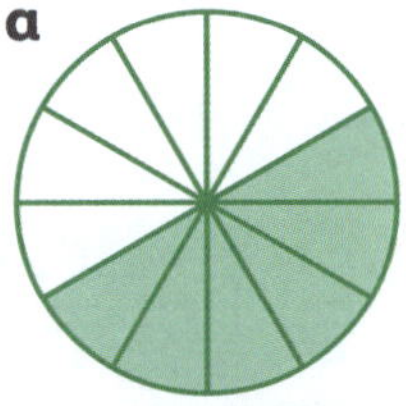

b

c

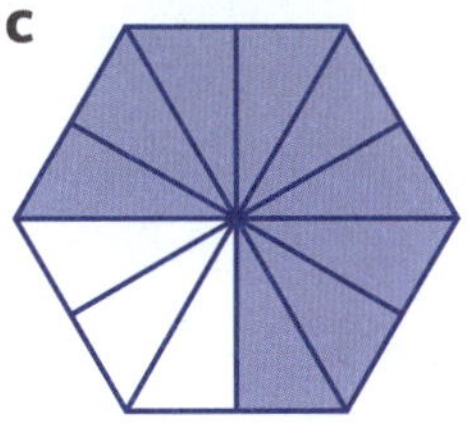

d

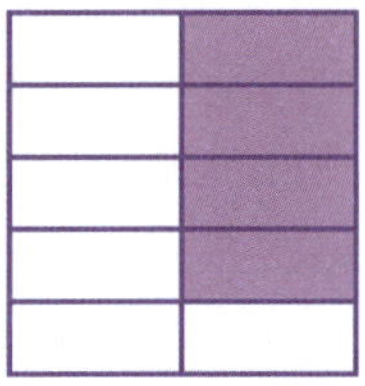

e

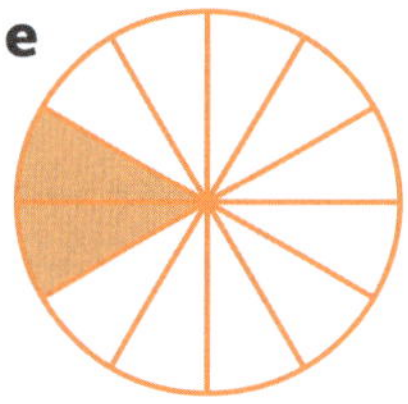

f

2 Look again at the diagrams in Exercise 1. With a partner, work out how we would write the following fractions as decimals:

a $\frac{1}{2}$ **b** $\frac{1}{5}$

3 These thermometers took readings from patient A, patient B and patient C. Work out the new readings on each thermometer if:

a patient A's temperature goes up by 0.1 degrees

A

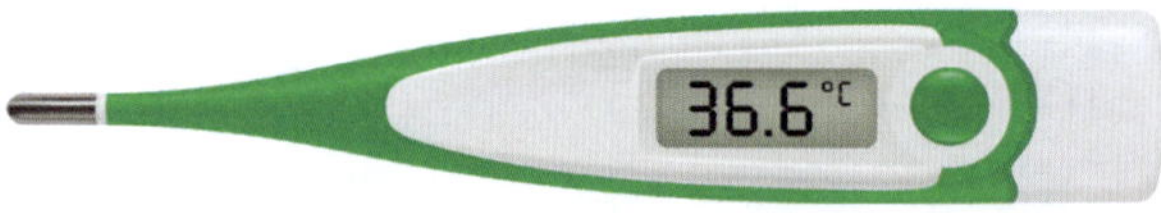

b patient B's temperature drops by 0.4 degrees

B

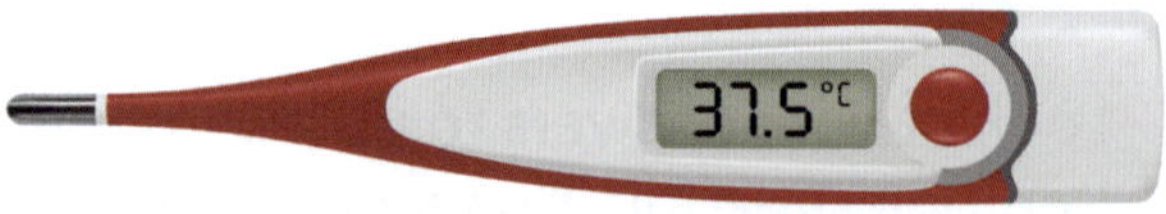

c patient C's temperature drops by half a degree.

C

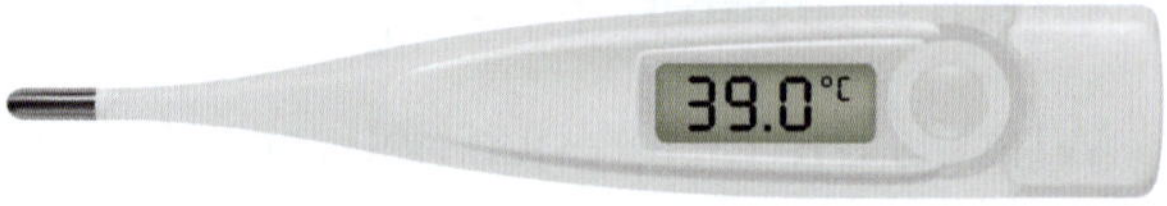

Hundredths

Key maths idea

Look at the ring on the scale in the picture. It weighs more than 5.3 g, but less than 5.4 g.

Here is the measurement in a place value table:

Tth	Th	H	T	O	.	t	h
				5	.	3	7

The *t* stands for tenths. The *h* stands for **hundredths**.

$5.37 = (5 \times 1) + (3 \times \frac{1}{10}) + (7 \times \frac{1}{100})$

Each of these blocks represents 1 whole.
The shaded area represents a fraction of the whole.

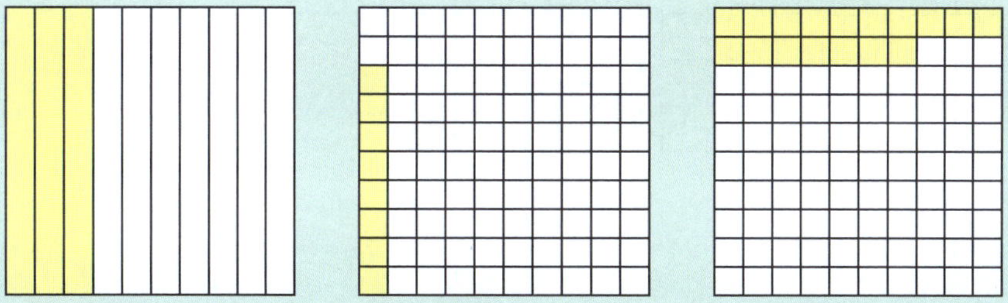

$\frac{3}{10} = 0.3$ $\frac{8}{100} = 0.08$ $\frac{17}{100} = 0.17$

0.3 has one **decimal place**. That means there is one digit after the decimal point.

0.08 and 0.17 each have two decimal places. The digits show us tenths and hundredths.

The zero before the decimal point tells us there are no whole parts in these decimals. Some decimals have a whole and fractional part.

For example, Shana's bag weighs 1.39 kg.

1 whole kilogram + $\frac{39}{100}$ parts of a kilogram

= 1.39 kg

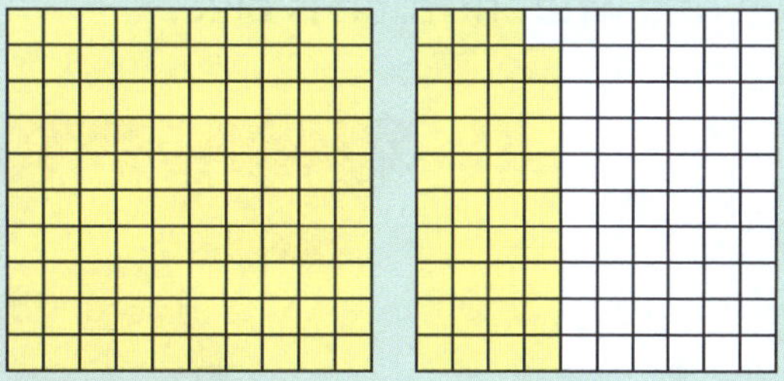

In expanded notation we write: 1.39 = 1 + 0.3 + 0.09.

Tth	Th	H	T	O	.	t	h
				1	.	3	9

Key words
hundredths
decimal place

In a place value table, each column's value is ten times more than the column to its right. That means each column's value is $\frac{1}{10}$ of the column to its left.

We say 'one point three nine kilograms'.

1 Write these money amounts using decimal fractions:

a forty dollars and twenty-five cents

b seventy dollars and fifty cents

c ninety dollars and ninety-nine cents

d twelve dollars and ten cents

e one hundred and sixty-seven dollars and five cents.

2 Write the number shown in each diagram as a decimal number.

a

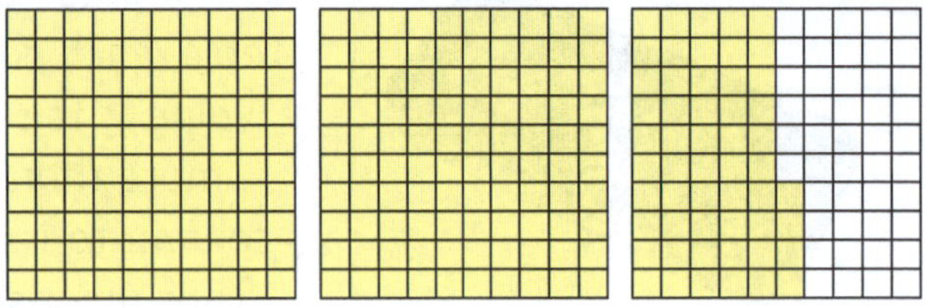

b

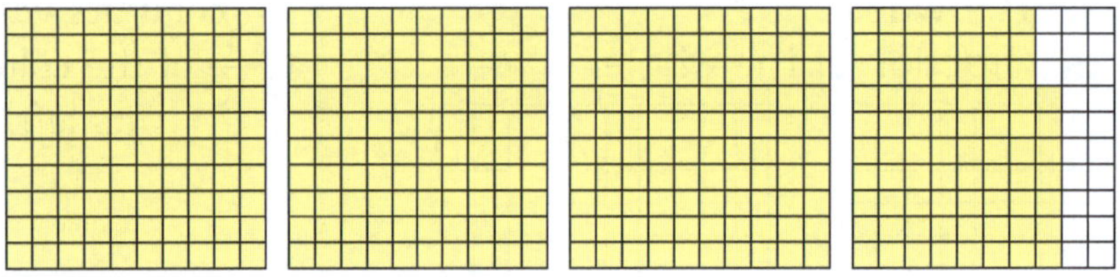

c

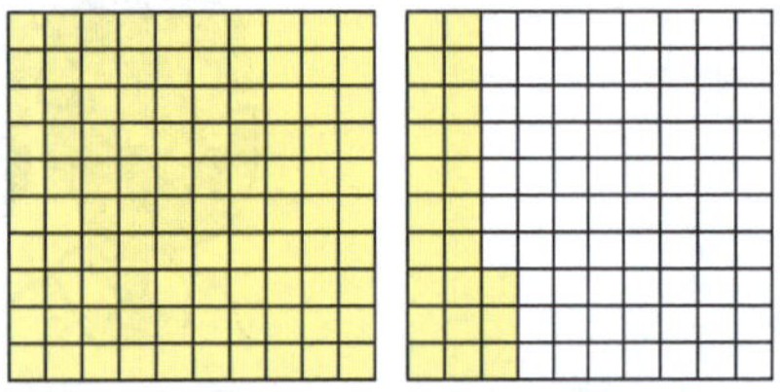

Maths detective: Investigating fractions of shapes

In each of these diagrams, a yellow triangle represents one unit fraction of the whole. For each shape, work out how many yellow triangles it would take to cover the whole shape. Use this number to work out what fraction of the shape is yellow, what fraction is red, what fraction is green and what fraction is blue.

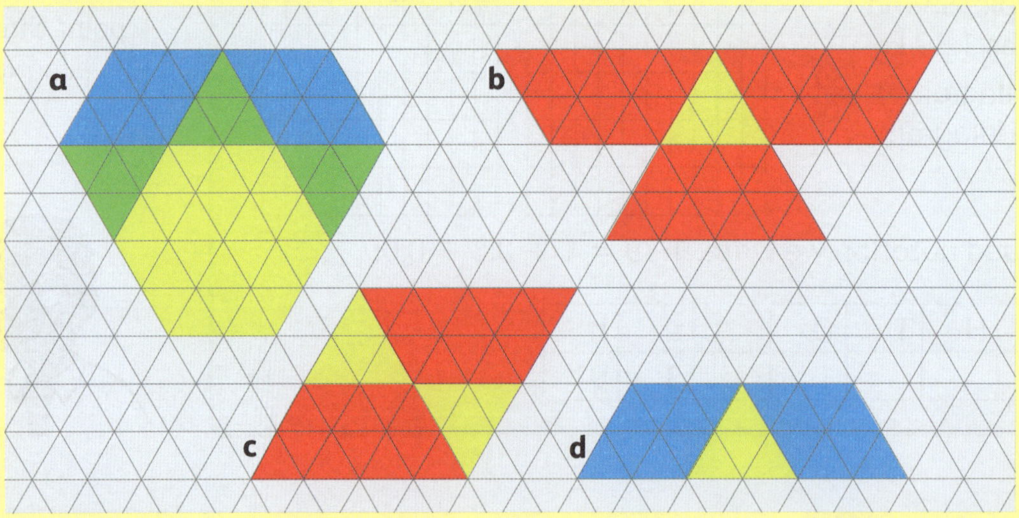

Decimals in grids and charts

Key maths idea

We can show **fractional numbers** in different ways.
A fractional number is any number that has parts that are a fraction of a whole. Proper fractions, improper fractions, mixed numbers and decimals are all types of fractional numbers.
What is the value of the shaded area in the blocks below?

Example 1

1 whole block + $\frac{23}{100}$ of a block

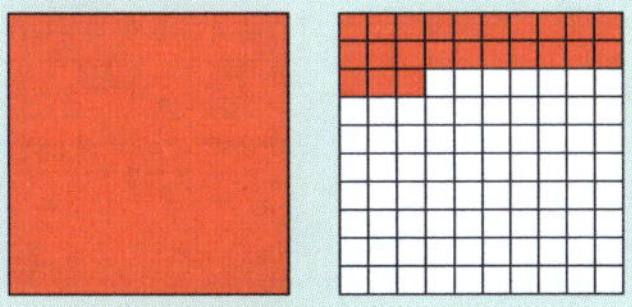

$1\frac{23}{100} = 1.23$
one point two three

H	T	O	.	t	h
		1	.	2	3

Example 2

2 whole blocks and + $\frac{57}{100}$ of a block

$2\frac{57}{100} = 2.57$
two point five seven

H	T	O	.	t	h
		2	.	5	7

Key word

fractional numbers

1 In this grid, 3 out of 100 blocks are shaded. The shaded area is equal to $\frac{3}{100}$ of the whole grid. In decimal form, we write 0.03.

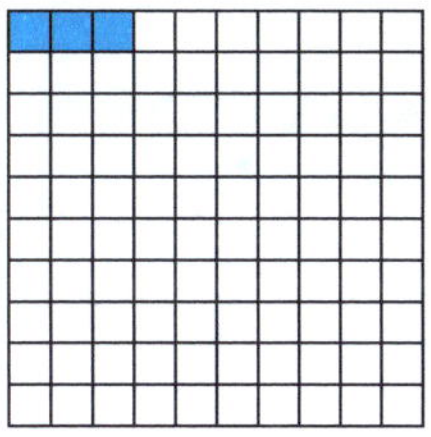

For each of the following grids, write what fraction is shaded. Write it as a proper fraction and then as a decimal.

a

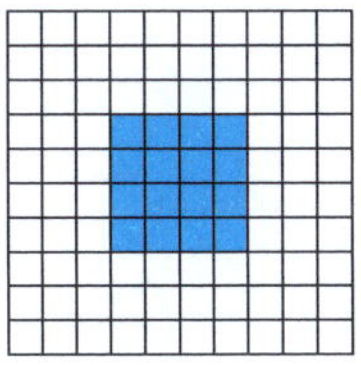

b

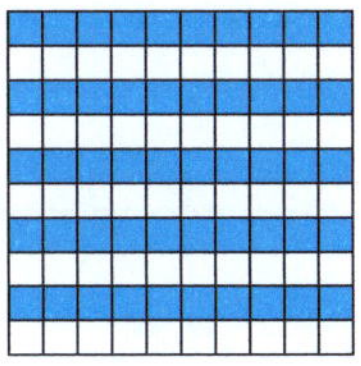

c

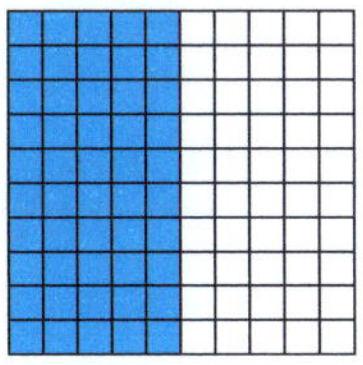

d

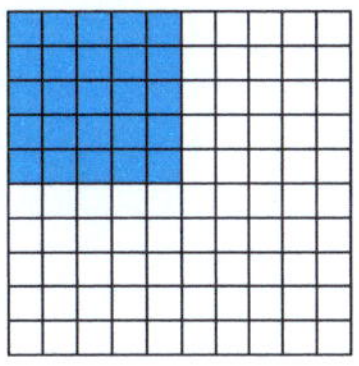

e

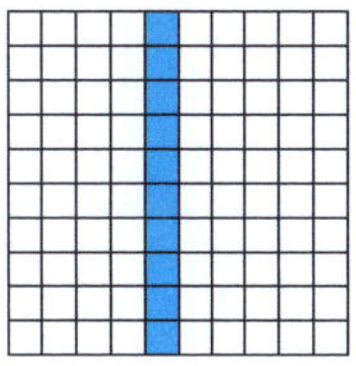

f

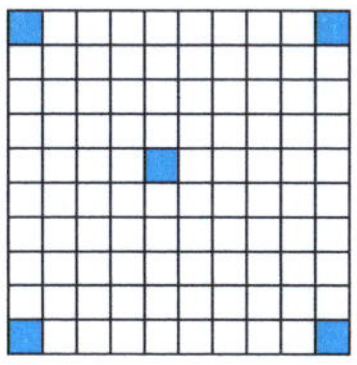

2 Answer these questions about the number in the place value chart.

Hth	Tth	Th	H	T	O	.	t	h	th
2	7	0	2	3	4	.	5	0	7

a What is the value of the 2 in the first column on the left?
b What is the value of the 7 in the last column on the right?
c What is the value of the digit in the tenths column?
d What is the value of the digit in the tens column?
e In this number, two of the digits are zero. Why do we have to put zeros in these places?
f Write the number using expanded notation.

3 Show these numbers in place value charts. You can use the empty chart below to help you. Once you have written in all the decimals, read them aloud.

Hth	Tth	Th	H	T	O	.	t	h	th

a 550.233 **b** 0.06 **c** 10.32 **d** 56 301.008

e 405 670.201 **f** 548.03 **g** 6.004 **h** 8 906.105

4 In each diagram, the whole shaded block is equal to 1. Write the value of the shaded area as a fraction (or mixed number) and as a decimal.

a

b

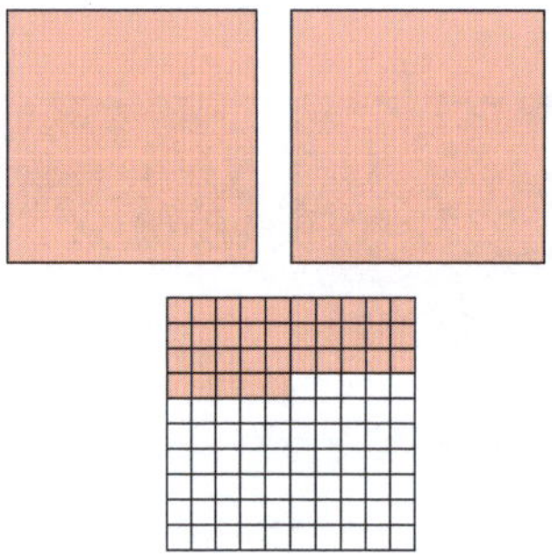

c

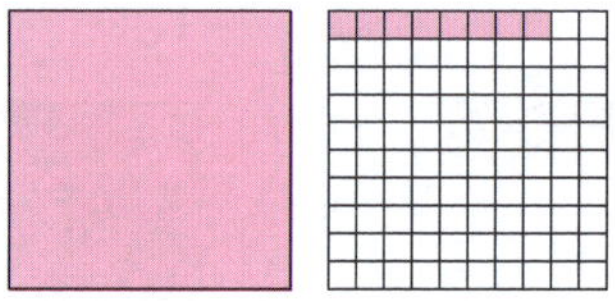

d

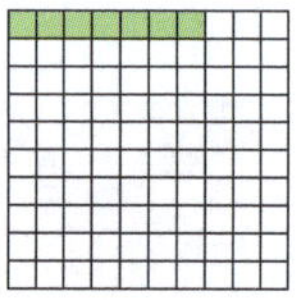

Decimals on number lines

Key maths idea

This number line is marked in tenths. It shows the fractional numbers between 125 and 127. You can see them as decimals and the equivalent fractions.

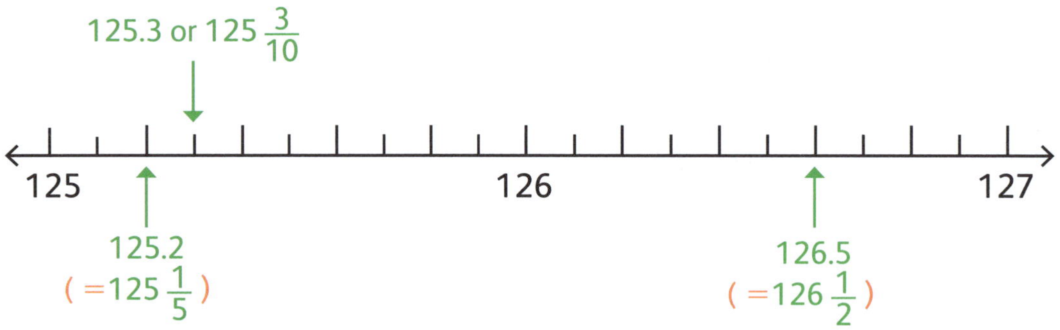

Now imagine that you can zoom in on the number line to see the section between 125.1 and 125.2.

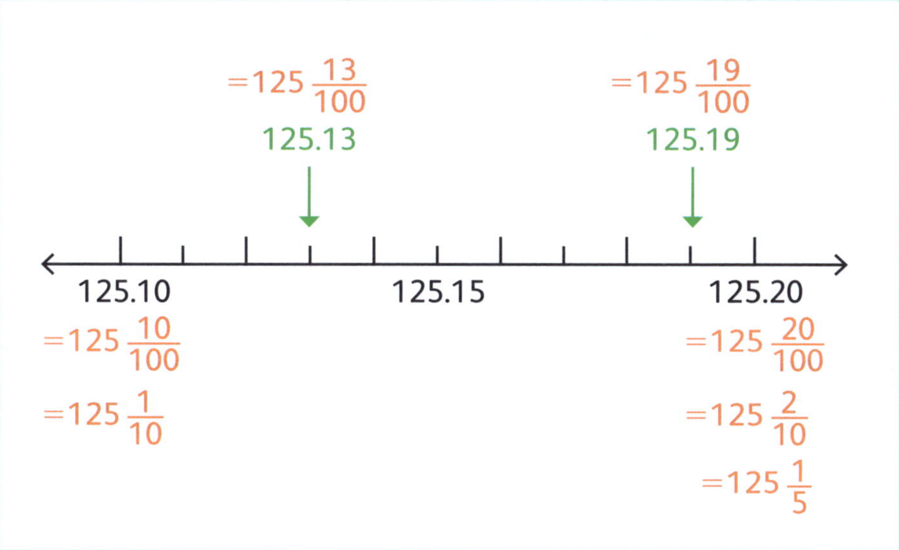

Remember: 125.1 is the same as 125.10.

1 This number line is marked in tenths. Write the value of the marked numbers as a decimal and as a mixed number.

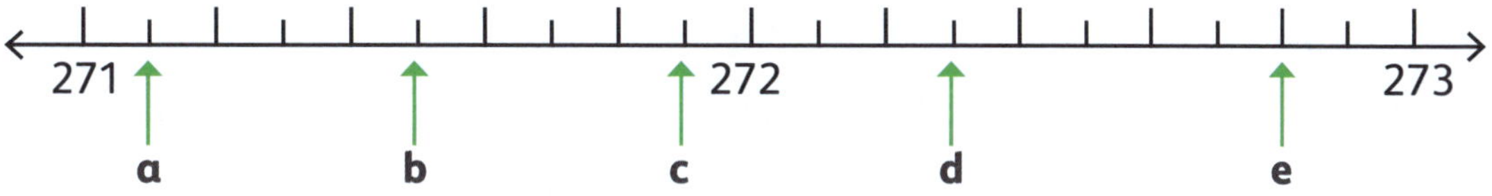

2 The next number line is marked in hundredths. Write the value of the marked numbers as a decimal and as a mixed number.

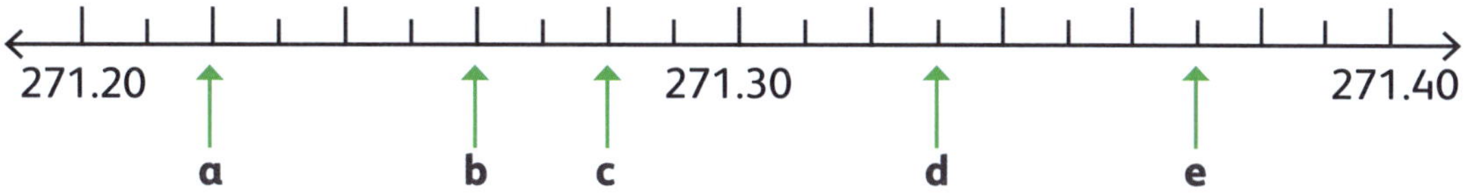

3 Complete these number sequences.

a 4.8, 4.9, 5.0, ____, ____, 5.3, ____, ____, ____

b 0.01, 0.02, ____, 0.04, 0.05, ____, ____

c 7.12, ____, ____, ____, 7.16, 7.17, ____

d 15.95, 15.96, 15.97, ____, ____, ____, 16.01, 16.02

4 In the number 145.97, the 9 has a value of 0.9 or $\frac{9}{10}$.

H	T	O	.	t	h
1	4	0	.	9	7

Write each of the numbers that follow in a place value chart like the one above.

a 12.54 **b** 3.08 **c** 0.09 **d** 3.45

e 100.98 **f** 10.07 **g** 563.25 **h** 49.99

5 For each number in Exercise 4, write the value of the red digit.

6 Write each set of decimal fractions in ascending order.

a	0.5	1.0	0.7	0.2	1.2	1.9	0.9			
b	5.01	5.1	1.5	1.05	1.11	5.11	5.11			
c	0.2	0.19	0.5	1.01	0.13	0.1				
d	0.7	10.4	22.2	111.3	1.5	2.3	0.5	12.9	1.8	52.8

7 Use decimal fractions to make the number 1 in at least ten different ways. Use a calculator if you want to. For example, 0.3 + 0.7 = 1.

Adding and subtracting decimals

Key maths idea

When you add or subtract decimals, take care to add up the digits in each place correctly. If you work in columns, you need to line up the columns and make sure the decimal point is in the correct place.

Example 1

```
  15      1.5
+ 21    + 2.1
----    -----
  36      3.6
```

Example 2

```
  58      5.8
- 24    - 2.4
----    -----
  34      3.4
```

What is similar about working with whole numbers and working with decimals?

If the numbers have different numbers of decimal places, you can use zeros as placeholders.

Example 3

14.5 − 3.02

```
    4 1
  14.50
 - 3.02
 ------
  11.48
```

Note that 14.5 is equal to 14.50 but 145 is not equal to 1 450. Why?

1 Work out the answers mentally. Add the tenths first. Then add the ones and write the total.

a	0.2 + 0.7	b	0.1 + 0.5	c	0.5 + 0.5
d	6.2 + 5.7	e	4.1 + 5.5	f	12.5 + 10.5
g	11.4 + 6.4	h	2.5 + 0.6	i	17.1 + 2.6

2 Work out the answers mentally. Make sure to subtract tenths from tenths and ones from ones.

a	0.5 − 0.4	b	1.5 − 0.2	c	8.5 − 0.5	d	9.9 − 8.3
e	0.9 − 0.8	f	1.6 − 0.6	g	0.9 − 0.3	h	1.5 − 0.4

3 Work out the answers mentally. Add hundredths together, then tenths, then ones.

a	0.44 + 1.21	b	3.62 + 0.34	c	0.05 + 0.06	d	1.12 + 1.18
e	1.5 + 2.09	f	11.5 + 10.05	g	12.99 + 1.11	h	4.04 + 6.06

4 Carmen is knitting a scarf.

- **a** Her knitting is 11.5 cm long and she knits another 5.7 cm. How long is it now?
- **b** She makes a mistake and unravels 4.2 cm of her work. What is the new length of her scarf?

5 Work in columns.

a	0.6 + 0.35	**b**	5.25 + 7.4
c	25.52 + 4.4	**d**	12.8 + 0.18
e	7.51 + 0.09	**f**	100.1 + 11.01
g	80.8 + 800.08	**h**	4.45 + 40.5

Remember to put extra zeros as placeholders if you need to.

6 Subtract. Work mentally or in columns.

a 14.7 − 0.4 **b** 12.55 − 2.4 **c** 65.56 − 1.6 **d** 1.88 − 0.9

7 Complete these decimal number sequences.

- **a** 0.5, 1.0, 1.5, ____, ____, ____, ____, ____
- **b** 0.2, 0.4, 0.8, ____, ____, ____, ____, ____
- **c** 0.6, 1.2, 1.8, ____, ____, ____, ____, ____
- **d** 2.5, 5, 7.5, 10, ____, ____, ____, ____, ____
- **e** 0.3, 0.6, 0.9, ____, ____, ____, ____, ____
- **f** 0.8, 1.6, 2.4, ____, ____, ____, ____, ____

Hint: you can use what you know about place value, adding decimals and even multiples to help you.

Real-world maths: Using decimals in everyday life

We often use decimals in real life. Here are some examples:

- **Money**: Many countries use dollars and cents. \$1 = 100 c.
  That means 1 c = 0.01 of one dollar.
  \$125.85 means 125 **dollars** and 85 **cents**.
- Mass: 1 kg = 1 000 g.
  One gram is one thousandth (0.001) of a kilogram.
  625 g = 0.625 kg
- Lengths and distances: 1 km = 1 000 m.
  One metre is one thousandth (0.001) of a kilometre.
  0.947 km = 947 m
  100 cm = 1 m
  58 cm = 0.58 m

Key words

money
dollars
cents

1 A family is on a beach holiday. At the store, the owner says that each item will cost \$65.50 less. Write the new price for each item.

2 Four athletes took part in a 50 m sprint. Their finishing times are shown in the table.

- **a** Who finished first and who finished last?
- **b** How much faster was Leon than Keisha?
- **c** What was the difference in seconds between Leon's time and Nikki's time?
- **d** Which of these is the closest estimate for the time these runners would take to run 100 m? Give a reason for your choice. (10 seconds; 2 minutes; 20 seconds)

Athlete	Finish time
Keisha	8.8 seconds
Devone	9.0 seconds
Leon	8.6 seconds
Nikki	9.1 seconds

3 **a** Write the capacity of each of these five bottles from smallest to largest.

b Express the capacities in millilitres.

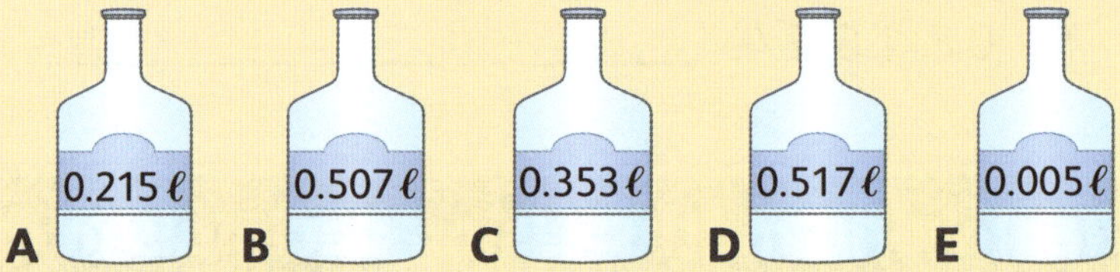

4 Write the value of the 7 in each of these numbers.

1.75 litres | 7 000 kg | 0.7 m | 75 ml | 7 g | 17 mm

5 Amari lives 545 m from school. Akiel lives 0.65 km from the school and Michael lives 0.1 km closer to school than Akiel.

- **a** How far does Amari live from school in km?
- **b** Who lives closer to the school: Michael or Amari?
- **c** How many metres does Michael live from the school?

6 Solve these problems. Write your answer as a mixed number.

- **a** Brianna uses $\frac{1}{5}$ m of wire to make a pair of earrings. How many pairs of earrings can she make using 1 m of wire?
- **b** Your mother makes face cloths. She uses $\frac{1}{3}$ of a metre to make 3 face cloths. How many metres of cloth does she need to make 12 face cloths?

7 Solve these problems. Write your answer as an improper fraction.

- **a** Ruth made a dozen cookies. She uses $\frac{2}{6}$ of a tub of butter for her recipe. How much butter will she need for 5 dozen cookies?
- **b** Devon makes wire animals. He uses $\frac{7}{8}$ m of wire to make one wire animal. How much wire will he need if he makes 6 wire animals?

Real-world maths: Planning an aquarium

A Grade 4 class is planning to set up a classroom aquarium. Read this information about setting up an **aquarium** to help them plan.

Key word
aquarium

How to set up an aquarium

When you set up an aquarium, you will need to think about:

- the size of the tank
- the number of fish
- the types of fish
- equipment you need to keep your aquarium safe, clean and healthy.

The number of fish you can get depends on the size of your tank. Experts say that you can fit 1 inch of fish per 1 gallon of water. So a tank that holds 10 gallons of water can house 10 inches of fish. You need to consider the length that the fish will be when they are fully grown. 10 inches of fish = 5 fish that are each 2 inches long at their fully grown length, or 2 fish that are each 5 inches long.

Some fish, such as guppies, platy fish and tetras, like to move in groups we call schools. This means you should have at least six or more of one type of fish. Guppies grow up to 2 inches in length. Tetras are 2.5 to 3.5 inches long, and platy fish grow up to 2.8 inches.

Your equipment might include:

- a water filter – to keep the water clean
- a UV steriliser – to help control algae and infections
- a heater – to regulate the temperature (for tropical fish such as beta or platy fish)
- drops to remove chlorine from tap water
- gravel for the bottom surface
- plants to decorate and oxygenate the tank
- an electric air pump
- a stand, if you do not have a big enough surface for the tank
- a hood (if you have jumping fish)
- ornaments or toys for decoration.

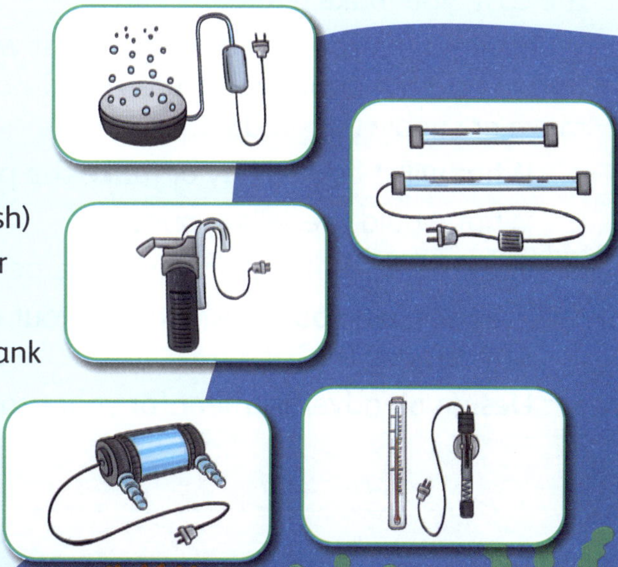

> 1 gallon = 3.8 litres
> 1 inch = 2.54 cm

1 How many litres of water fit into a tank that holds:
   a 10 gallons?
   b 18 gallons?
   c 25 gallons?

2 Work out the centimetre length of a fully grown:
   a guppy   b tetra   c platy

3 Michael wants to put a school of 8 tetras and 10 guppies in his tank. What tank size does he need?

4 In groups, plan what you would need to set up an aquarium in your class.
   a Find out:
   - What materials do you need?
   - What size tank can you fit in your classroom?
   - What kinds of fish will you get and how many of each?
   - What would the basic supplies and the fish cost?

   b Find out which supply stores sell aquarium equipment and fish. Find out the prices of the items you need. Draw up a budget for a simple classroom aquarium.
   c Are there any products that are not available in your area? Make a list of these.

5 An entrepreneur is someone that starts a new business. **Entrepreneurship** includes all the activities of planning, starting and running a business. Entrepreneurs have to identify a need and work out how to supply their **goods** or **services**. They also need to draw up a **budget**. **Profit** is the difference between the **cost** and the selling **price** when you sell something for more than it cost to produce. If the price is lower than the cost, you make a **loss**.

   Work in groups. Identify a product that was difficult to find for your aquarium. Think about how you could make or supply the product.
   a What will it cost to buy or make the product?
   b What would the selling price be?
   c How much profit would you make per item?
   d Where could you sell it? Explain your choice of location.
   e Design an advertisement for your product.

Key words
entrepreneurship
goods
services
budget
profit
cost
price
loss

What I have learned

1 Use these fractions to explain each term. You can use a sketch, a number line, a diagram or write a few sentences for each term.

$\frac{1}{2}$	0.5	0.01	$\frac{1}{100}$

a decimal fraction
b decimal point
c decimal places
d tenths
e hundredths

2 Write your own example to show how adding or subtracting decimals works in the same way as adding or subtracting whole numbers.

Practice questions

1 Draw your own diagrams to show these mixed numbers.
a $1\frac{1}{2}$ b $2\frac{1}{4}$ c $4\frac{1}{3}$ d $1\frac{3}{4}$ e $1\frac{7}{10}$

2 Write each improper fraction as a mixed number.
a $\frac{5}{2}$ b $\frac{7}{4}$ c $\frac{9}{5}$ d $\frac{8}{6}$ e $\frac{13}{4}$

3 Match the improper fraction with its equivalent mixed number. Write the pairs in your notebook.

$\frac{5}{4}$	$\frac{6}{5}$	$\frac{8}{7}$	$\frac{4}{3}$	$\frac{5}{2}$	$\frac{9}{7}$	$\frac{7}{2}$

$1\frac{1}{3}$	$2\frac{1}{2}$	$1\frac{1}{4}$	$1\frac{2}{7}$	$3\frac{1}{2}$	$1\frac{1}{5}$	$1\frac{1}{7}$

4 Look at these rectangles. What fraction of each rectangle will you be left with if you remove one piece?

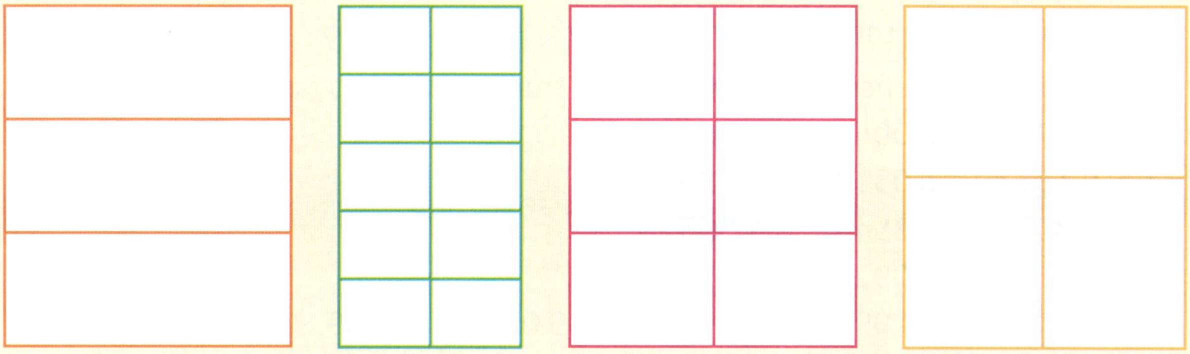

5 Five people take part in a fun run. This is how long it took them to finish the course.

Romano	Ayla	Jalisia	Devon	Ruth
$\frac{2}{5}$ of an hour	$\frac{5}{12}$ of an hour	$\frac{7}{10}$ of an hour	$\frac{1}{4}$ of an hour	$\frac{3}{6}$ of an hour

a Who ran for more than half an hour?
b What fraction of an hour did they have left when they completed their run?
c Work out how many minutes each person ran.
d How many minutes of an hour did the fastest person have left?

6 Calculate the answers. Show your working out.

a $\frac{2}{5} + \frac{2}{15}$ b $\frac{5}{9} + \frac{1}{3}$ c $4 - \frac{2}{3}$

d $\frac{10}{12} + \frac{1}{4}$ e $\frac{3}{4} + \frac{1}{12}$ f $\frac{7}{8} + \frac{1}{2}$

7 In this table the sum of the fractions in each row and each column is the same.

A	$\frac{1}{4}$	$\frac{2}{3}$
B	$\frac{1}{2}$	$\frac{7}{12}$
C	$\frac{3}{4}$	D

a Work out the value of A, B, C, D.
b Write each fraction from A to D in its simplest form.

8 Jasmine helps her mom cook jerk chicken and rice to sell at a market. She mixes $2\frac{1}{2}$ kg of chicken with $\frac{1}{4}$ kg of vegetables. Later she cooks $1\frac{1}{3}$ kg of rice.

a What is the total mass of the food she cooked?
b They sell 15 portions of the chicken dish at lunchtime. Each container costs \$220. How much money did they make from the sale of the 15 portions?

9 Benjamin has $\frac{2}{3}$ m of rope and Lloyd has $\frac{5}{7}$ m of rope.

a Whose piece of rope is longer?
b What is the total length of rope they have?
c If they cut $\frac{1}{3}$ m from Lloyd's rope, how much does he have left?
d How much do they have left together?

10 Write each set of measurements in order from greatest to smallest.

a 2.04 km 2.004 km 2.104 km
b 0.131 ℓ 0.031 ℓ 0.13 ℓ
c 1.305 kg 1.35 kg 1.503 kg
d 1.92 kg 1.092 kg 1.29 kg
e \$9.00 \$0.99 \$9.90 \$9.09 \$0.09 \$9.99

10 Self check

Check you can do this:	If you have difficulty, try this:
1 Identify proper fractions, improper fractions and mixed numbers.	Go over the work on page 143 to 145. Practise modelling fractions in different ways.
2 Express fractions as decimals and vice versa.	Revise the work on tenths and hundredths on pages 146 and 147 and the exercises you completed from page 148. Also look at the grids and charts on pages 148 and 149.
3 Complete sequences of decimal numbers.	Revise the decimal number lines on pages 150 and 151 and Exercise 7 on page 153.
4 Convert between different types of fractions (proper, improper, mixed numbers, decimals).	Go back over the modelling work on pages 144 and 145. Remind yourself how you model an improper fraction and convert between improper fractions and mixed numbers. For converting between decimals and fractions, look at the grids and charts on pages 149 and 150.
5 Add and subtract using decimals and other fractions.	Revise pages 152 and 153.
6 Solve problems involving decimals and other fractions.	Revise the activities on pages 153 to 156.

Unit 2: Measurement

Chapter 11 Length and area

In this chapter you will explore this focus question:

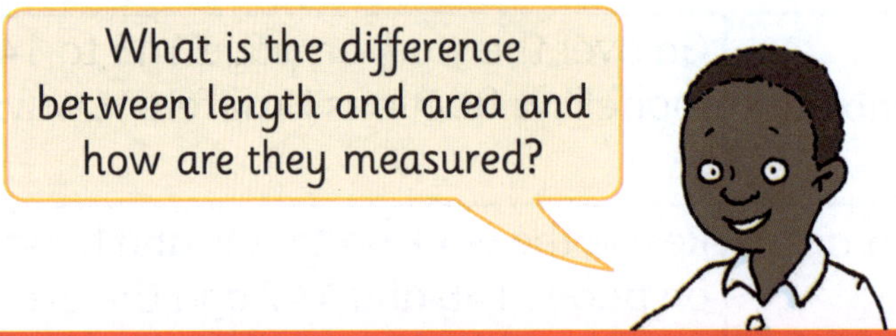

To answer this question, you will:

- calculate perimeters of regular and irregular polygons
- find the area of various objects and figures
- distinguish between units of length and units of area
- calculate area in square units and using grids.

Starting point

Here is a **plan** of a classroom. A plan is a drawing or diagram of a place or object that shows its **dimensions**. On this plan, 1 cm on the square **grid** represents 1 m in real life. A grid is an arrangement of lines that form square units. Grid paper makes it easier to draw plans.

1 **Perimeter** is the distance around a shape or the **length** of its boundary.

 a Explain in your own words how you can work out the perimeter of this classroom.

 b Work out the perimeter.

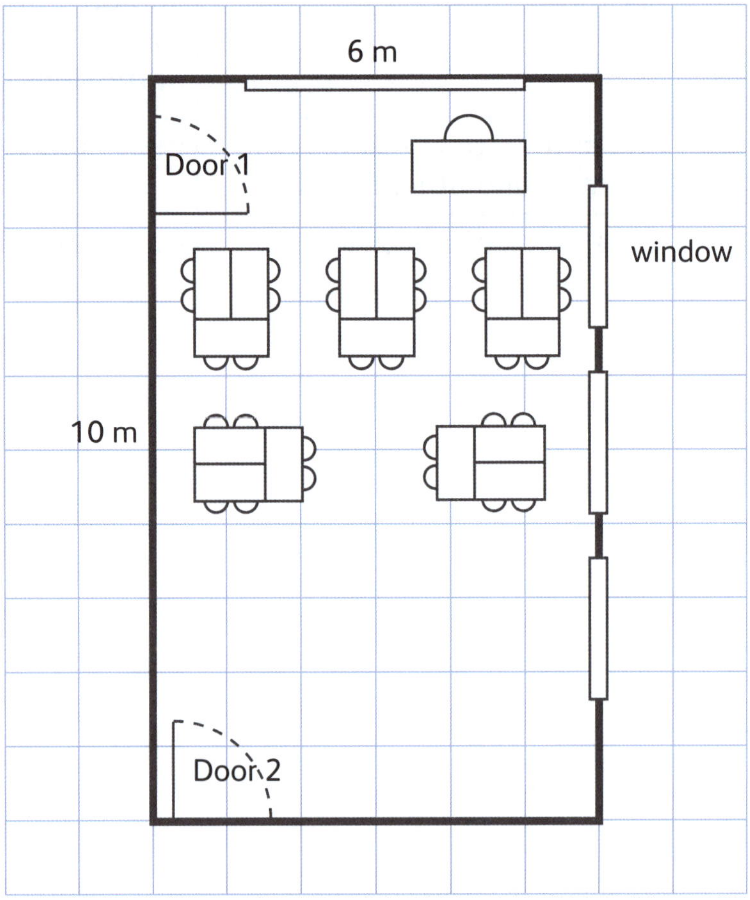

Teacher's desk – found in top right corner.
Students' desk – any of the remaining five.

2 The teacher sits at the front desk and the students sit at the grouped desks. Each group is made up of two or three desks.
  a How many groups of desks are there in this classroom?
  b Each group of desks seats six students. How many grid squares make up each group? Count the grid squares for each group, including the squares for the chairs.
  c Do you think that the classroom could fit desks for 12 more students?
  d Do you think that the classroom could fit desks for 18 more students? Why or why not?

Key words
plan
dimensions
grid
perimeter
length

3 Use a ruler to measure the sides of these shapes. Add them up to find the perimeters.

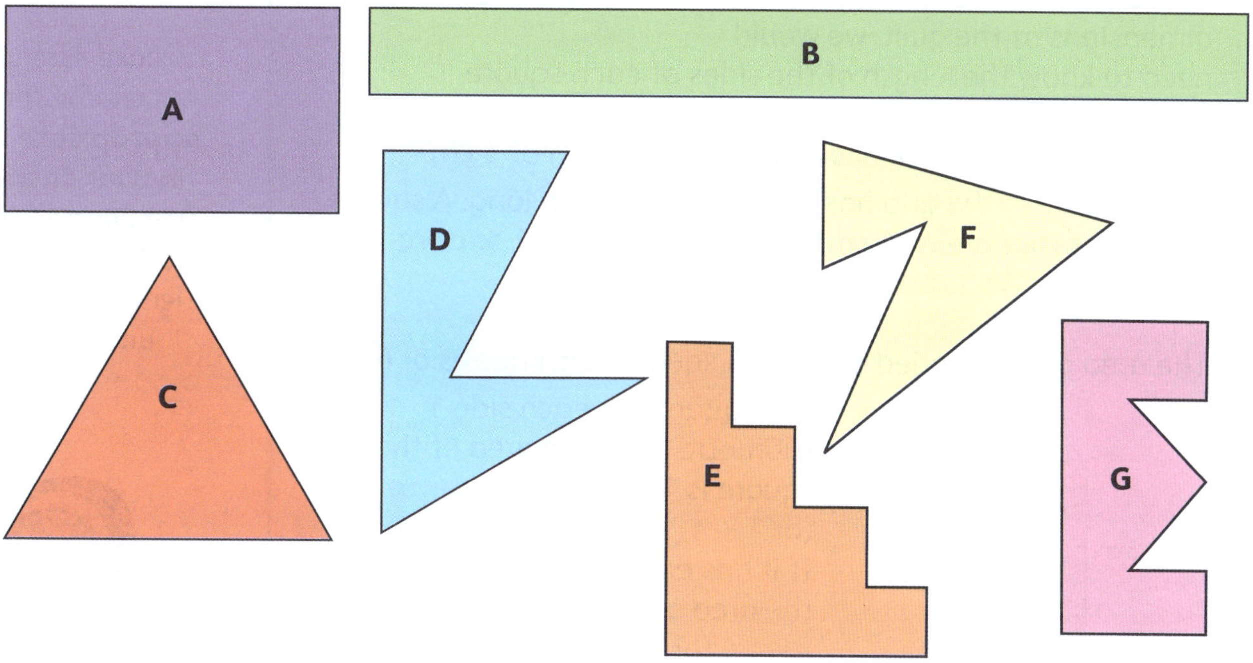

4 Work out the perimeter of these shapes:
  a a square with each side 22 cm long
  b a triangle with two sides of 13 cm and one side of 5 cm
  c a regular pentagon with all its sides 1.2 cm long.

5 A five-sided piece of lawn has two sides of 2.75 metres long and two sides of 6.5 metres long. Its perimeter is 23.25 metres. How long is the fifth side?

6 Find some circular objects, such as a plate, a mug or lids of jars. Use string to help you measure the perimeter of each object.

Area

Key maths idea

Area is the amount of space a 2D shape takes up on a flat **surface**. Look at this quilt. It has 6 squares in each row and 7 squares in each column from top to bottom.

We can work out the total number of squares:

$6 \times 7 = 42$ squares

In order to know the actual dimensions of the quilt, we would need to know the length of the sides of each square.

In this grid, each block has a length and width of 1 cm. Each square in the grid has sides that are 1 cm long. A square with sides that are all 1 cm long has an area of 1 **square centimetre**. We write it like this: 1 **cm^2**.

The area of the shaded shape is 6 square centimetres or 6 cm^2.

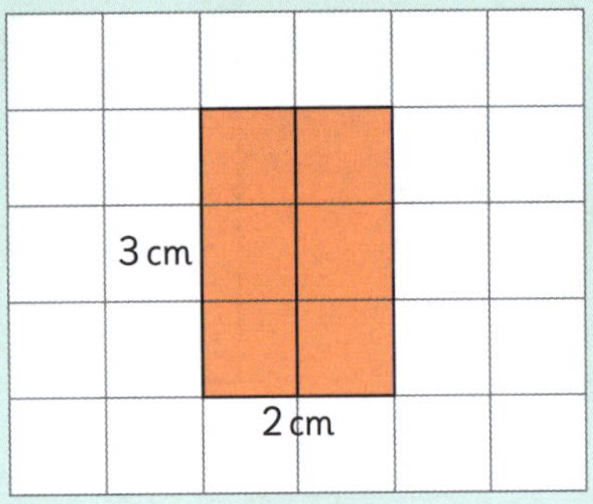

If a square has each side 1 kilometre long, the area of the square is 1 **square kilometre** (or 1 **km^2**).
If it has each side 1 metre long, the area of the square is 1 **square metre** (or 1 **m^2**).

Key words

area
surface
square centimetre (cm^2)
square kilometre (km^2)
square metre (m^2)
appropriate

Make sure you choose the **appropriate** units in your answers. Ask yourself: Am I working with units (length) or square units (area)?

1 In this grid, each square of the grid represents 1 cm^2. Count the squares to work out the area of each shaded shape on the grid. Write it using cm^2.

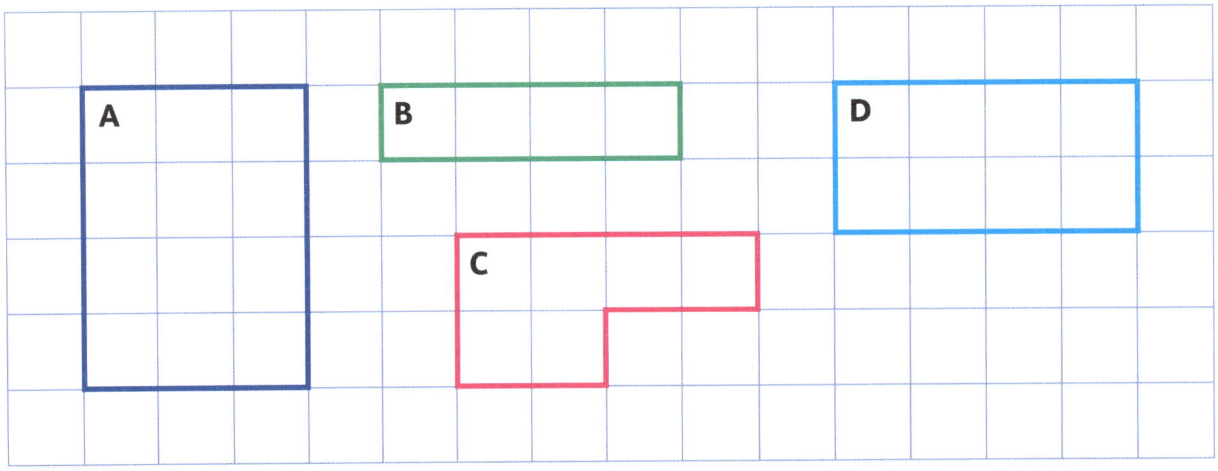

Measurement

2 Look at these shapes.

a Which shape has the greatest perimeter?

b Which shape has the greatest area?

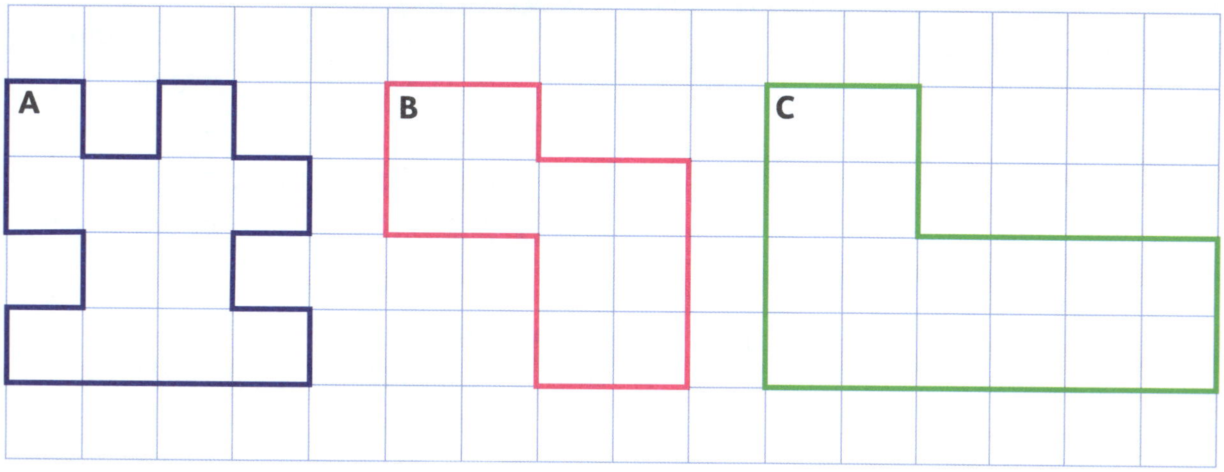

3 Use a piece of grid paper. How many different shapes can you draw that have an area of 24 cm^2?

4 On grid paper, draw shapes with these perimeters:

a 8 cm b 15 cm c 120 mm d 21 cm e 225 mm

Calculate the area of each shape. Compare with a partner. Were your areas the same or different? Why?

Maths detective: Doubling dimensions

Use your maths detective skills to investigate what happens when you double the dimensions of a shape.

Use a spreadsheet or other software to draw a rectangle that has a length of 3 cm and a width of 2 cm.

1 Investigate what happens to the area of the rectangle if you:

a double the length but keep the width at 2 cm

b double the width but keep the length at 3 cm

c double both the width and the length.

2 Do the same investigation again, for rectangles of 4×5 cm and 6×6 cm.

3 What pattern do you notice? Describe the pattern in your own words.

Finding areas of irregular shapes

Key maths idea

Look at these shapes.

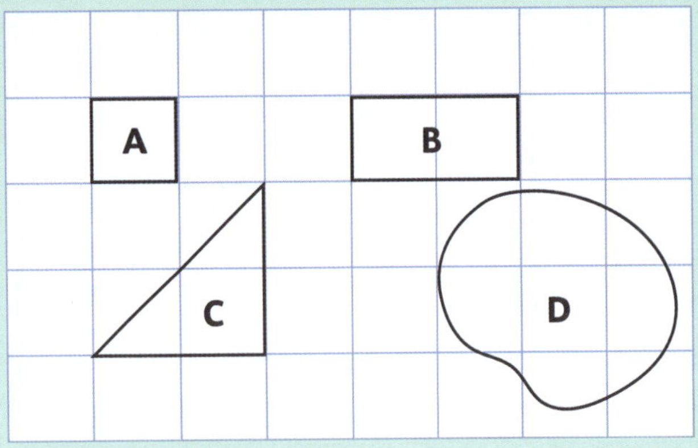

Shape A has an area of 1 **square unit**. Shape B has an area of 2 square units. What about **shapes** C and D?

Sometimes you can't count an **exact** number of whole squares. But you can **estimate** the area of the shape:

Step 1: Count all the whole squares.
Step 2: Combine pairs of half squares to make whole squares.
Step 3: Parts that are greater than half a square – round up to 1 square.
Step 4: Parts that are less than half a square – round down to 0.
Step 5: Add up the squares to get the total estimate of the area.

Key words

square unit
shape
exact
estimate
approximate

When we estimate the area of a shape, we work out its **approximate** area.

1 A class of students collected some big leaves and traced them onto grid paper. Each block of the grid = 1 cm^2.

a Estimate the area of each leaf.

b Collect some leaves of your own. Trace them onto grid paper. Estimate and work out the area of each leaf.

2 The word TOTAL has been spelled out using square tiles.

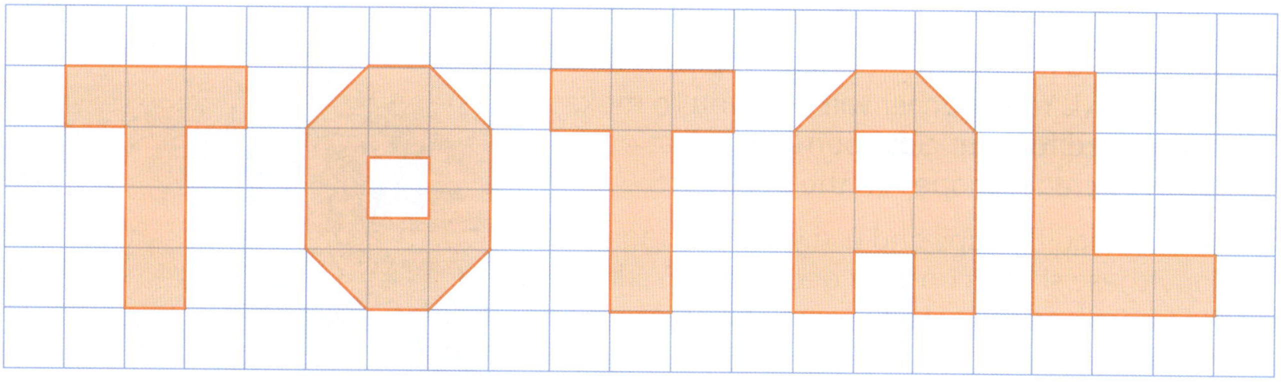

a What is the area of each letter in square units?

b What is the total area of the letters in square units?

c Why did you only need to find the area of four of the letters?

d If each square has a length of 1 cm, what is the perimeter of the letter L?

3 Write your own short word on square grid paper. Swap with someone and answer parts *a* and *b* of Exercise 1 for your new word.

4 Use square tiles to find the area of two different flat surfaces in your classroom. The surfaces should have a square or rectangular shape.

a First estimate how many tiles you will need to cover each surface.

b Then use the tiles to cover the surface.

c Count the tiles.

d Check your answers using multiplication.

5 Your teacher will give you square grid paper. Work with a partner and explore these ideas.

a Draw outlines of circular lids and other odd-shaped objects onto the square grid paper. Count the squares and compare the areas of the shapes.

b Estimate how many squares will be covered if you put one hand, with fingers together, on the square grid paper. Next, trace around one hand with fingers together. Count how many squares your hand covers and see how close you are to your estimate.

c Trace all six faces of a box on 1 cm square grid paper. Find the area of each face and add them to find the surface area.

6 Look at the four shapes on this grid. They have been drawn on a square grid to show the size of the area of each shape. Look at the shapes and the grid.

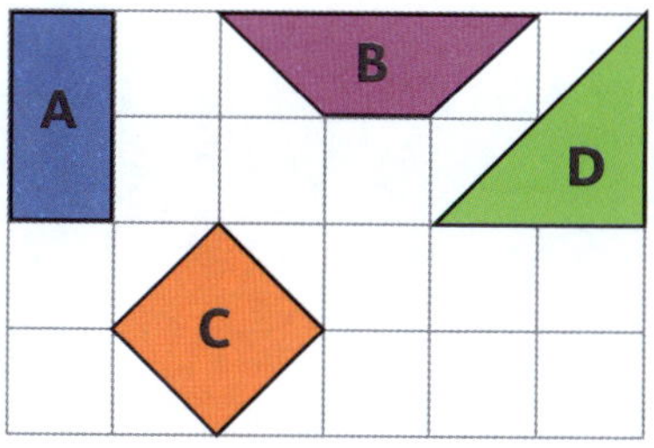

a What similarities and differences do you notice about the shapes?

b If each square on the grid measures 1 cm^2, then what is the area of each shape?

c Now measure the perimeter of all four shapes. Which perimeter is the shortest? Which perimeter is the longest?

Measurement

7 Use grid paper to measure the area of:

a a slice of bread **b** a cookie **c** a money note.

8 A geoboard is a board with pegs or pins that form a grid. Your teacher will help you to make a geoboard for your classroom. Use elastic bands to create shapes on the geoboard.

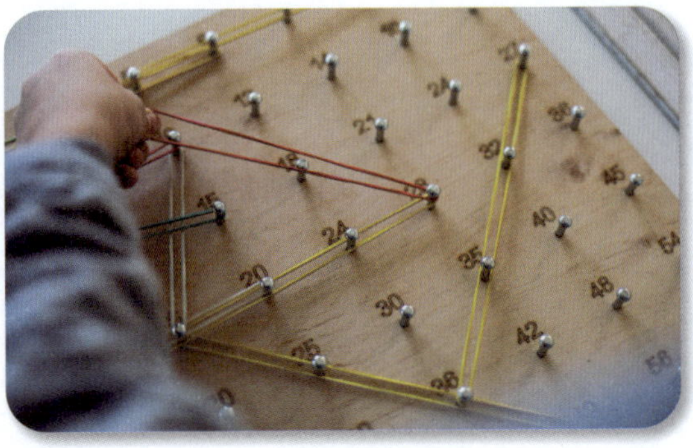

Work in small groups and use your geoboards. Copy these shapes on your geoboard. Work out the perimeter and area for each shape.

a

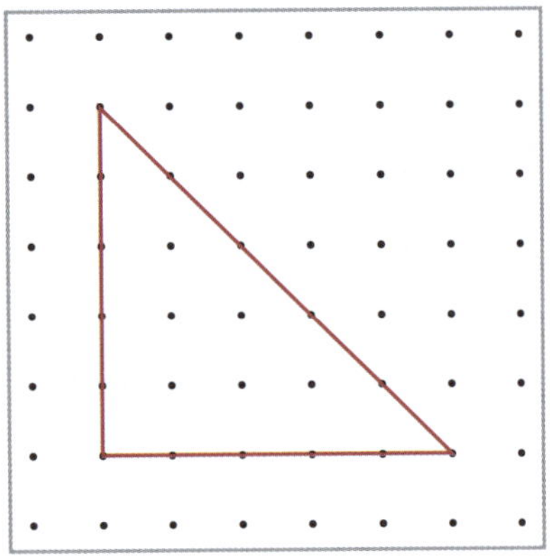

b

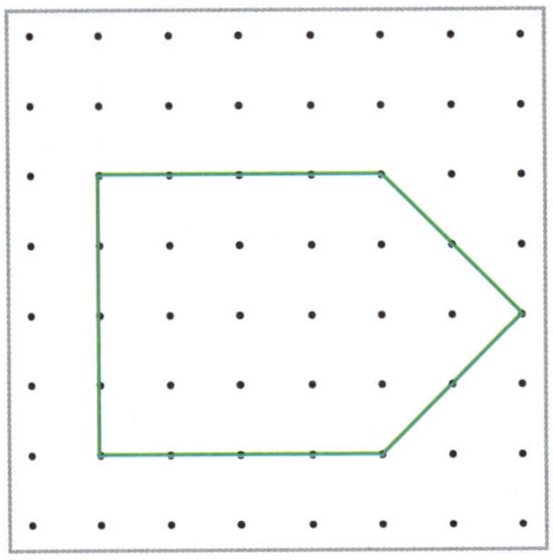

c

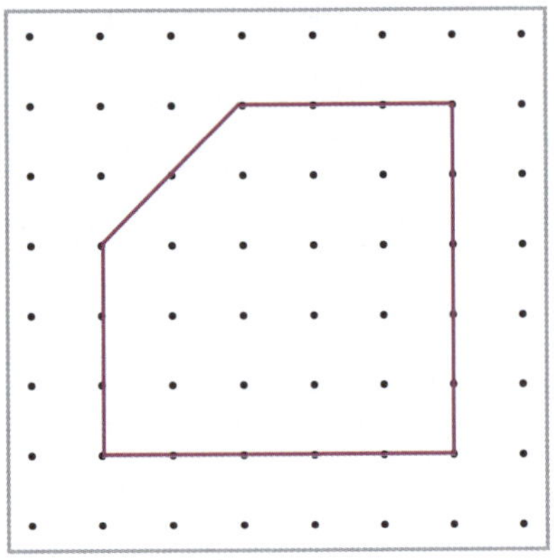

d

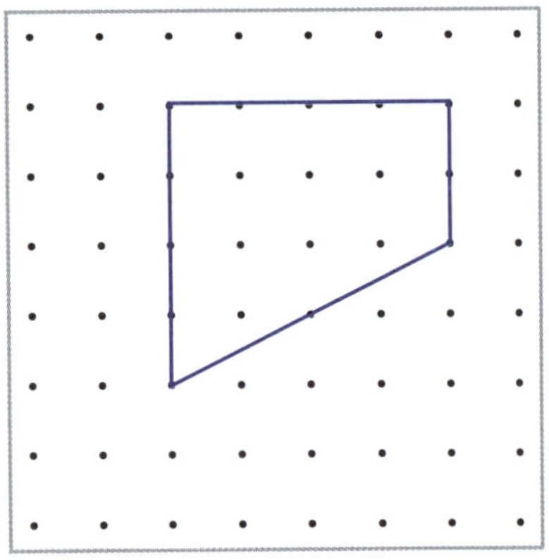

e

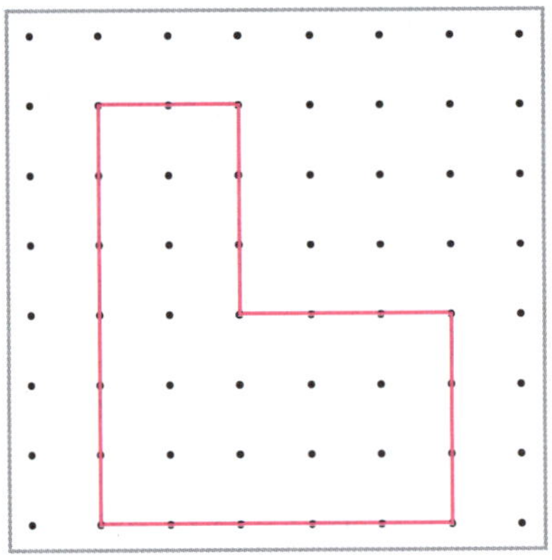

f

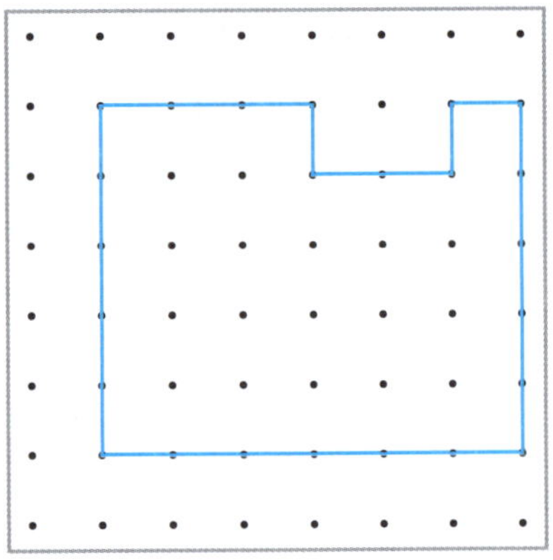

9 Read the dimensions and make the shapes on your geoboard. On grid paper, draw each shape and label the perimeters and areas. (P = perimeter and A = area)

a P = 10 lengths
b P = 14 lengths
c P = 12 lengths
d P = 8 lengths
e A = 13 squares
f A = 16 squares
g A = 24 squares
h A = 19 squares

Measurement

Distances between places

Key maths idea

You learned in earlier chapters about the relationship between kilometres and metres. If a distance is not an exact number of kilometres, we can write it in different ways, for example:

Distance in km	Distance in m	Distance in km and m
5.5 km	5 500 m	5 km 500 m
6.25 km	6 250 m	6 km 250 m
8.739 km	8 739 m	8 km 739 m

Remember:
1 000 m = 1 km

1 Write the following in km.

a 400 m **b** 1 200 m **c** 4 380 m **d** 8 102 m
e 50 m **f** 7 m **g** 706 m **h** 11 295 m

2 Write the following in m.

a 20 km **b** 7.5 km **c** 0.25 km **d** 0.001 km
e 1.4 km **f** 0.06 km **g** 11.511 km **h** 9.80 km

Real-world maths: Parishes of Jamaica

This map shows the parishes of Jamaica. The scale of the map is 1 cm: 15 km. That means 1 cm on the map represents 15 km in real life.

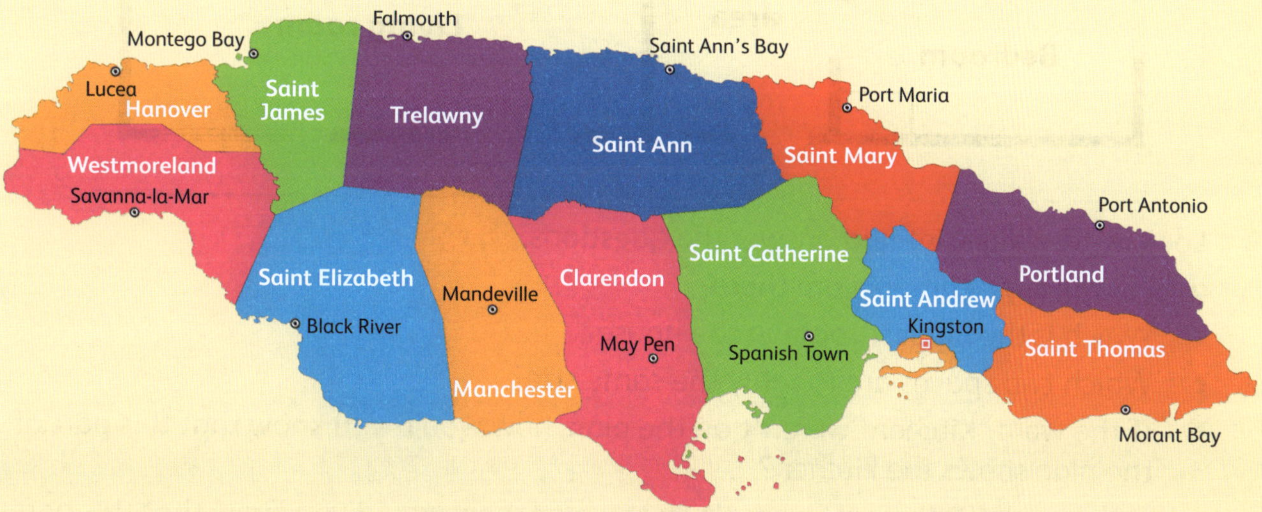

1 Use a piece of string to measure the perimeter of each parish on the map.

2 Use the string to work out the direct distance from:
  a Port Maria to Kingston
  b Kingston to Montego Bay.

3 When you travel between these places, you may travel a longer distance to get there. Why is that?

4 Use an atlas or an online map of your area (on a computer, tablet or smartphone). Find an example of places that are:
  a 1 km away from the church
  b 1 km away from the school
  c 1 km away from the beach.

Real-world maths: Careers that use perimeter and area

Designers, builders and **architects** are just some of the people who use perimeter and area in their daily work. A **blueprint** is a copy of an architect's drawing with the details and dimensions of the building.

In the past, the copy was printed on blue paper, but today most plans are on white paper. A blueprint is also called a **floor plan**.

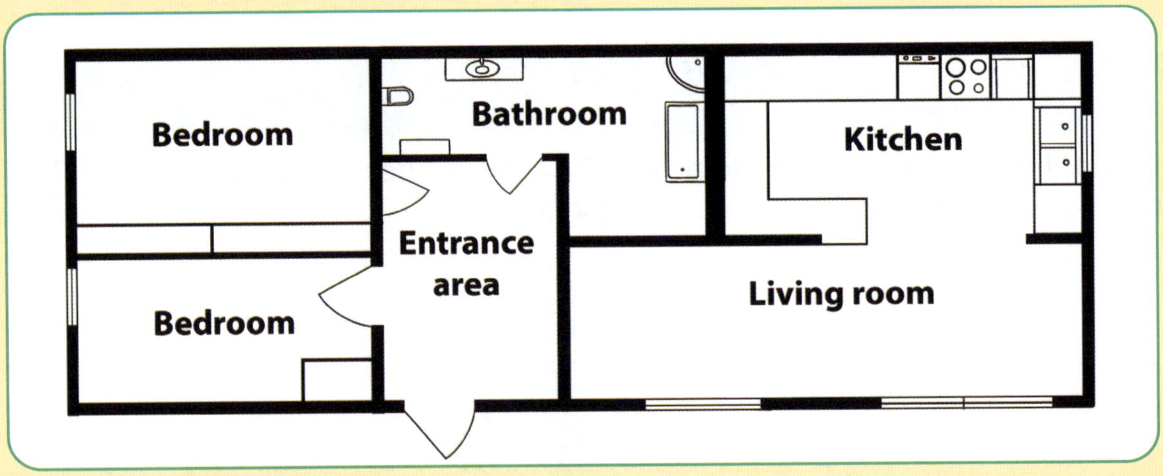

1 Look at this blueprint and answer the questions.
  a How many bedrooms are there?
  b Which is the biggest room in the house?
  c Which two spaces are roughly the same size?
  d If the word 'Kitchen' weren't on the plan, how would you know that this part of the plan shows the kitchen?
  e If the word 'Bathroom' weren't on the plan, how would you know that this part of the plan shows the bathroom?
  f Draw the symbols that show where the windows will be built and where the doors will be built.

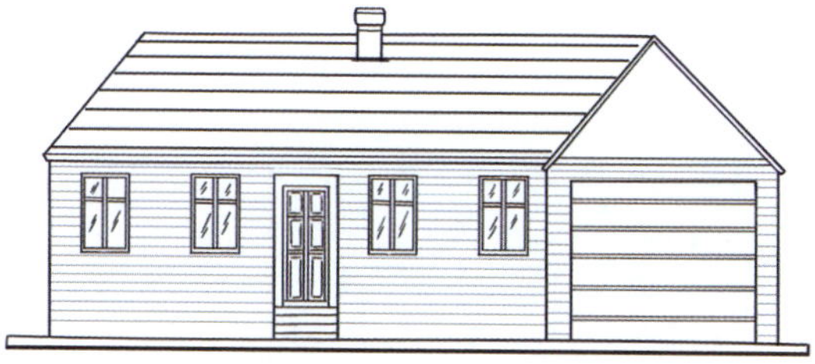

Front view

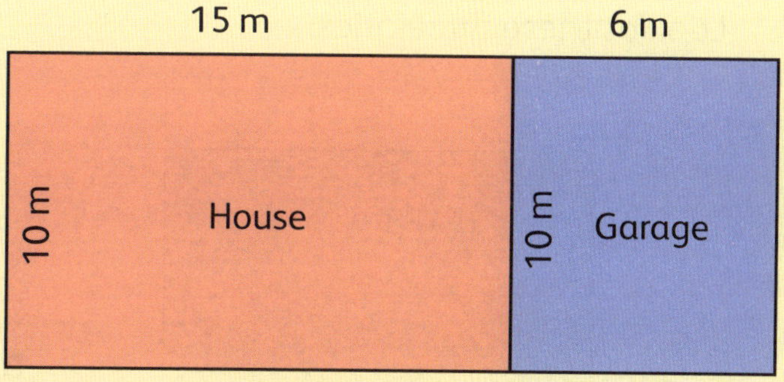

Simple floor plan

2 a Calculate the perimeter of the house only.
  b Calculate the perimeter of the garage only.
  c What is the total perimeter measurement of the house and garage together?

3 a Now use grid paper to create a drawing or plan of an imaginary building that you would like to have. Get creative and think of buildings like a dog house, doll house, tree house or a home you might want to live in one day.
  b Once your drawing is complete, count the squares and work out the perimeters and areas of the walls and rooms in your house. Write these in your notebook.

Key words
- architect
- blueprint
- floor plan

What I have learned

Draw a sketch of three different shapes that have the same area. Label it with sentences that explain the meanings of these terms:

perimeter	area	length	square	unit

Practice questions

1 Write the missing words that will complete each sentence.
  a The _____ is the distance around the outside of a shape.
  b To calculate the perimeter you use the length of each _____ and then _____ the measurements.
  c The area of a shape is the number of _____ that you need to _____ the surface.
  d You can find the area by _____ the number of squares in the shape.

2 Draw any shapes with a perimeter of:
  a 32 cm b 19 cm c 200 mm d 29 cm

3 Make up two mathematics problems about perimeters and swap your problems with a partner. Solve each other's problems.

4 Look at these three grids.

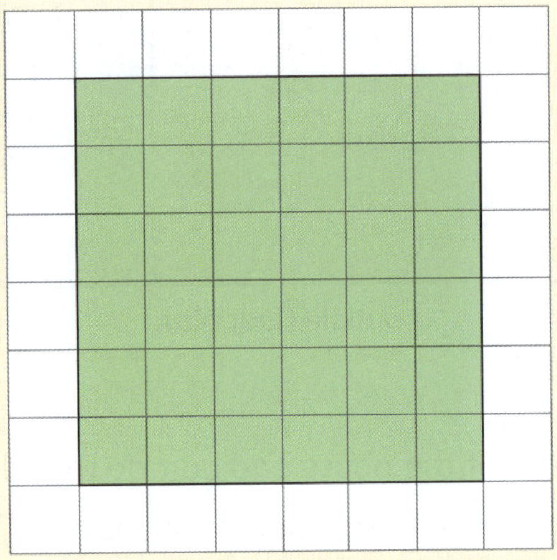

A

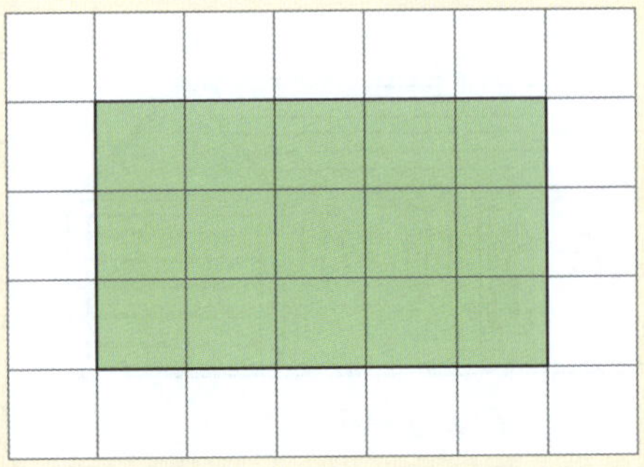

B

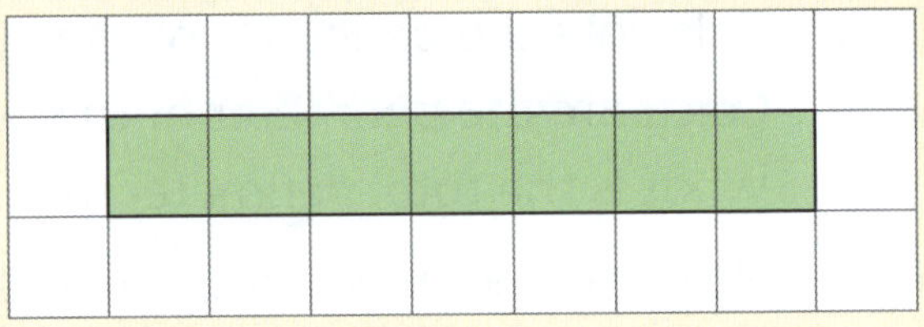

C

- **a** What is the area of each of the green areas in shapes A to C?
- **b** What is the area of the white areas in each shape, A to C?
- **c** Work out the total area of each shape.
- **d** Now work out the perimeter of each shape.

5 Your aunt is knitting a blanket for a doll.

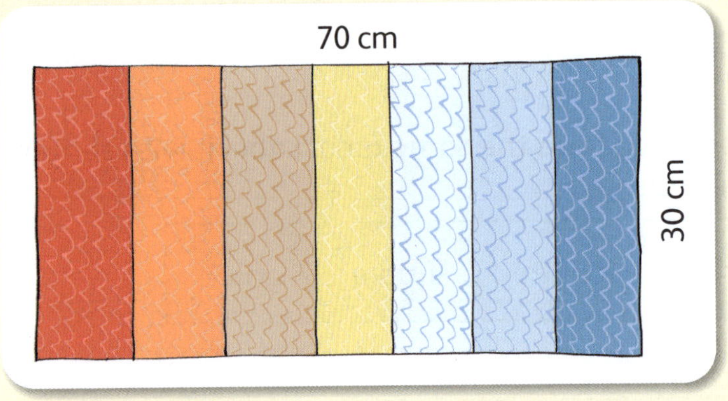

A 200 cm

B 2 100 cm^2

- **a** Write down the calculation for the area of the blanket.
- **b** Is A or B the measurement for the area of the blanket?
- **c** Now write down the calculation for the perimeter of the blanket.
- **d** Explain the differences between perimeter and area in your own words.

6 Shawna has 9 chickens and 3 pigs. She needs to plan the layout of her farmyard so that all the animals have enough space. Use these guidelines to help you plan:

- each pig needs around 0.7 m^2 of outside space
- each chicken needs about 0.9 m^2 of outside space and 0.3 m^2 inside the coop.

Use square grid paper. Draw and label a plan for the chicken coop and the outside yard. Write down the perimeter and square unit measurements for each of the spaces.

11 Self check

	Check you can do this:	If you have difficulty, try this:
1	Calculate perimeters of regular and irregular polygons.	Go over the activities on pages 160 and 161.
2	Find areas of various objects and figures.	Revise pages 162 and 163 for areas of regular shapes, and pages 164 to 166 for areas of irregular shapes.
3	Distinguish between units of length and units of area.	Go back to page 162 and go over the relationship between square units, such as cm^2, and units of length, such as cm.
4	Calculate area in square units and using grids.	Revise the work on pages 164 to 166.

Unit 3: Geometry

Chapter 12 Symmetry and position

In this chapter you will explore these focus questions:

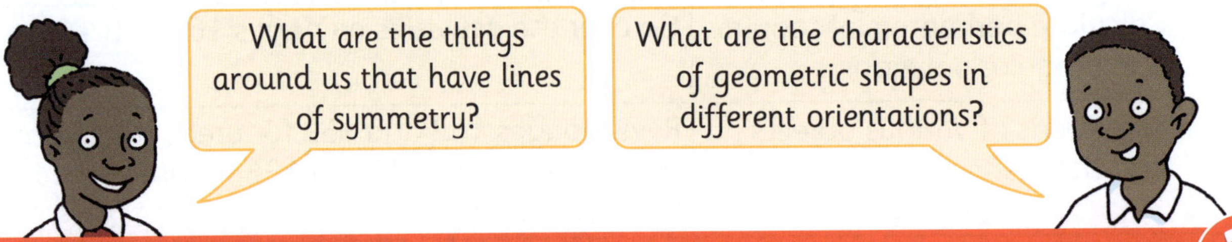

To answer these questions, you will:

- explain what symmetry is
- identify the mirror line of a reflection
- show possible lines of symmetry in different shapes and objects
- describe locations on a grid
- describe flips, turns and slides.

Starting point

1 Look at the Jamaican flag. It has a line drawn down the centre.

a Identify any polygons that you can see on the flag.

b What do you notice about the way the line divides the flag?

c What other line could we draw to divide the flag in this way?

2 This is a diagram of a circle. The line across it is a diameter. A diameter is a line that intersects two points on the circumference of the circle and passes through the centre.
What similarity do you notice between the way the diameter cuts the circle and the way the line cuts the flag in Exercise 1?

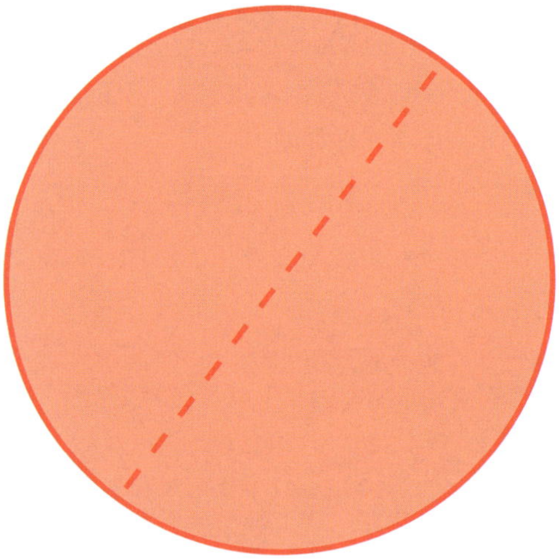

Lines of symmetry

Key maths idea

We say a shape is **symmetrical** or has **symmetry** if you can draw (or imagine) a line down the middle that divides it into two parts that match exactly.

Each red dotted line shows a **line of symmetry**. It is also called a **mirror line**, because the two sides mirror each other. The mirror line divides a shape into two **congruent** parts.

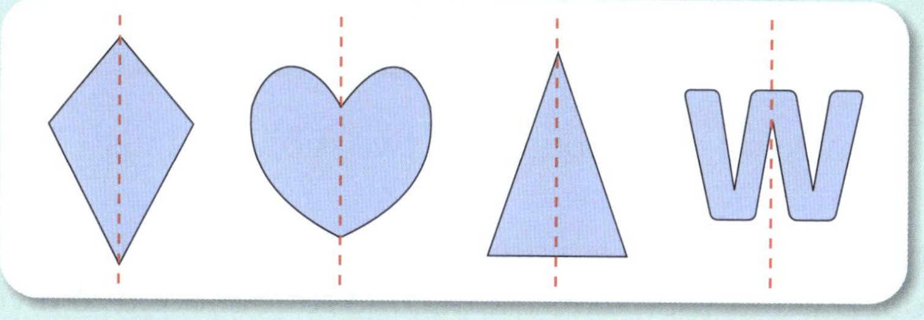

Key words

symmetrical
symmetry
line of symmetry
mirror line
congruent

1 Does the dotted line show a line of symmetry? Say 'yes' or 'no' for each shape.

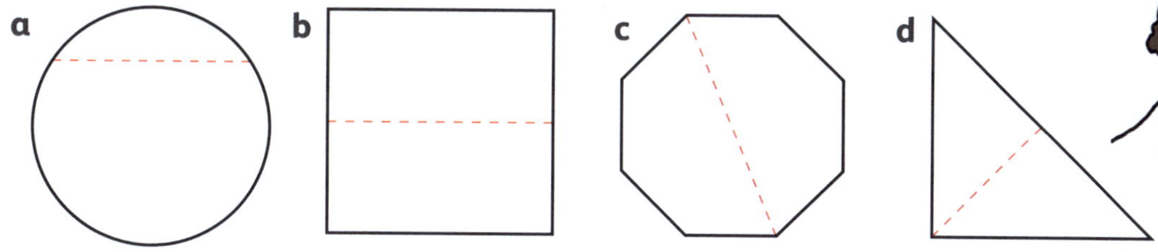

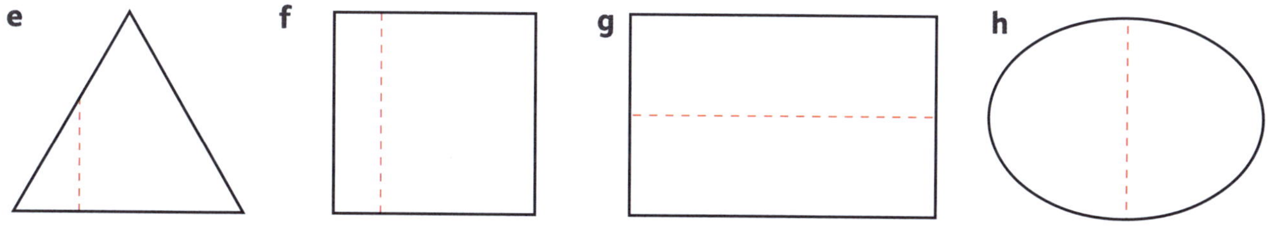

2 Look at each shape. Say whether it is symmetrical or not. Check by placing a ruler or pencil along the mirror line.

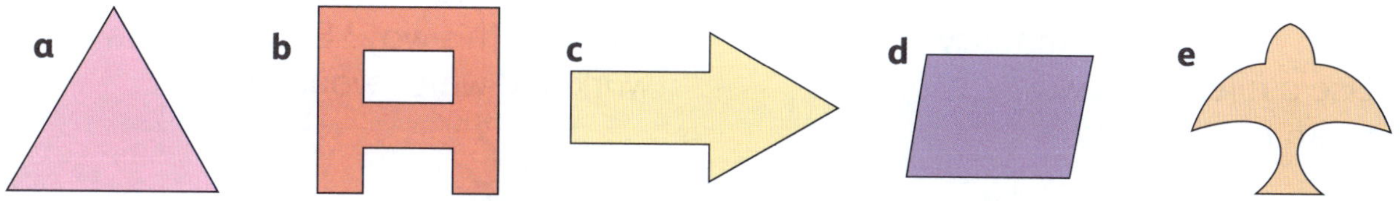

Exploring symmetry

Key maths idea

You can explore symmetry in polygons, such as triangles and rectangles, and other 2D shapes, such as circles, ovals and irregular closed shapes. You can also see symmetry in your environment.

Remember that we say a figure is symmetrical, or has symmetry, if we can fold it in half and the two parts match exactly.

1 Work in pairs. Your teacher will give you a geoboard and two rubber bands to work with.

- **a** Take turns to use one rubber band to make a figure on your geoboard and then give it to your partner. Your partner then uses the second rubber band to divide the figure into two equal parts.
- **b** Draw each figure and its line of symmetry in your notebook.

2 Work in pairs. Your teacher will give you shapes to fold. Explore whether each of these shapes has a line of symmetry. If it does, how many lines of symmetry does it have?

a 
rectangle

b 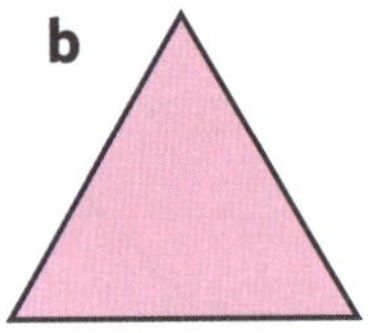
equilateral triangle

c 
square

d 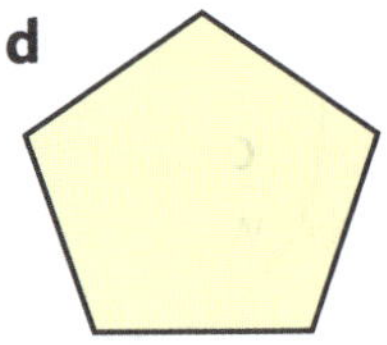
pentagon

e 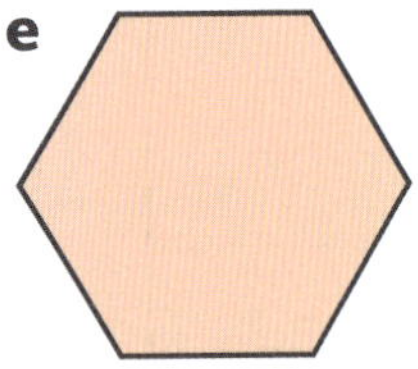
hexagon

f 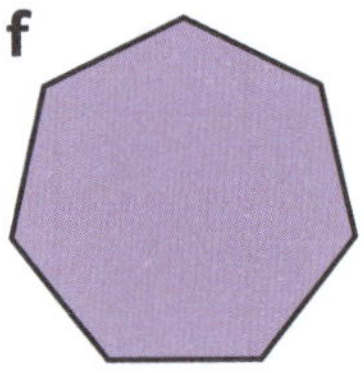
heptagon

g 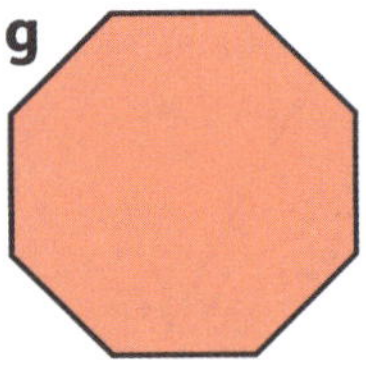
octagon

3 Look at the pictures of the fruits below.

- **a** This star fruit has more than one line of symmetry. Use a ruler to show them to a partner.

- **b** Which of the following fruits could you divide in this way? Explain which would work and which would not.

A 

B 

C 

D 

4 You can use ink blobbing to make symmetrical designs on paper.

You will need:
- paper
- ink or paint
- brushes or droppers.

Fold a sheet of paper in half. Place some paint or ink on one side. Fold it over to create your symmetrical design.
Compare your designs with your classmates' designs. What happens if your ink does not touch the mirror line?

Figures with more than one line of symmetry

Key maths idea

A symmetrical figure can also have more than one line of symmetry.

Look at the letter H. You can see it has two lines of symmetry, a **vertical line of symmetry** and a **horizontal line of symmetry**.

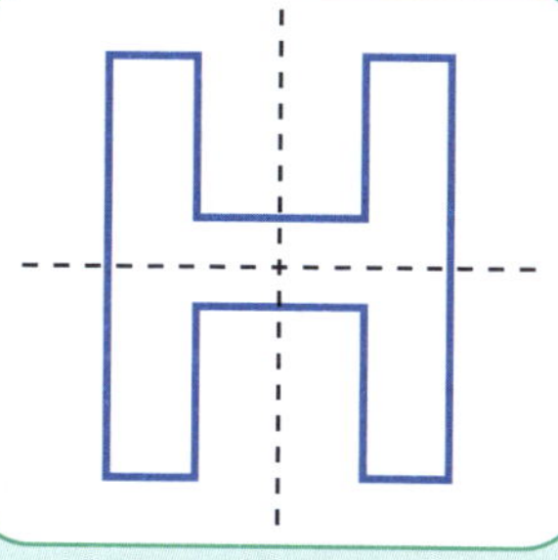

Some figures also have a **diagonal line of symmetry**. The square below has four lines of symmetry. Two are diagonals. In the triangle, all the lines of symmetry are diagonals.

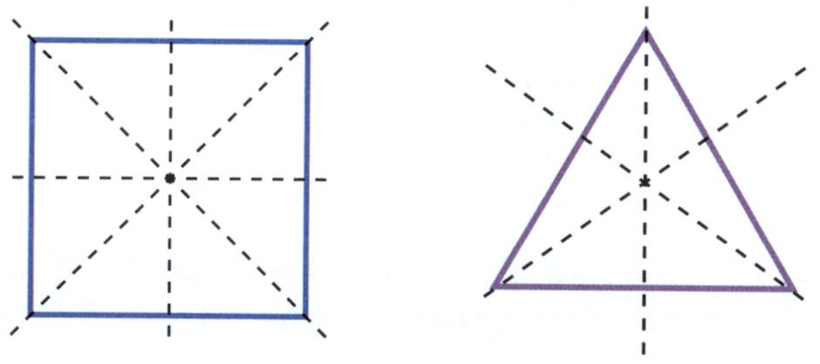

Key words

vertical line of symmetry
horizontal line of symmetry
diagonal line of symmetry

1 Use grid paper to draw your own shapes with more than one line of symmetry.
  a Draw two shapes that have two lines of symmetry.
  b Draw two shapes that have three lines of symmetry.

2 Copy or trace these shapes. Draw the mirror image to complete each shape.

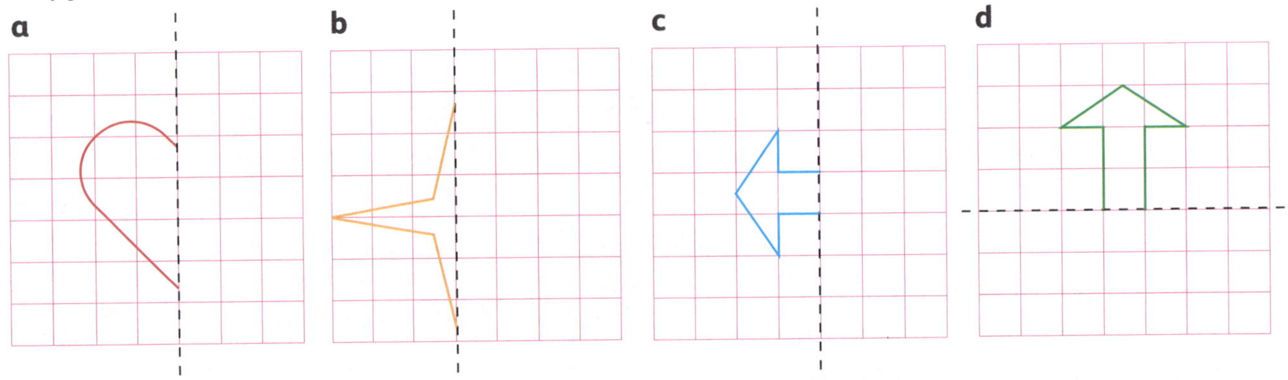

3 Use symmetry to help you work out what these words are. Write the words in your notebook.

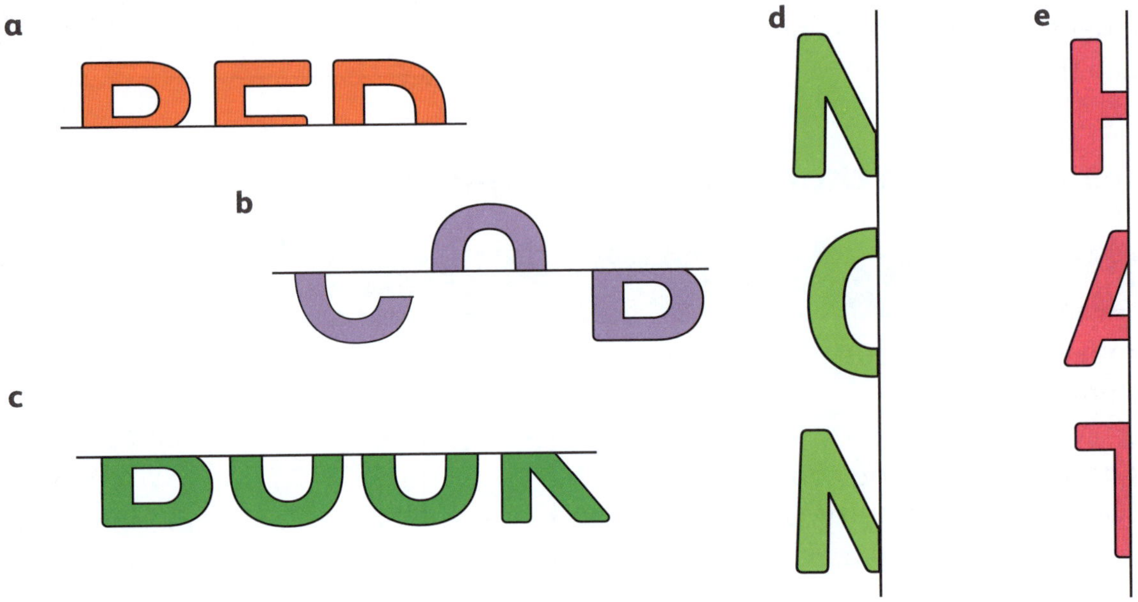

4 Follow the instructions to find lines of symmetry in numbers and in letters of the alphabet.

a Write the word 'JAMAICA' in capital letters. Look at each letter to see if it has symmetry. Draw the lines of symmetry on each letter.

b Now write your name and the names of three other students in your class. Write the names in capital letters and show the lines of symmetry in the letters.

c Choose seven more letters of the alphabet, write them in capitals and draw the lines of symmetry.

Real-world maths: Symmetry in my environment

1 Look at the numerals from 0 to 9.

a Which of the numerals have line symmetry? Write them down.

b Write down the numerals that don't have line symmetry.

0 1 2 3 4 5 6 7 8 9

2 Work with a partner. Look at each of these objects. Describe any lines of symmetry that you notice.

a

b

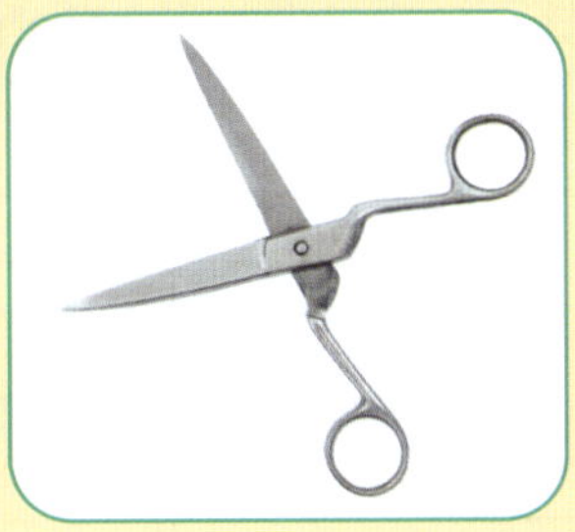

c

d

e

f

g

3 Take a walk around at home or at your school. Identify ten different objects that have line symmetry. Draw a sketch of each object to show the line of symmetry.

Position on a grid

Key maths idea

A **grid** is an arrangement of parallel **horizontal** and **vertical** lines. They create an array of blocks called **cells**, in **rows** and **columns**. Rows run horizontally (from left to right) and columns run vertically (up and down). We use letters and numbers to give **coordinates** that describe the **location** of an object on the grid. We usually name each column with a letter and each row with a number.

For example, on this grid, the sun is located in the top row, which is row 5, and the fourth column, which is column D. The coordinates of this position are D5.

Key words

grid
horizontal
vertical
cell
row
column
coordinates
location

1 Write the coordinates that describe the position of:

a the bee **b** the fish **c** the owl **d** the rabbit.

2 **a** What is in the same column as the fish?

b What is in the same row as the rabbit?

3 Draw a grid with five columns and six rows. Label the columns A to E and the rows 1 to 6. Draw the following pictures in the correct positions:

a a flower at D4 **b** a car at A5 **c** a ball at E6

d an apple at B2 **e** a butterfly at C1 **f** a cat at C5.

4 Draw two more pictures in any position on your grid. In your notebook write the positions on the grid of where your drawings can be found.

5 The numbers 1 to 100 are arranged in this chart. Identify the numbers at the following positions:

a A9 **b** H7 **c** D6 **d** J1

e E10 **f** C7 **g** I9 **h** J4

10	1	2	3	4	5	6	7	8	9	10
9	11	12	13	14	15	16	17	18	19	20
8	21	22	23	24	25	26	27	28	29	30
7	31	32	33	34	35	36	37	38	39	40
6	41	42	43	44	45	46	47	48	49	50
5	51	52	53	54	55	56	57	58	59	60
4	61	62	63	64	65	66	67	68	69	70
3	71	72	73	74	75	76	77	78	79	80
2	81	82	83	84	85	86	87	88	89	90
1	91	92	93	94	95	96	97	98	99	100
	A	**B**	**C**	**D**	**E**	**F**	**G**	**H**	**I**	**J**

6 Work with a partner and use the hundred chart again. Take turns to ask and answer questions about where to find numbers on the chart. For example, where would you find number 48?

Geometry

Slides, flips and turns

Key maths idea

The word **transformation** means 'change'. In maths, we use it to describe a change in the size, position or **orientation** of a shape. The shape that you start with is called the **source object**. The transformed shape is the **image**. There are three main kinds of transformations:

- A **slide** is also called a **translation**. The source image moves horizontally and/or vertically to a new position.
- A **flip** is also called a **reflection**. The source image is reflected in a mirror line.
- A **turn** is also called a **rotation**. The source image rotates about a point, to a given angle and in a given direction.

The image and the source object are **congruent**. That means they have the same size and shape. Only the position or orientation has changed.

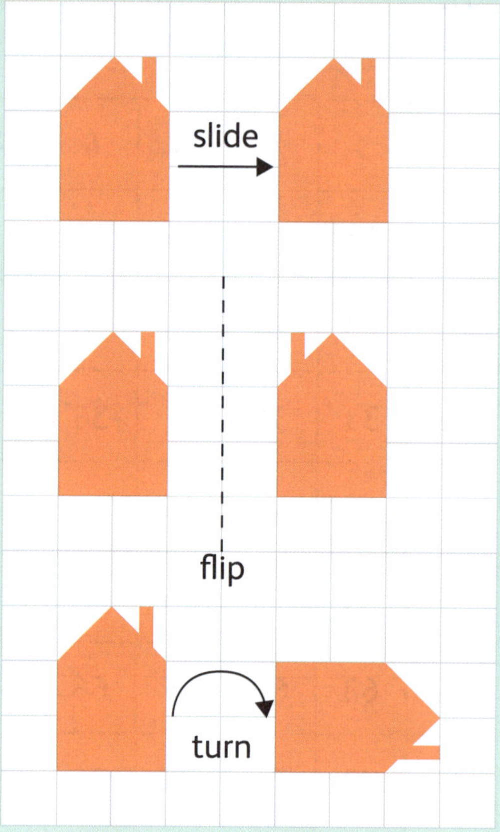

Key words

transformation
orientation
source object
image
slide
translation
flip
reflection
turn
rotation
congruent

Discuss the photograph of this floor pattern with a partner.

1 Which lines of symmetry can you see?

2 Which shapes are congruent?

3 How have flips, slides and turns been used to create this pattern?

Slide

Key maths idea

To slide or move a shape the whole shape must move the same distance and in the same direction. It may move horizontally forwards or backwards (or left or right), or vertically up or down, or both.

In this diagram, the orange shape shows the image (the shape in its new position). Look at one of the corners of the shape. Count down and along to see how many units it has moved.

This source object has moved 5 units down and 3 units to the right. The arrows show you how it has moved.

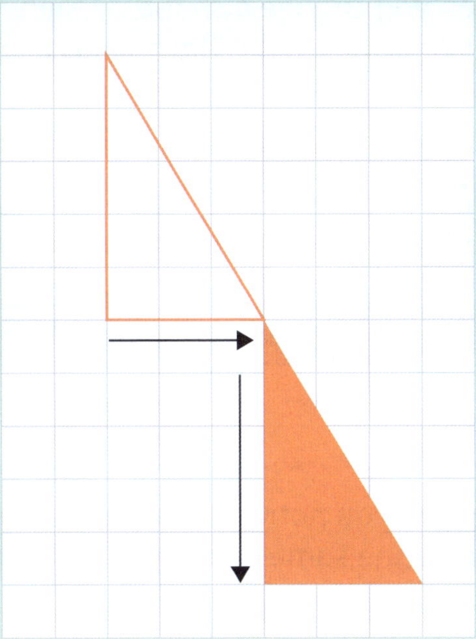

A is the source object and B is the image in each pair. Describe the slide. Give the direction and the number of units in each direction.

a

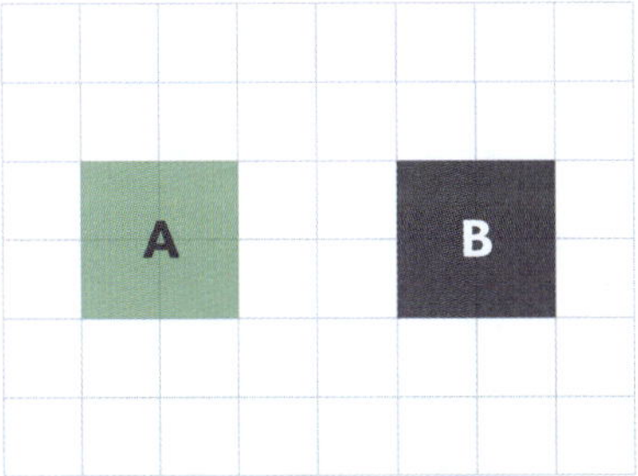

b

c

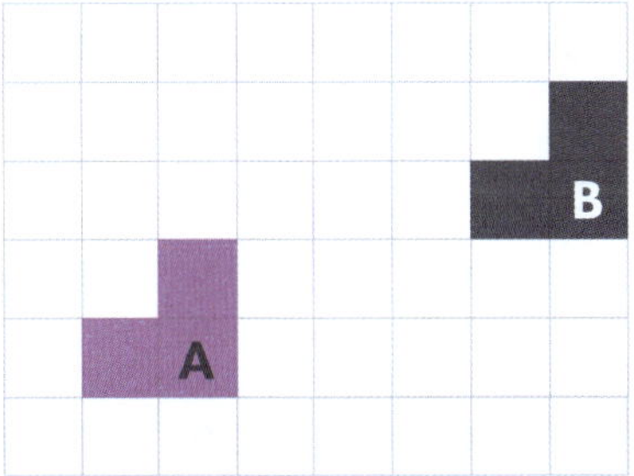

d

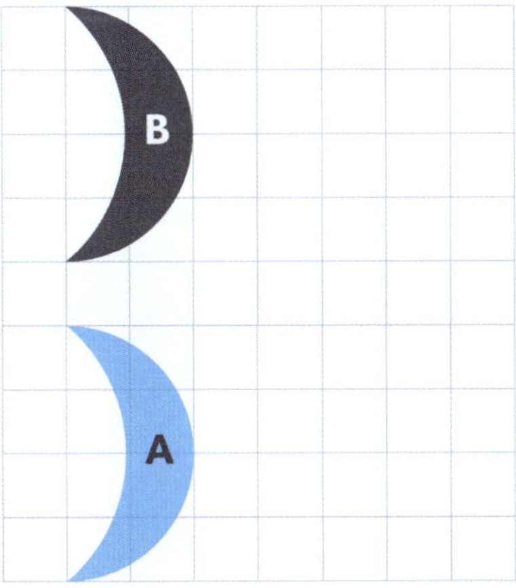

e

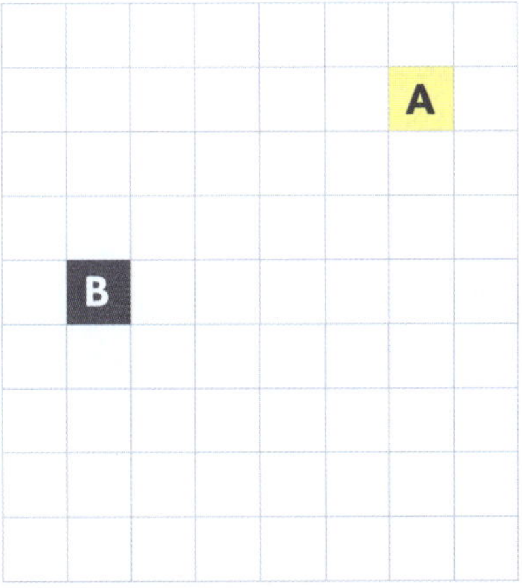

f

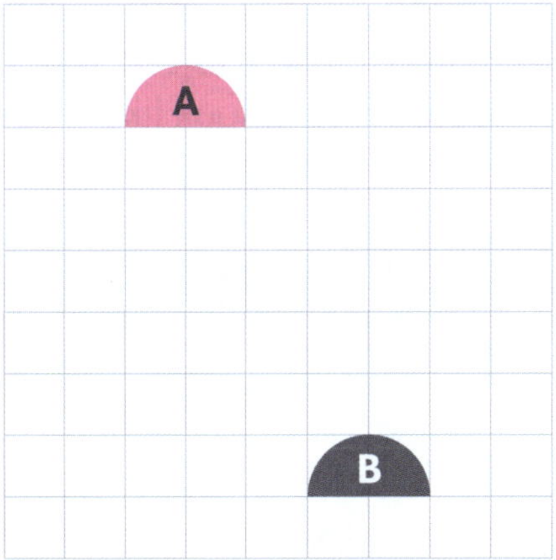

Flip

Key maths idea

When you change a shape by flipping it, remember what you learned about mirror lines of symmetry. A flip is also called a **reflection**.

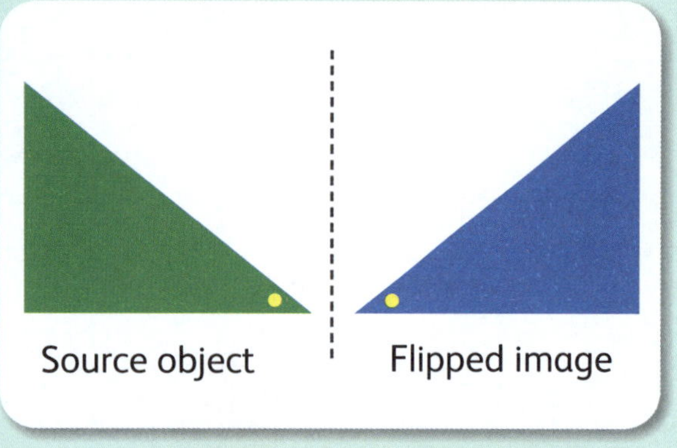

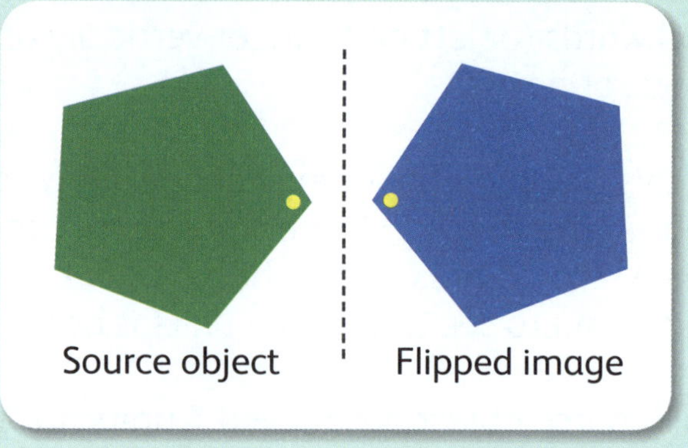

What do you notice about the size and shape of these figures after they were flipped? What is the same and what is different?

1 **a** On grid paper, draw a triangle.

b Choose one of the sides of the triangle to use as the mirror line. Draw a reflection of the triangle in the mirror line.

c Describe the new shape.

d Now add reflections on each of the other sides.

e Describe what new shape you have created.

2 Fold a piece of paper. Cut out a shape along the fold line. Open it out. Describe the shape you have made. With a partner, explore how you can use this technique to make cutouts that look like flowers, leaves, snowflakes or spider webs.

Turn

Key maths idea

A turn changes a shape by turning it around a given point.

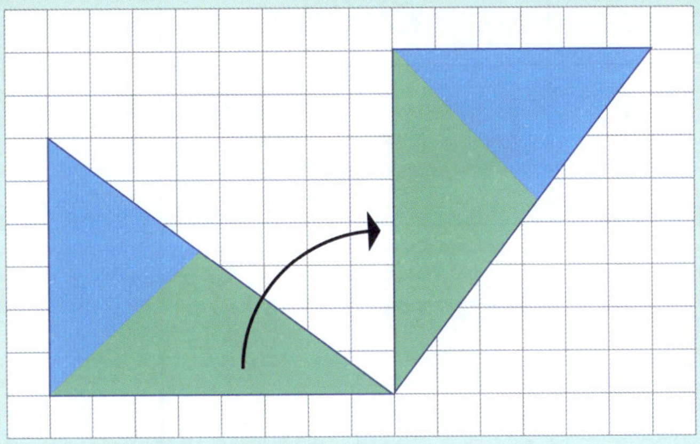

$\frac{1}{4}$ turn clockwise

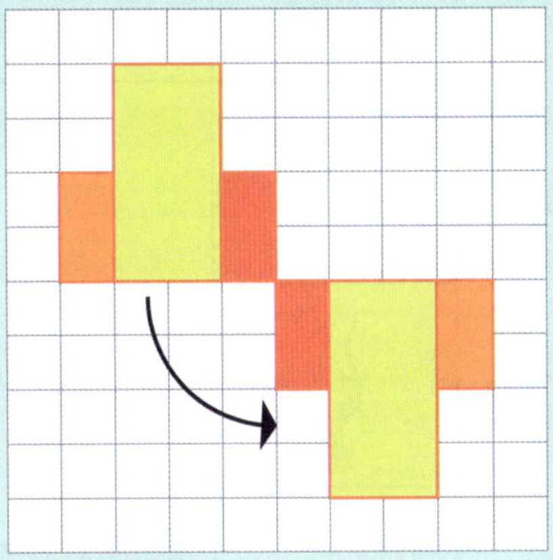

$\frac{1}{2}$ turn anticlockwise

1 Look at each of the turns shown above. Can you think of a turn in a different direction that would produce the same image?

2 For each pair of figures, say whether it shows a slide, flip or turn.

a

b

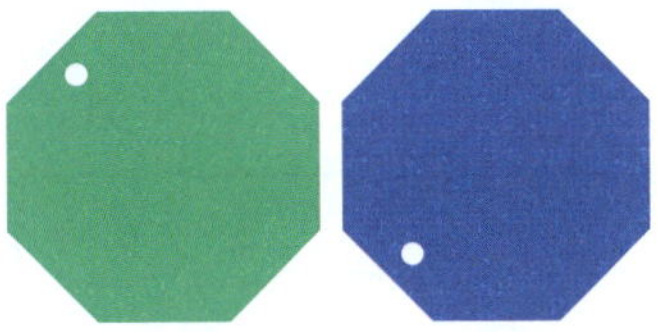

c

d

e

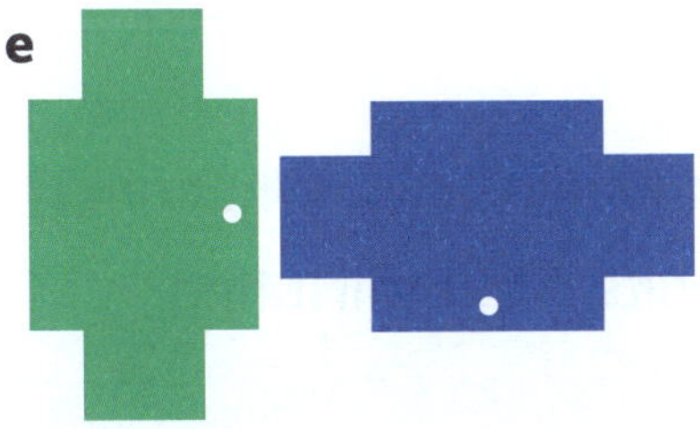

f

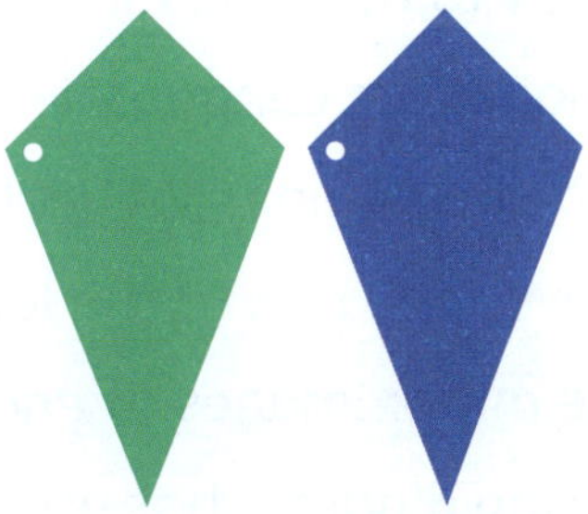

g

h

Maths detective: Exploring transformations and location

Use this grid to answer the questions.

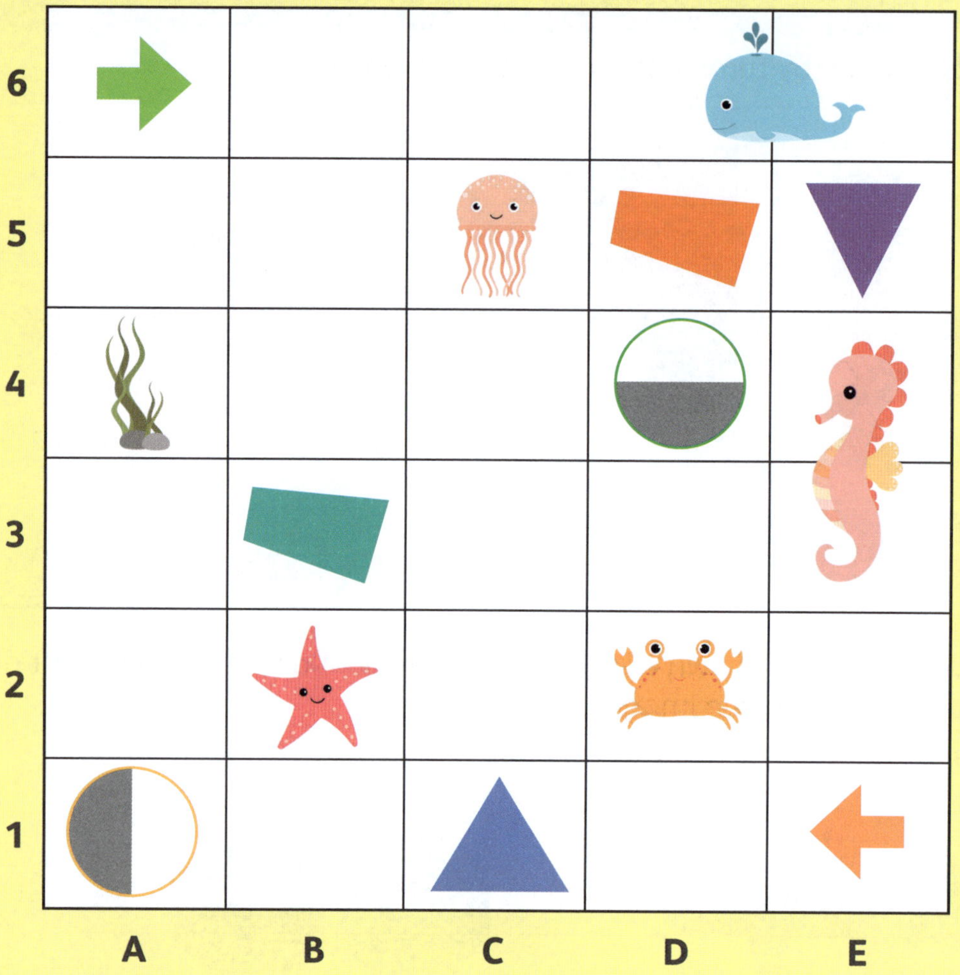

1 Which animals are in these cells?

a B2 b D2 c D6 and E6

2 Describe the transformation that happened when these shapes were moved.

a The shape in B3 to the shape in D5.

b The shape in A6 to the shape in E1.

c The shape in C1 to the shape in E5.

3 Which pairs of shapes are congruent?

4 I am an animal located 4 units vertically up from the triangle in C1. What am I?

5 If the starfish moves one unit to the left and up two units, what will it find?

6 Which shapes are mirror images of each other?

7 Work in pairs. Design a game that uses pictures on a coordinates grid.

What I have learned

Make a small booklet with your own examples of flips, slides and turns. You can cut out pictures from newspapers or magazines, or use found shapes, such as leaves or coins. Draw or stick the shape in the booklet and then draw it in a new position. Describe the transformation you have made.

Practice questions

1 a Draw a rectangle and draw the lines of symmetry.

b Use your drawing to explain these terms:
mirror line   reflection   symmetry

c Write the names of three people you know, in capital letters. Draw the lines of symmetry for each of the letters.

d Design a table that shows which letters have the same number of lines of symmetry.

2 Design and draw two flags. One of the designs should have symmetry, the other should not be symmetrical.

3 Copy the grid into your notebook. Follow the instructions to find the hidden shape.

a Colour in these cells:
F1, A3, D5, B5, C1, F4, A5, E5, B3, C2, D1, F2, A4, C5, F3, E1, C3, F5

b What hidden shape do you find?

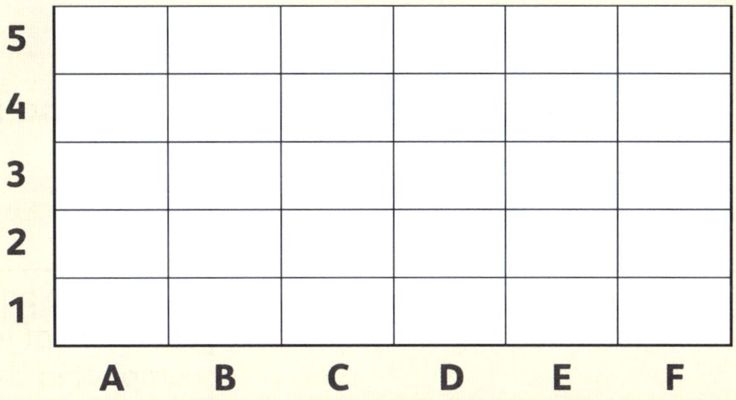

4 Look at these diagrams. The first figure in each row is the source object. Copy the other figures in the row and label each one slide, turn or flip.

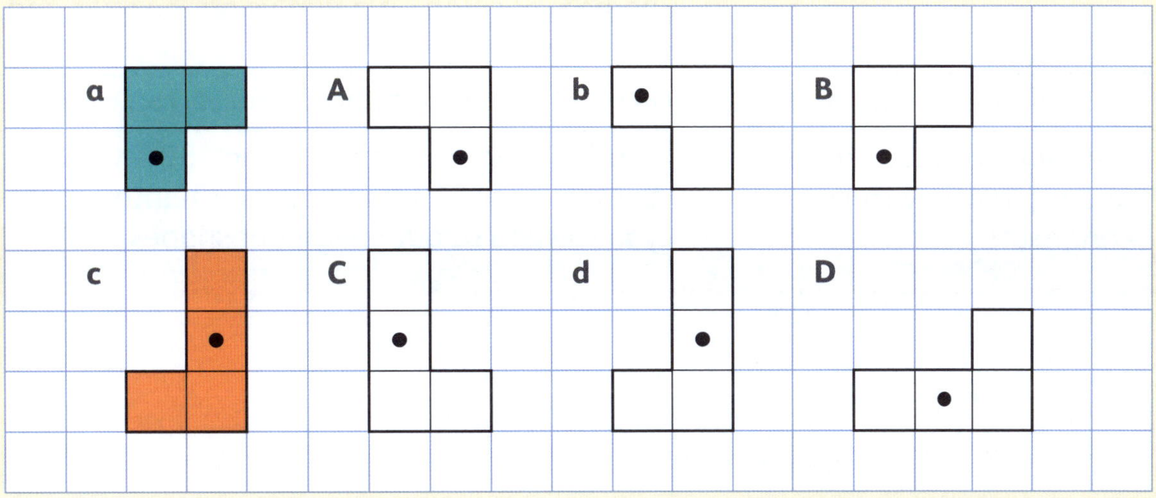

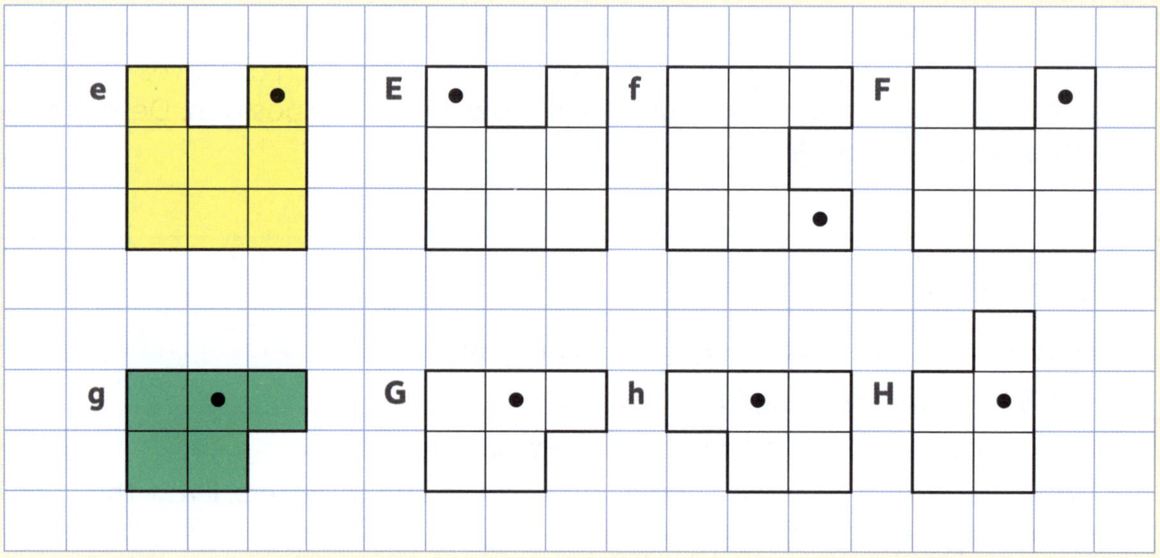

5 Draw a sketch to explain each of these terms:

a flip b turn c slide

12 Self check

Check you can do this:	If you have difficulty, try this:
1 Identify lines of symmetry.	Go back to pages 173 to 176. Go over the definition of line symmetry and practise completing shapes along a mirror line.
2 Describe locations on a grid using columns and rows.	Go back to pages 178 and 179. Revise what we mean by grids, rows and columns, and how to name a position on a grid using the rows and columns.
3 Describe flips, slides and turns.	Revise the work on pages 180 to 183.
4 Identify congruent shapes and explain why they are congruent.	Read the information on page 180 and also go back to pages 102 and 103 in Chapter 7, where you learned about congruent shapes.

Unit 4: Algebra

Chapter 13 Patterns and variables

In this chapter you will explore these focus questions:

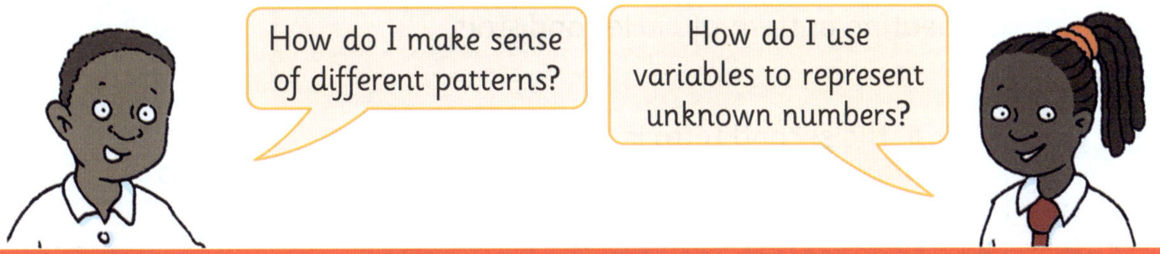

To answer these questions, you will:

- describe and create geometric patterns
- use rules to describe and create number patterns
- describe shape and number patterns, and predict how they will continue
- use letters to represent variables
- write algebraic expressions to describe and solve problems.

Starting point

1 Look at each number **pattern**. Work with a partner. Find two different ways to describe each pattern in words. Say what you think should come next in each pattern. Work out what **operations** you can use to work it out.

   a 1 112, 2 223, 3 334, 4 445, 5 556, …

   b 10 × 10, 100 × 100, 1 000 × 1 000, …

   c 12 + 34, 23 + 45, 34 + 56, …

> The four main operations are addition, subtraction, division and multiplication.

> **Key words**
> pattern
> operations

2 Tell your partner how you think each pattern grid continues. Choose one of the patterns. Copy it onto grid paper and continue it to fill 6 more columns of blocks.

a
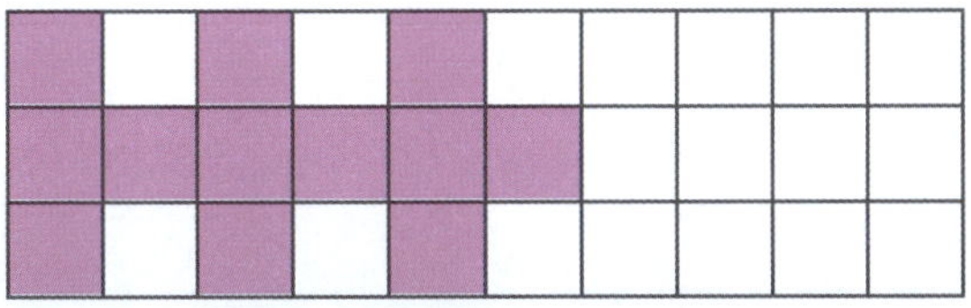

b
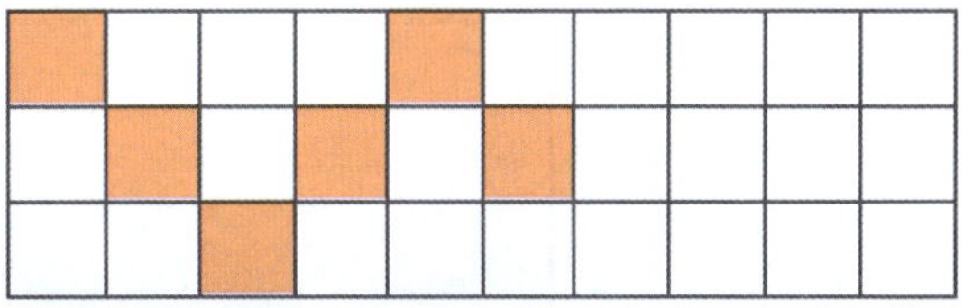

c
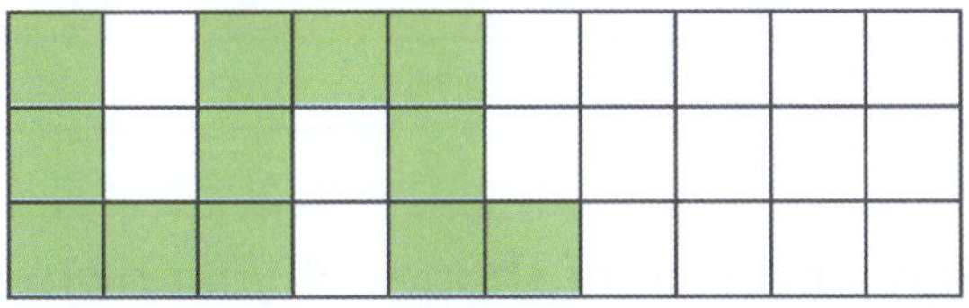

d
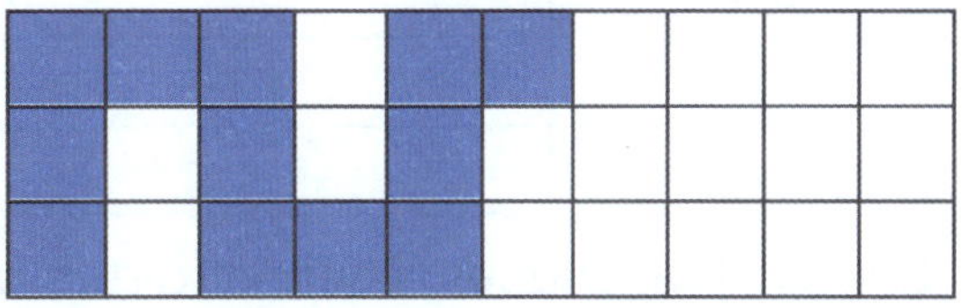

3 Use grid paper to create your own block pattern.

Geometric patterns

Key maths idea

A **geometric pattern** is a set of shapes arranged in a **sequence** that follows a particular rule. The **terms** are the shapes (or groups of shapes) at each **position** in the pattern.

Here is a simple repeating pattern of circles and stars:

The first term is a star, the second term is a circle and so on.

In the next pattern, the first term has two shapes – a circle and a star. Each term has one more shape than the term before.

Term 1 Term 2 Term 3 Term 4

When you know the rule for a pattern, you can make a **prediction** about what comes next in the sequence. Start by checking whether each term has **more** or **less** than the previous one.

Key words

geometric pattern
sequence
terms
position
prediction
more
less

1 Use the second pattern above. Copy and complete this table.

- **a** Write the number of circles and stars in each of the four terms. Then write the total number of shapes in each term.
- **b** Draw terms 5 and 6 of the pattern.
- **c** Complete the next two rows in the table.
- **d** Look at the number patterns you have made in each column. Predict how many circles, stars and total shapes will occur in terms 7 and 8. Write your prediction in the table.
- **e** Draw the last two terms and check whether your prediction was correct.

Term	Number of circles	Number of stars	Total number of shapes
1st			
2nd			
3rd			
4th			
5th			
6th			
7th			
8th			

2 Choose two shapes. Think of a rule and design a pattern of shapes of your own. Swap patterns with a partner. Work out the next two terms in your partner's pattern.

Maths detective: Growing patterns

Case 1

Look at the pattern of coloured pegs.

1 Describe the shape of each term and how it has been made.

2 How many pegs will the next shape need?

3 Why do you think this is called a 'growing pattern'?

4 Describe what the next two shapes will look like.

5 Copy and complete this table about the terms of the pattern.

Term	How many pegs?
1	1
2	4
3	9
4	
5	

6 What do you notice about the relationship between the term numbers and the number of pegs?

7 If the shape pattern continued in the same way, which term would require 100 pegs?

Hint: think about the shapes as arrays or think of what you know about factors and multiplication.

Case 2

Choose one of these peg shapes. Use it to design your own growing pattern. Draw the first three terms in the pattern, then give it to a partner to continue.

Algebra

Real-world maths: Patterns around us

Patterns are an important part of **design**. Patterns can be **functional** – this means they help objects to work properly. For example, a staircase needs repeated steps of even height so that people can walk easily up and down. The spokes of a bicycle make it possible for the wheel to roll evenly, but they also create a pattern of triangles. Some patterns are **decorative**. They make objects interesting or attractive.

1 The picture below shows a room inside a house.

- **a** Identify examples of the following shapes in the picture: squares, rectangles, circles and triangles.
- **b** Identify which shapes are repeated to make patterns.
- **c** Give two examples of patterns that help an object to perform its function. Explain your answer.
- **d** Give two examples of decorative patterns.

2 Look around at home and at school. Identify five different examples of patterns that help a design to perform its function. You could look at clothing, floor tiling patterns, furniture or any other objects.

3 Write a paragraph to explain why patterns are an important part of design work.

Key words
design
functional
decorative

Patterns and rules

Key maths idea

A pattern is a group of numbers or objects that appear in a certain order. Another word for a pattern is a 'sequence'. Each number or object in a pattern is called a 'term'. Each term in a pattern has a fixed position. In the pattern 3, 9, 15, 21, the number 3 is the first term, the number 9 is the second term and so on.

Key word

tabular form

1 Look at this pattern made from matchsticks:

Pattern A

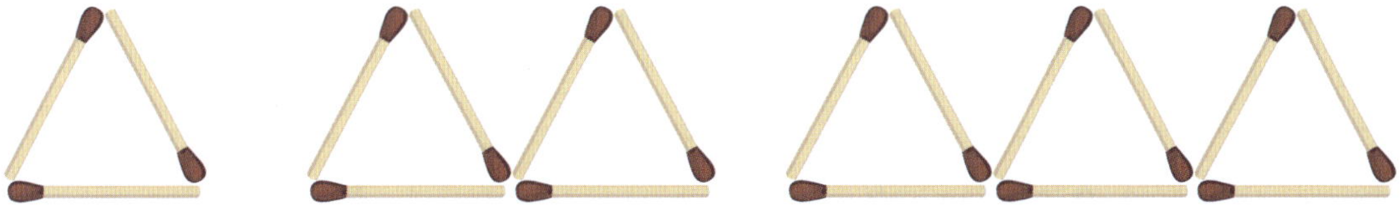

Term 1 Term 2 Term 3

a Draw the fourth and fifth terms of the pattern.

b Without drawing the sixth term, write the number of matchsticks you would need to make it.

c Copy and complete a table like this about pattern A:

Term	1	2	3	4
How many matchsticks?				

2 Now look at this pattern made from matchsticks:

Pattern B

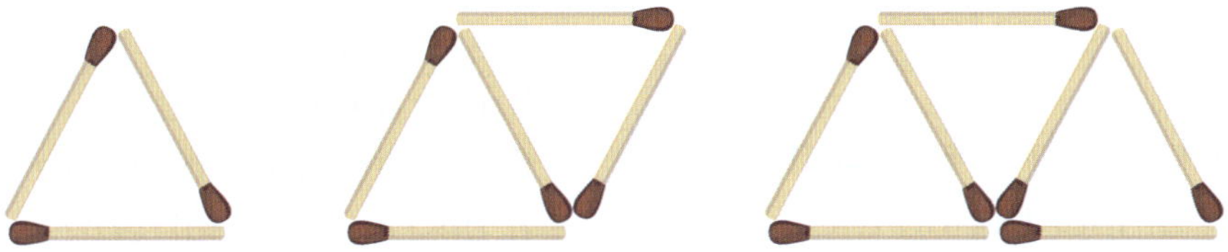

Term 1 Term 2 Term 3

Remember: when we express a pattern using numbers and set it out in a table, we call this showing the pattern in **tabular form**.

a Talk to a partner. What do you notice is similar or different between pattern A and pattern B?

b Draw the fourth and fifth terms of this pattern.

c Without drawing the sixth term, write the number of matchsticks you would need to make it.

d Work in a small group. Discuss how you could explain, in numbers or words, how to find the next terms in each pattern. Then share your ideas with the class.

e Express pattern B as a number pattern in tabular form.

Key maths idea

Numbers in patterns can be in **ascending** order or **descending** order, for example:

3, 9, 15, 21 or 21, 15, 9, 3

We can use the numbers we know in a pattern to work out what will come next.

When we work out what will come next in a number pattern, we are finding a **rule**.

We can use addition, subtraction, multiplication or division in rules for number patterns. Number patterns can continue indefinitely by following the rule to work out the next term.

When you see three dots at the end of a pattern this means that the pattern continues.

Example

Here is a number pattern: 3, 9, 15, 21, 27, ...

There is a difference of 6 between the terms.

$3 + 6 = 9, 9 + 6 = 15, 15 + 6 = 21, 21 + 6 = 27$

The rule is to add six each time.

The next two numbers in the pattern are 33 and 39.

3, 9, 15, 21, 27, 33, 39, ...

Key words

ascending
descending
rule

1 Write the next three terms in each pattern.

- **a** 1, 8, 15, 22, ...
- **b** 84, 72, 60, 48, ...
- **c** 4, 8, 16, 32, ...

2 Read the rule. Then write the first five terms of each pattern.

- **a** Start at 19. Subtract 2 to find the next term.
- **b** Start at 10. Double the number to find the next term.
- **c** Start at 101. Add 11 to find the next term.

3 In a number pattern, the first term is 100.

- **a** Make up a rule that would create an ascending pattern. You may use any number operation (addition, subtraction, multiplication or division). Write the first five terms.
- **b** Make up another rule to create a descending pattern. Which operation did you use? Explain why you used this one.

Talk about

Discuss with a partner: how could you use multiplication to create a descending pattern? Think about the different kinds of numbers you have learned about this year.

Exploring number sentences

Key maths idea

A **number sentence** is a combination of numbers and mathematical symbols.

For example, 5 + 2 = 7 is a simple number sentence.

Number sentences can also contain letters or other symbols.

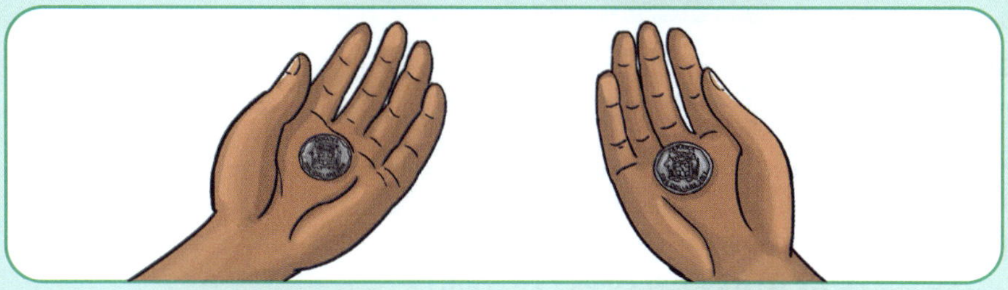

Example

I have \$10 in one hand and \$5 in the other. How much do I have altogether?

\$10 + \$5 = \$15

Number sentences are also called **equations**, because they contain an equal sign. The **expressions** on each side of the = sign are equal to each other.

Key words

number sentence

equation

expression

\$1

\$5

\$10

\$20

1 You have a set of the four coins in the picture.

- **a** What is the total value of the coins?
- **b** Use two or more coins. What sums can you make that have an odd number as the answer? Write number sentences to show your answers.
- **c** Use two or more coins. What sums can you make that have an even number as the answer? Write the number sentences.
- **d** Is it possible to add up the coins in a different order to get a different total? Explain your answer.

2 Use real or pretend coins in the values shown above, but this time you have more coins of each value. How many different ways can you make \$25 using only \$1, \$5 and \$10 coins?

3 Is it possible to make an odd total using only even coins (such as \$10 and \$20 coins)? Explain your answer.

4 Is it possible to make an even total using only odd coins (such as \$5 and \$1)? Why?

5 Work in pairs. Find five different combinations of coins that make a total of \$50.

Maths detective: Finding out about money

You may have heard the expression 'Money doesn't grow on trees'! Have you ever wondered where money comes from? Where is it made and what is it made out of?

The process of making money is called **minting**. Coins are made from different kinds of metals and paper notes are printed on special paper. The paper has to be strong enough so that it does not tear easily when people fold it or handle it. It also has to be produced in a way that makes it difficult to **forge**. 'Forging' means making fake money. This is illegal. Forged notes are also called **counterfeit** notes.

Do some detective work to find out more. If possible, your teacher will arrange for your class to visit the Bank of Jamaica's Money Museum. If you do this, you can use your visit to gather your answers to the questions that follow. If not, you will need access to the internet, on a computer, tablet or smartphone, or you will need to find information in reference books in a library.

Key words

minting
forge
counterfeit

1. What materials are used to make money? Find out:
   - **a** What kind of metals are used to make coins? How are the patterns on the coins made?
   - **b** What information is printed on each coin?
   - **c** Which symbols and images are on the coins? What do they represent?
   - **d** What kind of paper is used to make notes?
   - **e** How is the paper strengthened so that it does not tear easily?
   - **f** What materials are used to make sure that the notes are not easily forged?
2. Find out a little more about the history of the Jamaican dollar.
   - **a** When did we start using Jamaican dollars?
   - **b** What currency was used before that?
   - **c** Which denominations were first introduced? How has this changed over time?
3. Write a letter to a friend describing your visit to the Jamaican Money Museum. If it was not possible to visit the museum, imagine you are writing to a friend overseas to explain what currency is used in Jamaica and some typical prices of things in our country.

Arithmagons

Key maths idea

An **arithmagon** is a kind of a missing number puzzle in the shape of a polygon. Each number in a square is the sum of the two numbers in the adjacent circles. Adjacent means 'next to'.

You need to start with the numbers that are given.

You can add two circle numbers to get the number for the square between them.

You can subtract a circle number from the adjacent square to get the number for the circle on the other side.

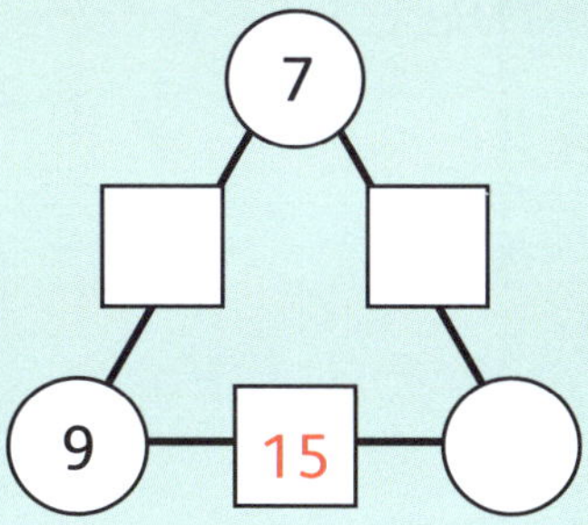

7 + 9 = 16
15 − 9 = 6
6 + 7 = 13

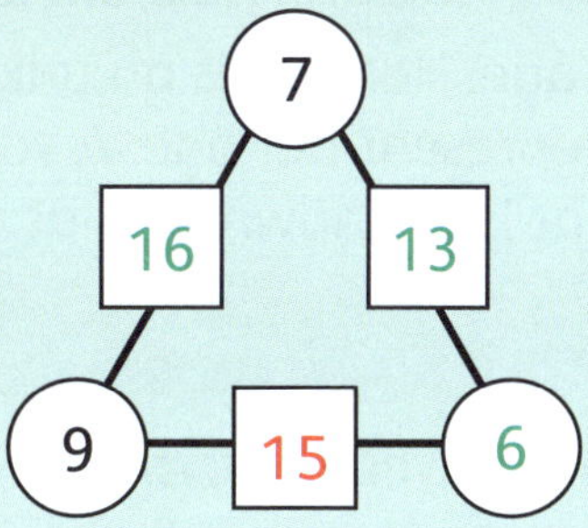

Key word
arithmagon

Copy the arithmagons and work out the missing numbers.

a

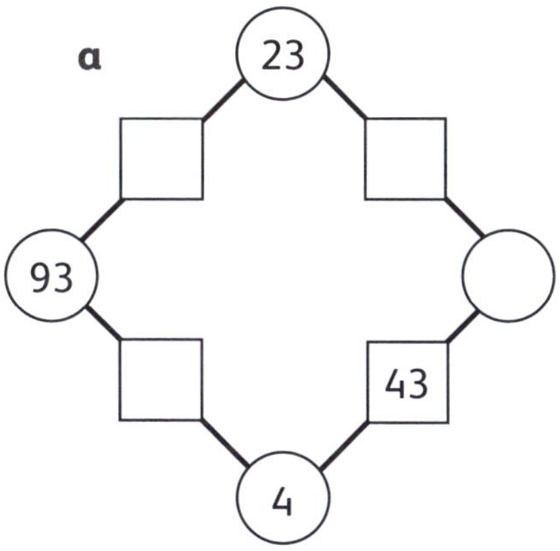

b

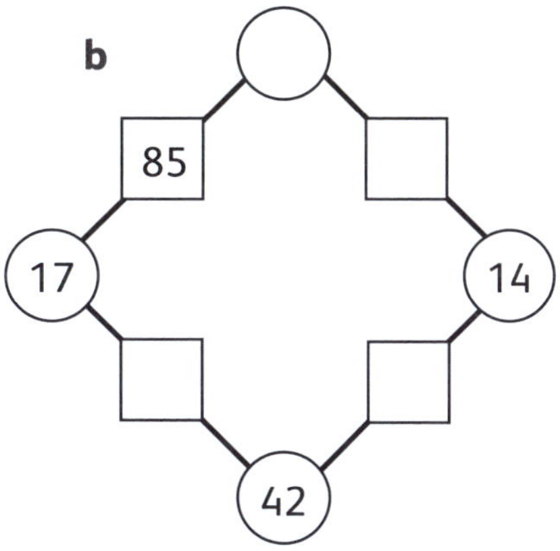

c

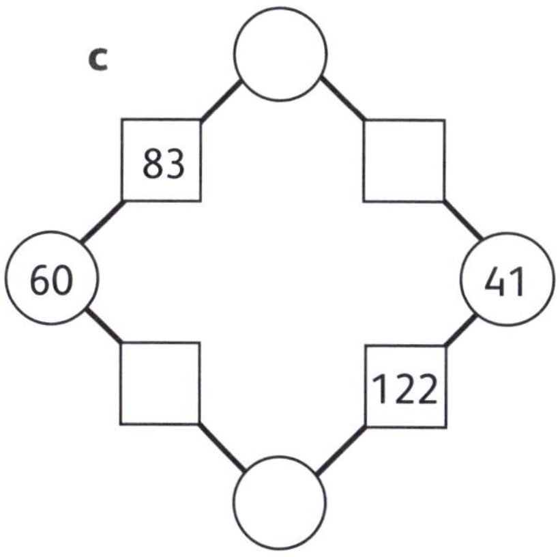

d

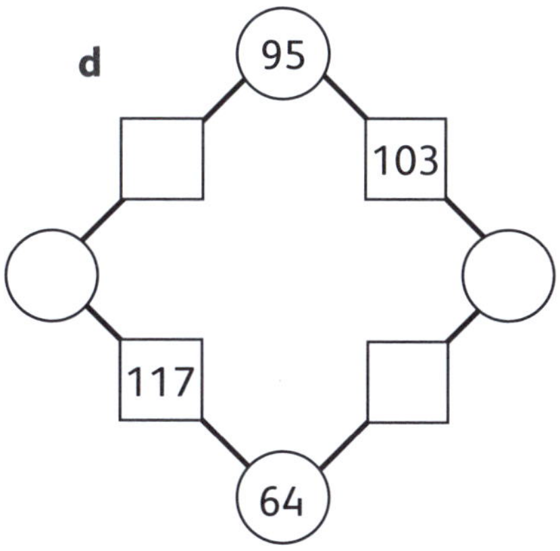

e

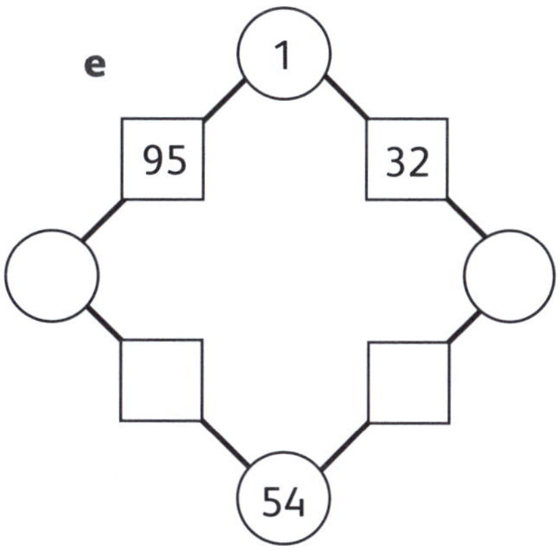

f

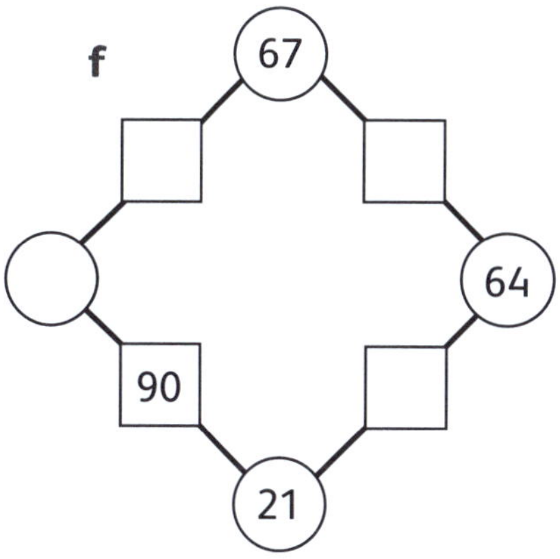

Algebra

Using variables to write expressions

Key maths idea

A **number sentence** can help you to find the **solution** to a problem. But sometimes you do not know all the numbers in a number sentence. You can use letters of the alphabet to represent the numbers you don't know in the number sentence. The letter that represents an **unknown** value is called a **variable**.

Example

Sandy is two years older than Melani. How old is Sandy?

We do not know Melani's age. Her age is an unknown value. We can use a variable to represent her age. A variable is a letter that represents a value that is unknown or that can change.

Let Melani's age = m

Whatever age Melani is (whatever value m has), Sandy is 2 years older. So this **expression** describes Sandy's age: $m + 2$

Algebraic expressions do not have an equal sign. Here are some more algebraic expressions and what they mean in words:

Algebraic expression	What it means in words
$n - 2$	two less than the number n
$15 + x$	15 more than the number x
$40 \div f$	40 divided by the number f or how many times does f go into 40
$12w$	the product of 12 and w (or 12 multiplied by the number w)

Notice that when you multiply a number by a variable, you always put the known number before the variable. It is called the **coefficient** or **constant**.

$a \times 5$ is written as $5a$ (5 is the coefficient)
$q \times 25$ is written as $25q$ (25 is the coefficient)

Key words

number sentence
solution
unknown
variable
expression
coefficient
constant

1 Write the algebraic expression for each word phrase. The first one has been done for you.

- **a** four less than the number a $\qquad a - 4$
- **b** 100 divided by the number b
- **c** the product of 9 and c
- **d** 25 more than k
- **e** 18 less than the number p
- **f** q less than 100

2 Write words to explain what each of these expressions mean.

a $8p$ **b** $y + 6$ **c** $k \div 6$ **d** $t + 40$

Equations

Key maths idea

Now, when we know what age Melani or Sandy is, we can create an equation.
Let Sandy's age = s
We know $s = m + 2$
When Melani is 12 years old, we can make this equation:

$$s = 12 + 2$$
$$s = 14$$

An expression <u>does not</u> contain an equal sign.
An equation uses an equal sign to show expressions that are equal to each other.
$n - 2$ is an expression,
but $n - 2 = 5$ is an equation.
You can write an equation to help you solve problems.

Example

Jay has n number of pencils. His friend then gives him 5 more pencils. If Jay now has 11 pencils, how many pencils did he have to begin with?

We know that: $n + 5 = 11$
In other words, something plus 5 makes 11.
How can we find out the missing number?
We take 5 away from 11!
This is how you write it:

$$n + 5 = 11$$
$$n + 5 - 5 = 11 - 5$$
$$n = 6$$

Think: 5 more than n is equal to 11.

1 Decide if the following are expressions or equations:

a $c \div 3 = 8$ **b** $9y + 7c$ **c** $4p + 4 = 32$ **d** $t + 9 = 20$

e $3f$ **f** $100 - a$ **g** $14 + g = 20$ **h** $9x + 5 = 23$

2 Read the word problems. Solve each problem using algebra. Don't forget to set out your working like the example above.

a Mark is catching fish. By lunchtime he has caught *k* fish. In the afternoon he catches another 6 fish. If he caught 14 fish altogether, how many did he catch before lunchtime?

b Alex has collected some money for charity. She has raised *d* dollars so far. Her mom then gives her \$800, which gives her a total of \$4 600. How much money did Alex have before the extra donation?

c Tori is counting her steps. By the end of the school day, she has walked *s* number of steps. She went for a walk after school which was 3 500 steps. Now she has a total of 7 800 steps. How many steps had she walked by the end of the school day?

d Sally baked *b* number of cupcakes. She gave seven to her friends and has nine left. How many cupcakes did Sally have to begin with?

e Jack is playing marbles with his brother. He has *m* number of marbles and gives twelve to his brother to play with. Jack still has 21 marbles left. How many marbles did he have to begin with?

3 These students have each made mistakes in their working. Work in pairs. Explain the mistakes in their working. Correct their work.

a

$4p = 16$

$4p - 4 = 16 - 4$

$p = 12$

b

$7x + 3 = 24$

So $10x = 24$

$10x \div 10 = 24 \div 10$

$x = 2.4$

4 Write your own equations to work out the value of the variable in each bar model.

a

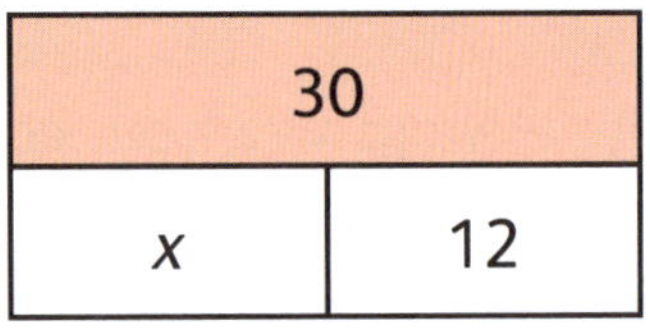

b

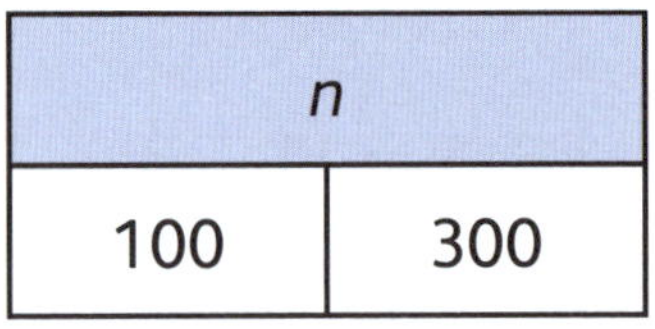

c

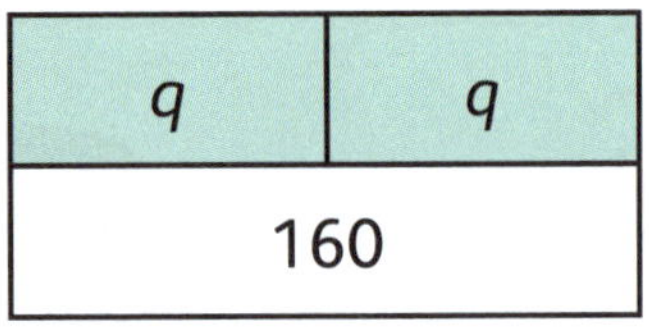

d

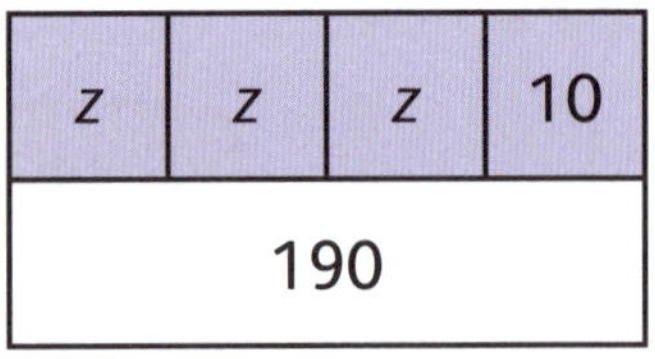

What I have learned

Create a poster with examples and definitions that show what you learned in this chapter. Include:

- an example of a geometric pattern, first as a shape pattern, then in tabular form
- three examples of different number patterns
- definitions and examples of these key words: term, rule, position, variable, unknown, constant, equation and expression.

Practice questions

1 Look at the pattern of shapes and complete the tasks that follow.

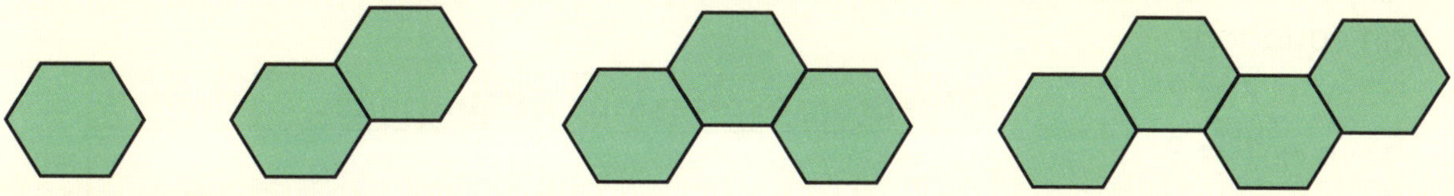

a Copy and complete the table.

Term	1	2	3	4
Number of sides	6			

b Draw the fifth term in the pattern.

c Write a sentence to describe how to find the next term in this pattern. You may use numbers or words.

2 One ant has 6 legs, two ants have 12 legs and so on. Show this pattern using a table. End your table after 6 ants.

3 Choose the pattern that has 10 as the third term and 22 as the sixth term.

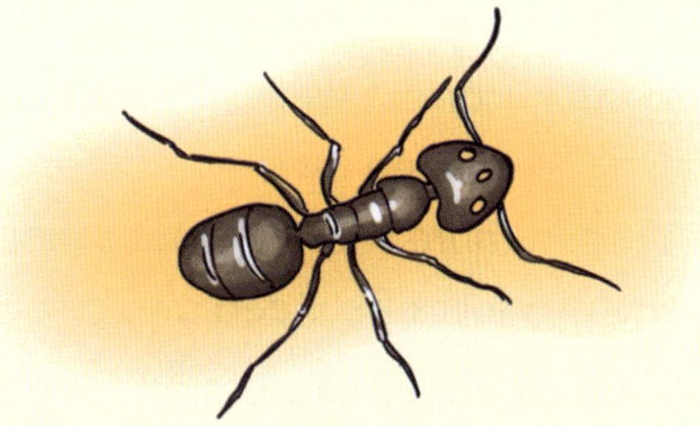

a 2, 6, 10, 14, 18, 22, 26, …

b 14, 12, 10, 8, 6, 4, …

c 4, 7, 10, 13, 16, 19, 22, …

d 23, 19, 15, 11, 7, 3, …

4 Look at the number pattern and answer the questions that follow.

☐, 18, 33, 48, 63, 78, 93, …

a What numbers are the third and fifth terms of this number pattern?

b What is the next term?

c True or false? The missing first term of the number pattern is 3.

5 Write the next two terms of each number pattern.

a 8, 14, 20, 26, 32, …

b 72, 63, 54, 45, 36, …

c 3, 12, 21, 30, 39, …

d 320, 280, 240, 200, 160, …

Algebra

6 Find the missing numbers in these arithmagons.

a

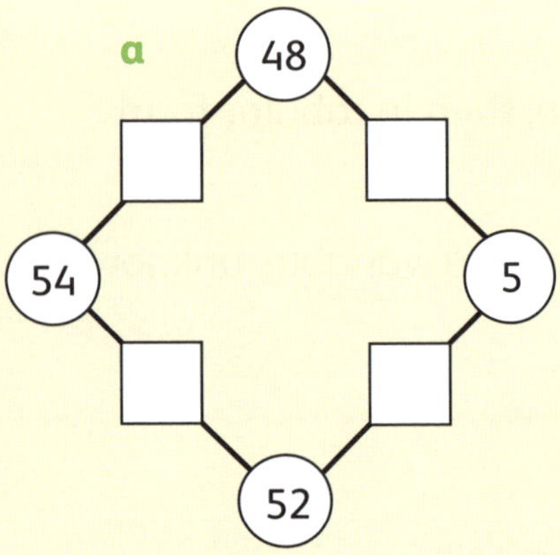

b

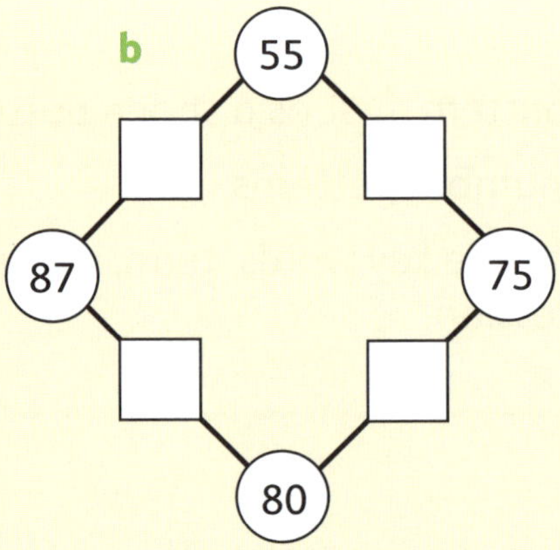

c

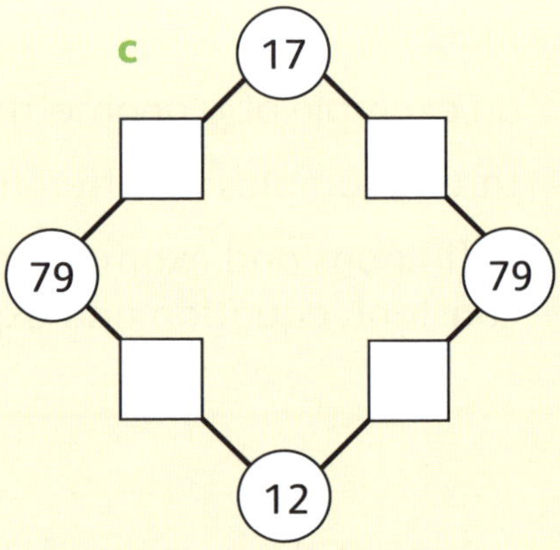

7 Copy the table. Categorise *a* to *f* according to whether they are an expression or an equation.

a $x + 7 = 25$

b $n - 1 = 9$

c $4a$

d $9 \div k$

e $15 + v = 40$

f $a + b - c$

Algebraic expressions	Equations

8 Solve the equations in Exercise 7.

9 Choose the correct word phrase to match this expression: $m + 3$.

a three less than the number *m*

b three more than the number *m*

c the product of three and *m*

10 Choose the correct word phrase to match this expression: $16b$.

a sixteen divided by the number *b*

b sixteen more than the number *b*

c the product of sixteen and *b*

11 Write an algebraic expression for each of these situations.

a six less than the number *y*

b the product of 10 and *w*

c nine more than the number *x*

12 Audrina is reading a book. On Tuesday, she had read *p* pages. On Thursday she read 25 pages more and finished the book which had 122 pages altogether. How many pages had Audrina read by Tuesday?

13 Yohan has *r* number of raffle tickets to sell. He has sold 23 tickets to family and neighbours and has 37 tickets left to sell. How many tickets did he have to begin with?

13 Self check

	Check you can do this:	If you have difficulty, try this:
1	Describe a geometric pattern in words and work out what comes next.	Go back to page 188 and revise the shape patterns. Revise what we mean by a term and practise saying which term comes next.
2	Express a geometric pattern in tabular form.	Revise the pattern work on pages 191 and 192, and look at how we write the terms of a pattern in a table.
3	Solve number puzzles.	Go back to page 195 and remind yourself how to solve an arithmagon.
4	Read algebraic sentences.	Revise what we mean by unknowns and variables on page 196. Remind yourself how to use a letter to express an unknown.
5	Explain the difference between an expression and an equation.	Revise expressions and equations on pages 196 and 197.
6	Decide whether to use +, −, × or ÷ to solve word problems.	Go over the word problems on page 198 and think about how you can decide which operation to use, based on what you are trying to work out.
7	Match algebraic sentences to corresponding words.	Look back at the examples on page 196, where you looked at algebraic expressions and explained what they mean in words.
8	Write algebraic sentences to explain word problems.	Look back at the examples on page 197, where you used variables in algebraic sentences to help solve word problems.

Unit 5: Statistics and probability

Chapter 14 Working with data

In this chapter you will explore these focus questions:

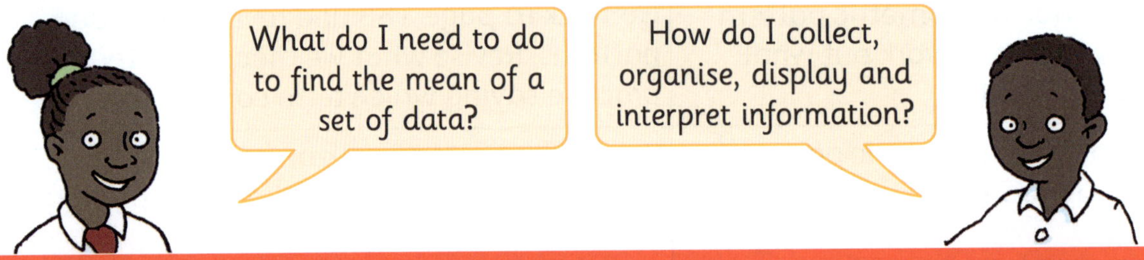

To answer these questions, you will:

- work out the mean, mode and median of a set of data
- solve problems involving mean, mode and median
- read and interpret bar graphs, line graphs, pictographs and pie charts
- draw your own pictographs and bar graphs to represent data
- convert data from a pictograph into a bar graph or vice versa.

Starting point

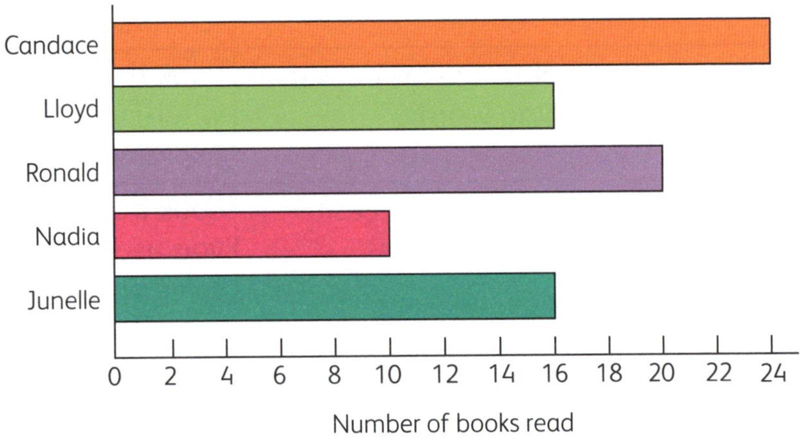

1 **a** What does the graph show?

b What do the numbers along the bottom tell us?

c What do the names down the side tell us?

d Why is it important for a graph to have a heading or title?

e What do you call this kind of graph?

2 Answer these questions about the data:

a Who read the most books?

b Who read the fewest?

c What was the total number of books that the five children read?

3 Redraw the graph so that the bars are vertical instead of horizontal.

Explore measures of central tendency: mean, median and mode

Key maths idea

Information is also known as **data**. Each number in a data set is also known as a **value**. There are different types of data. When you collect information about things that you can count (whole items), you are working with **discrete** data. You can represent discrete data using tallies, pictographs, block charts and bar graphs.

Example

Some students decided to open seven packets of sweets to investigate how many sweets are usually in a packet. This is how many sweets they found in the packets:

7, 9, 8, 9, 7, 8, 9

The students wrote the set in order: 7, 7, 8, 8, 9, 9, 9

The value in the middle of the set is the **median**. The median of this set is 8.

7, 7, 8, 8, 9, 9, 9

The value that occurs most often is called the **mode**. In this set, the mode is 9.

7, 7, 8, 8, 9, 9, 9

To calculate the **mean**, we add up all the values in the set, then divide by the number of **addends**.

$7 + 7 + 8 + 8 + 9 + 9 + 9 = 57$

$57 \div 7 = 8\frac{1}{7}$

The mean is a fractional number, because 57 is not divisible by 7. Another word for mean is **average**.

Mean, median and mode are called measures of **central tendency**. They tell us about patterns or trends in the data set.

Key words

information
data
value
discrete
median
mode
mean
addend
average
central tendency

Sometimes, there is more than one mode. This happens when two or more values occur the same number of times. If all of the values in a data set occur the same number of times, then there is no mode.

Work in groups. Your teacher will give each group a branch with some leaves.

1 **a** Each group should count the number of leaves on their branch.

b Write the totals of all the groups in a set. This is your data set.

c Write the data set in order from smallest to greatest.

d Is there a mode? Why or why not?

e What is the median value?

> Remember: the median is the value that occurs in the middle of the data set. If the data set has an even number of values, add the two middle values and divide by 2 to get the median.

f Calculate the mean.

2 This data set contains the ages of children who sing in the school choir.

7	8	10	9	12	11	9	9	10	11	12
12	12	11	9	10	10	10	8	10		

a What is the age of most children in the choir?

b How would you complete this statement?
'Most children who join the choir are between ___ and ___ years old.'

c Another child signs up for the choir. What do you think is their most likely age: 7 to 8, 9 to 10, or 11 to 12?

3 Aimee's mom has a store. She gets a fresh delivery of ginger cakes each Monday. This bar graph shows how many she sells in a week. Each block represents 1 ginger cake.

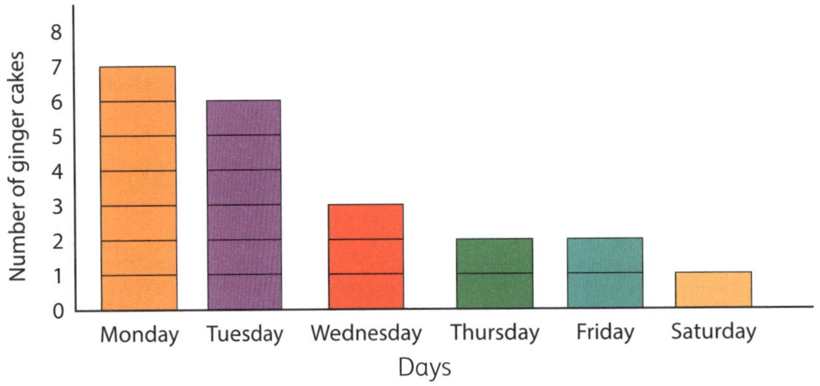

> Bar graphs are also called bar charts.

a What title would you give to the graph?

b Stack wooden cubes or construction blocks to model the graph shown.

c How could you rearrange your cubes to work out the mean? Discuss this with a partner and compare your ideas and solutions with another pair.

Collecting and understanding information

Key maths idea

Statistics is the area of mathematics concerned with gathering and understanding data. The term **survey** comes from the word meaning to look at something closely. When you perform a survey, you collect data from a **sample** to give you information about the whole **population**. Showing the data in a graph helps you to analyse and **interpret** it.

Key words

statistics
survey
sample
population
interpret

1 Some students recorded the number of cars in the school's parking area each day for five days.

Day	Number of cars
Monday	20
Tuesday	18
Wednesday	19
Thursday	18
Friday	18

Answer these questions about this data set:

a Which number best represents the number of cars in the parking area on any day?

b Would the following statement be correct?
'There are usually about 20 cars in the parking area.'
Give reasons for your answer.

c Johannes predicts that next Monday there will be 20 cars in the parking area, but Karl predicts that there will be 18. Explain who you think has made the best prediction and give a reason why.

2 Find the mean for each set of data.

a 12, 9, 7, 11, 6, 9 **b** 72, 66, 74, 89, 90, 89

3 The following daytime temperatures in degrees Celsius are recorded each day in Falmouth for one week in June: 31, 29, 30, 32, 31, 29, 28.

a What is the mode for this set of data?

b What is the median for this set of data?

c What is the mean daily temperature?

4 Ten students achieve the following percentage scores in a science test:
76, 54, 78, 63, 68, 54, 70, 66, 71, 60.

a What is the mode for this set of data?

b What is the median for this set of data?

c What is the mean percentage score?

Real-world maths: Mean, median and mode in real life

1 We use mean, median and mode in real life to help us in different situations. Think of ways the following people might use mean, median and mode in their work. Discuss your ideas in your groups.

a a restaurant owner
b a school teacher
c a climate scientist
d a fisherman

2 Carry out a survey of shoe sizes in your class. Record your results in a table. Then find the mean, median and mode for the data you have collected.

Using the mean to find the total set

Key maths idea

When we know the mean of a set of data, we can find the total set.

When we know the total set, we can find the mean.

Examples

15 watermelons have a mean weight of 9 kg. What is the total weight of the 15 watermelons?

When we find the mean, we divide the total by the number of values in the set. Division is the inverse of multiplication. So, to find the total, we multiply the mean by the number of values in the set.

$15 \times 9 = 120$ kg

The total weight of 15 watermelons is 120 kg.

1 A group of 20 students in a class record their height in centimetres. They work out that the mean height of a student is 138 cm. What is the total height of all students?

2 A farmer counts the number of peas in 85 pea pods. He has 680 peas. What is the mean number of peas in a pod?

3 The mean amount of sugar found in 8 candy bars is 18 grams. What is the total amount of sugar in the 8 candy bars?

4 A cinema records the ticket sales for a movie each day for one week. In total, 2 030 tickets were sold. What is the mean number of ticket sales per day?

5 The mean number of apples produced by an apple tree is 90. Darnel has 52 trees in his orchard. How many apples should he expect his trees to produce?

6 A café is open six days a week. Last week they sold 138 portions of fries. Approximately how many portions did they sell per day?

Solving problems using averages

Key maths idea

Make sure you remember these terms and what they mean:

- The **mode** is the value that appears most often in a set of data.
- The **median** is the value in the middle of a set of data, when it is arranged in order.
- The **mean** is found by carrying out two steps: first you add all of the values in the data set and then you divide the total by the number of values there are.

1 **a** In your class, carry out a survey to find out how many siblings each person has. Record your findings in a table.

b What is the mode number of siblings?

c What is the median number of siblings?

d Find the mean number of siblings.

2 This table compares the cost of a burger at five restaurants on one particular day:

Restaurant	Burger Shack	Just Burgers	Bob's Burgers	Surf 'n' Burger	BBQ House
Cost	\$220.00	\$180.00	\$200.00	\$230.00	\$220.00

a What is the mean cost of a burger?

b What are the median and mode of the burger prices?

c Explain how knowing the mean cost of items like this can help you to make decisions.

3 Work in groups. Choose one of the following surveys or create a survey of your own. Carry out the survey and record your findings in a table. Work out the mean, median and mode for the data you have collected. Present your findings to the class and talk about what the data tells us.

- The number of cars in the school's parking area during your maths lesson for a week.
- The number of students that take part in clubs or activities outside of school.
- The number of pets that students in your class have at home.

Statistics and probability

Maths detective: Games of chance and statistics

Case 1

Play this game with a partner or in a small group. You will need a die or you could use a random number generator instead of rolling a die.

- Decide who will go first by each rolling the die. Whoever gets the highest number goes first.
- Take turns to roll the die and record the scores for each player in a table. Do this until all players have rolled the die ten times.
- Find the mean score for each player. The student who has the highest mean score gets a point.
- To win the game, you need to be the first student to gain 3 points. Repeat until you find a winner!

Case 2

1 Work in pairs or in groups of three. You will need a deck of cards. Remove the jokers. The number cards are each worth their number value, the face cards are worth 10 and the aces are worth 1.
   - Draw 7 cards from the deck.
   - Sort them in order from smallest to greatest.
   - Find the mean of the set. Round it to the nearest whole number.
   - Find the median of the set.
   - Is there a mode? If there is, work it out.

2 Compare your results with the other groups. Write the following data sets:
   - the set of means found by all the groups
   - the set of medians found by all the groups.

3 Identify any trends you notice in your data.

Revisit what you know about graphs

Key maths idea

You have already practised collecting and recording data, and calculating the mean, median and mode of a set of data. Once you have collected data, there are different ways to display it. You can show data in **bar graphs**, **line graphs**, **pictographs** and **pie charts**. When you look at a graph, you can identify **trends** in the data. A trend is a pattern, for example, whether the numbers are going up or going down.

Here are some examples of graphs.

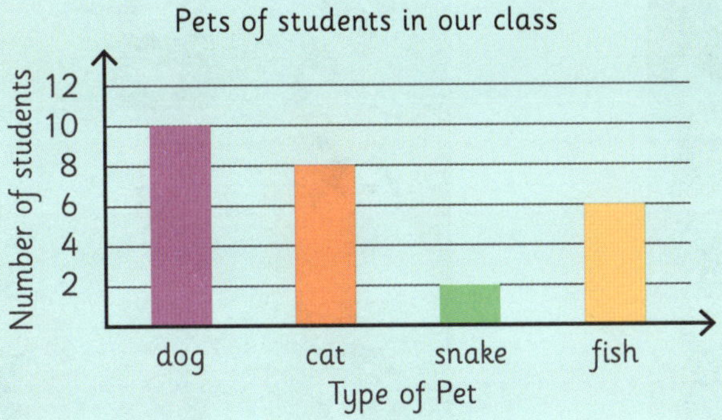

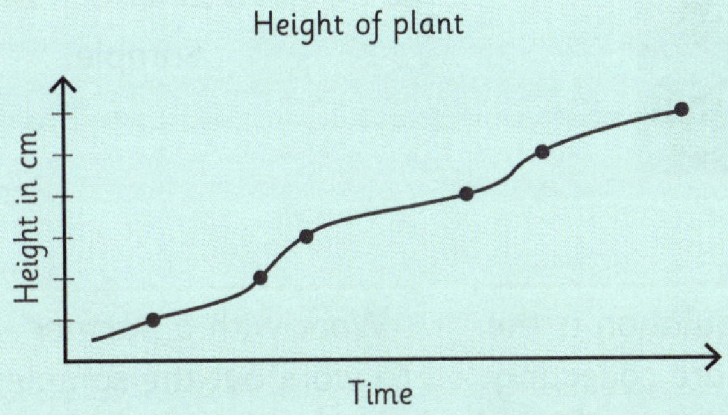

Key words

bar graph
line graph
pictograph
pie chart
trends

1 If possible, your teacher will show you a video or online tutorial about different types of graphs.

a Identify each graph as a pictograph, line graph, bar graph or pie chart.
Explain how you know which kind it is.

b Which two graphs represent the same data in different ways?
What is the same and what is different about the graphs?

c What similarity do you notice between the graph about pets and the graph showing the height of the plant? What difference do you notice between these two graphs?

2 With a partner, brainstorm why we need to use these different kinds of graphs. Think about these questions:

- Could you represent the data about the pets in a different way?
- Which other ways could you use?

3 You learned before about populations and samples.

Population

Sample

Remember that the population is the whole group that you are collecting data about. The sample is the number of individuals that you ask or study to get your information.

Work with a partner to work out the sample size for the pictograph and the bar graph on the previous page.

Using different graphs

Key maths idea

A Grade 4 class chose their favourite colours. They collected the information in a tally table.

Colour	Tally
Green	卌
Yellow	|||
Blue	卌 ||||
Red	卌 ||
Purple	卌	

Then they showed the data in a pictograph.

Favourite colours	
Green	★★½
Yellow	★½
Blue	★★★★½
Red	★★★½
Purple	★★★

Key:
★ = 2 students

Pictographs use symbols to represent data. They always have a key to tell you what the symbol represents. Pictographs can show a half or a quarter of a symbol, so here half a star is equal to 1 student.
Here is the same data shown in two different ways.

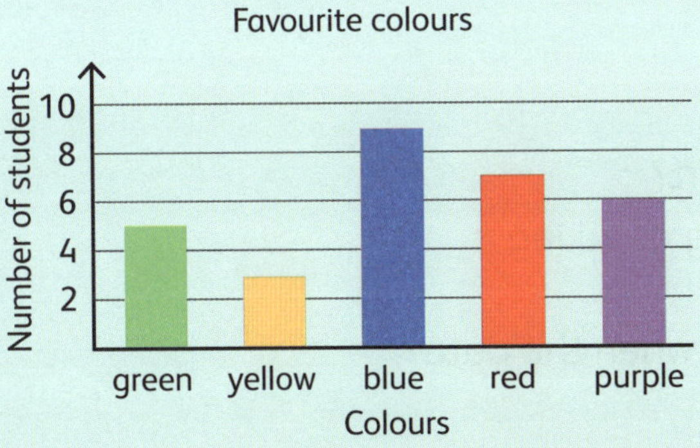

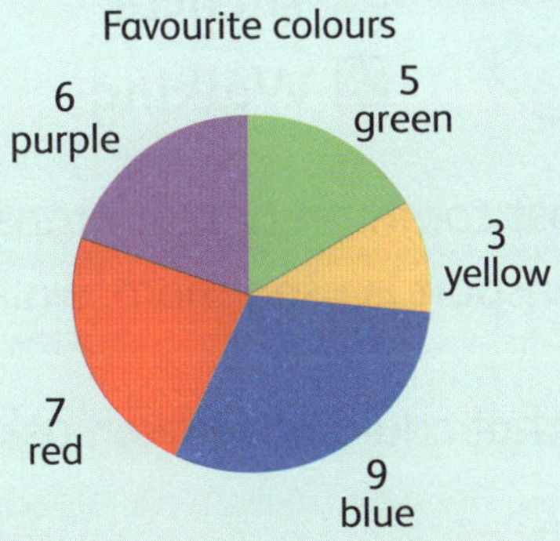

The first is a bar graph. A bar graph has an ***x*-axis** (horizontal line under the graph) and a ***y*-axis** (a vertical line, usually on the left of the graph). The bars are usually positioned vertically from the *x*-axis, or horizontally from the *y*-axis. In a bar graph, the width of the bars should be equal and the spaces between the bars should be equal.
The second chart is a pie chart. It is a circle that is divided into sectors. It shows the data as fractions of the whole sample.

Key words

x-axis
y-axis

1 Work in pairs and study the three charts in the Key maths ideas: the pictograph, the bar graph and the pie chart. Which one do you think shows the data most clearly? Which one is most helpful? Which do you prefer? Give reasons.

2 Look at the pictograph and then answer the questions.

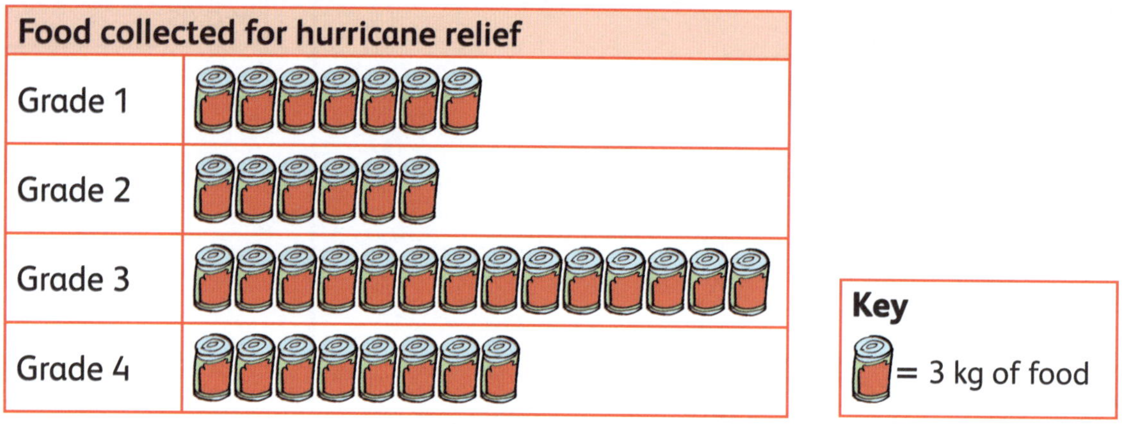

a What does 1 can on the pictograph represent?

b How much food did Grade 3 collect?

c How much less food did Grade 2 collect than Grade 1?

d Grade 5 collected 33 kg of food. Draw how this would be represented on the pictograph.

3 Look at the pie chart and answer the questions.

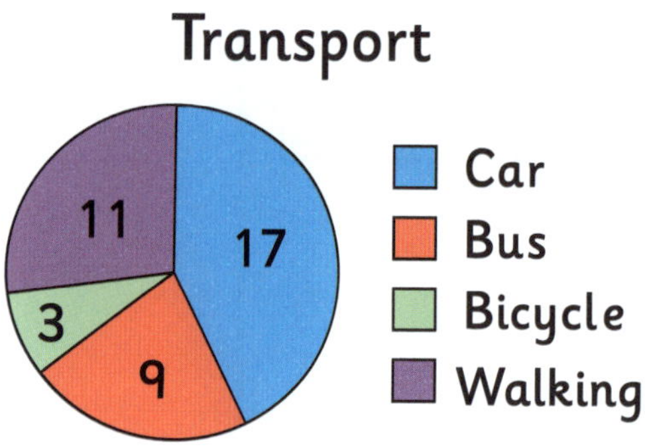

a What was the most common type of transport?

b Which type of transport is used more: walking or taking the bus?

c Write a question that could have been asked when the data was collected.

4 Carry out a survey in your class about the different types of transport students use to travel to school. Record your results in a table. Then organise and display your data in the form of either a bar graph or a pictograph.

Representing data in another way

Key maths idea

Sometimes, data is represented in one form and we want to convert it to a different form. For example, you have data in the form of a pictograph and want to **present** it in the form of a bar graph.

Example

Jay counted the number of butterflies he saw in his garden each day. This pictograph shows his results.

Butterflies counted

 = 12 butterflies

We can **convert** this information into a bar graph. To do this, think about how you will label the axes and what you will show in each bar. In this bar graph, the values are on the *y*-axis so it is a vertical bar graph. Each bar represents one day of the week.

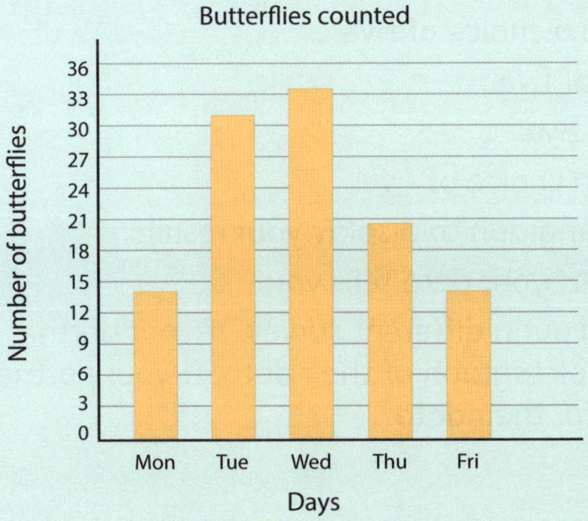

Key words

present

convert

Don't forget that half or quarter of a symbol represents a smaller amount!

Statistics and probability

1 The pictograph shows the number of pizzas three different pizzerias sell each week. Draw a bar graph to display the same data. Your bar graph may be horizontal or vertical.

2 The bar graph shows the ages of cats at an animal shelter. Draw a pictograph to display the same data. Don't forget to include a key.

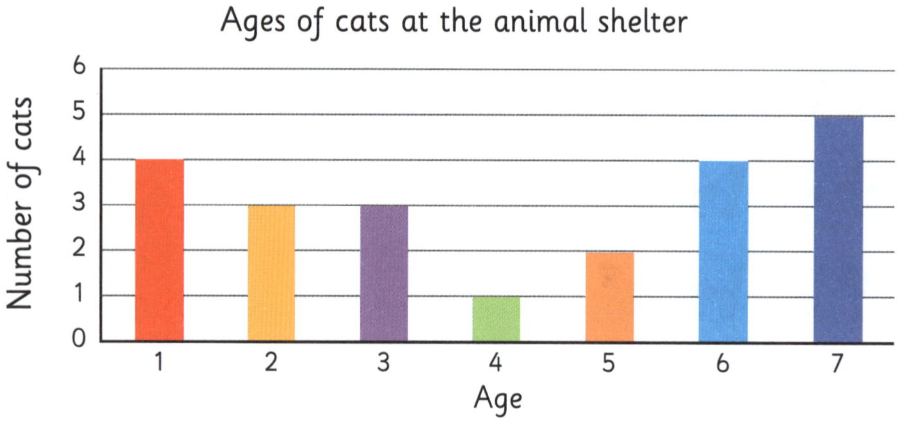

3 The table shows information about the food students bought at the school canteen on Friday.

Food	Burger	Pizza	Salad	Sandwich	Soup
Number of students	12	8	10	14	4

Choose to draw either a bar graph or a pictograph to represent this data set.

4 Choose one of the following surveys to carry out in your class. Record your results in a table.

- Favourite fruit or vegetable from a choice of five.
- Favourite animal from a choice of five.
- Favourite sport from a choice of five.
- Favourite type of weather from a choice of five.

a Draw either a pictograph or a bar graph to display your results.

b Write three sentences about what your data tells you.

c Swap with a partner who carried out a different survey. If your partner has drawn a bar graph for their data, draw a pictograph of their data. If your partner has drawn a pictograph, draw a bar graph for their data.

Line graphs

Key maths idea

Sometimes we want to collect data that shows us how things change over time. For example, how the height of a sunflower changes over time, or the temperature on each day of the week. We can display data like this in a line graph.

Example

The temperature is recorded at 12 noon each day for one week in September:

Day	Temperature (°C)
Monday	27
Tuesday	29
Wednesday	30
Thursday	28
Friday	32
Saturday	28
Sunday	26

The results are shown in a line graph:

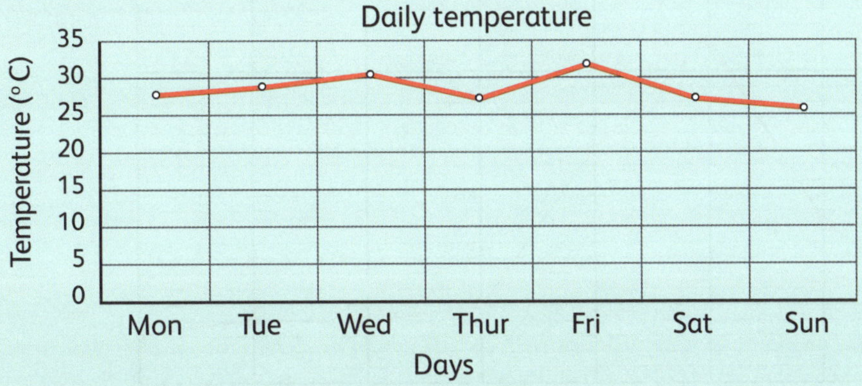

First we **plot** the **data points** on the graph. Then we draw a line to join the points. The line represents the changing temperature. A line that rises from left to right means the temperature is increasing. A line that falls from left to right means the temperature is decreasing.

We can see by looking at the line that the temperature is highest on Friday and it is lowest on Sunday.

We can describe the temperature over the week by looking at the line. This is called describing the **trend**. Here, the temperature rises steadily from Monday to Wednesday, before falling on Thursday and then rising to the maximum temperature on Friday. Over Saturday and Sunday, the temperature begins to fall.

Key words

plot

data points

trend

Statistics and probability

1 Look at the line graph and answer the questions.

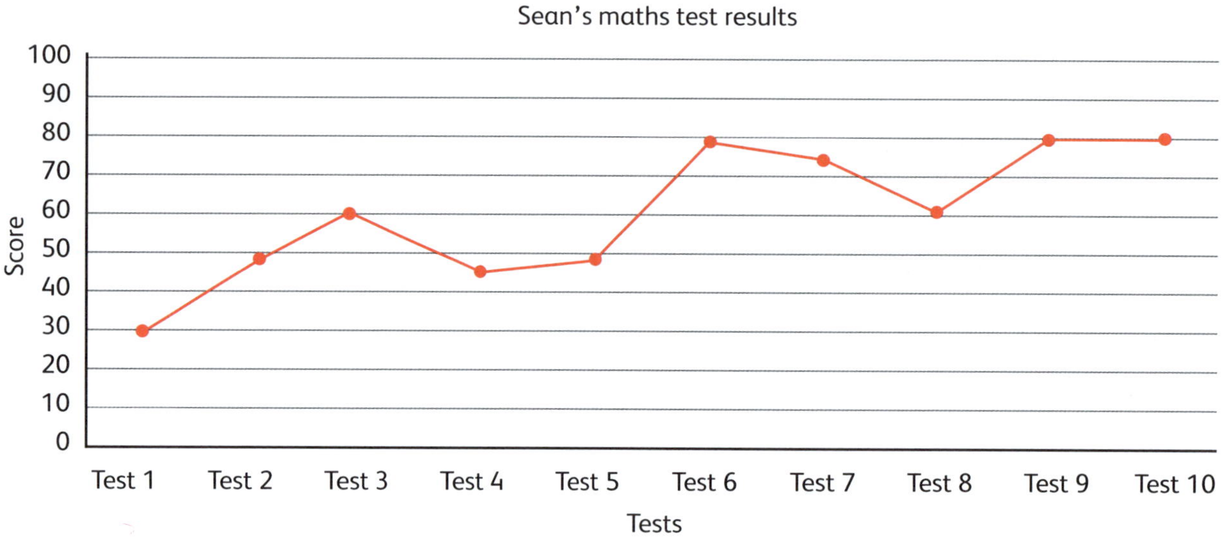

a What percentage did Sean score in Test 2?

b What percentage did Sean score in Test 9?

c Write a short description of Sean's score from Test 1 to Test 10.

d Is it true to say that in general Sean's score has improved?

2 Look at the line graph and answer the questions.

a What does this line graph tell us about?

b What method do you think was used to collect this data?

c Describe what happened between 12 noon and 3:00 p.m.

What I have learned

1 Look in newspapers or magazines, in books, or online, for more examples of each type of graph. Cut out, print out or copy some examples of the graphs you find.

2 Choose two of the graphs you found. Write two questions about the data presented in the graph. Give them to a partner to answer.

3 Make a journal entry about how people use graphs to present data about real-life situations.

4 Choose one topic and find out more about it:
- What are the popular after-school activities for students at your school?
- What time do most students go to bed at night?
- What do students like to bring to school for lunch?

Decide on a sample size. Collect your data and record your results. Then show your results in a graph.

Practice questions

1 Match each description to the correct type of average – mean, median or mode:

a The value in a data set that occurs the most often.

b The answer when you add together all the values in a data set, then divide by the number of addends.

c The middle value in a data set or the mean of the two middle values.

2 Twenty students take a spelling test. These are their scores.

20	19	11	18	12	18	15	17	10	18
10	19	16	18	19	20	18	14	12	18

Work out:

a the modal value

b the mean (rounded off to the nearest whole number)

c the median.

3 Ten students take a maths test. When the teacher adds together all the scores, the class total is 320. What is the mean test score?

4 A grocer records the bunches of bananas she sold each day for one week. If she sold 133 bunches altogether, what was the mean number of bunches sold per day?

Statistics and probability

5 The mean number of people who travel to work on the number 73 bus each day is 130. The bus company puts up posters to advertise its service. Do you think the claim in the advert is true? Explain why or why not.

6 A fisherman records the number of fish he catches each day in April. He didn't go fishing on nine days due to bad weather. The mean number of fish he caught each day was 16. How many fish did he catch in April?

7 Use the information in the pie chart to work out the answers to these questions.

a $\frac{1}{4}$ of all the patties Ginny sold were chicken. Work out the total number of patties represented in the pie chart.

b Work out how many ackee patties she sold. (Hint: chicken, ackee and pork made up half of all the patties.)

c If Ginny sold 8 shrimp patties, how many were beef?

Sales at Ginny's patty shop in one morning

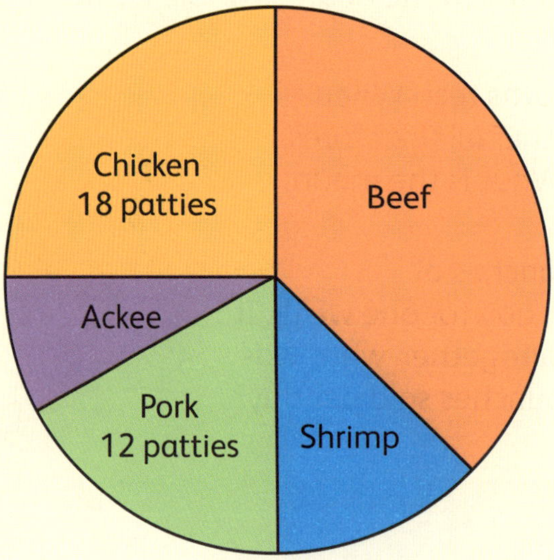

8 Look at the line graph and answer the questions.

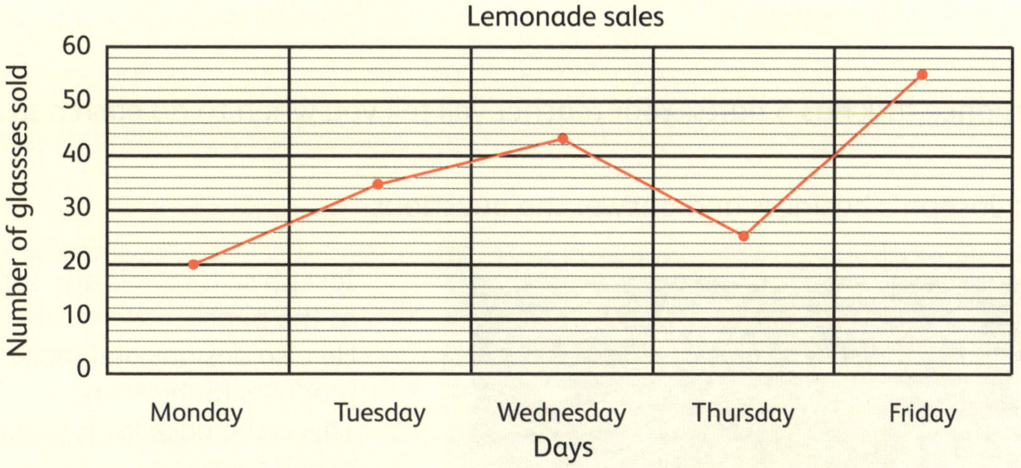

a State the days on which the most and least glasses of lemonade were sold.
b How many more glasses were sold on Wednesday than Monday?
c On which day were 34 glasses of lemonade sold?

14 Self check

Check you can do this: 	If you have difficulty, try this: 
1 Calculate the mean, mode and median of a data set.	Go back to the examples on pages 203 and 204. Practise ordering the values in the data set. Remember, the mode does not require any calculating. The median only requires a calculation if there is an odd number of values in the set.
2 Calculate the total set given the average and number of addends.	Revise the work on page 206.
3 Solve problems based on mean.	Revise the problems on pages 205 to 207.
4 Read and interpret bar graphs, line graphs, pictographs and pie charts.	Revise the different types of graphs on page 209 and the work on pages 210 to 216.
5 Present data using pictographs and bar graphs.	Revise pages 211 and 212.
6 Convert a pictograph to a bar graph or vice versa.	The work on pages 213 to 214 deals specifically with representing data from a pictograph on a bar graph or vice versa.

Statistics and probability

TERM 2

PEP task: Beach business

This performance task has 5 parts. Your teacher will tell you when to do each part.

Part 1

Read about Joseph's business and answer the questions.

Joseph sells towels and sarongs at the beach.
He also designs and makes the products for his store.
This is the price list for Joseph's beach towels:

TOWELS
Small – \$1 875 each
Medium – \$2 200 each
Large – \$2 600 each

1 Work out the cost of these sets of towels:

- **a** 10 small towels
- **b** 10 medium towels
- **c** 2 large towels, 1 small towel and 1 medium towel.

2 A customer spends \$13 125 on an order of small beach towels. How many did she buy?

3 Amari buys three towels of the same size. The Keeno family buy one large beach towel and a souvenir that costs \$4 000. Both customers pay the same amount. What size towels did Amari choose?

4 Malaki has \$16 000 to spend on towels. He wants to buy as many towels as he can, but he wants equal numbers of each size. How many can he buy of each size and how much change will he get?

Part 2

Work out these questions about the beach towels Joseph sells.

The dimensions of the beach towels are:

Small	70 × 150 cm
Medium	80 × 170 cm
Large	85 × 180 cm

1 **a** Work out the area of each towel in square centimetres.

b Convert the dimensions to metres and work out the area of each towel in square metres.

2 Work out the perimeter of each size towel.

3 Joseph displays the towels by pegging them to a clothes line as shown. His display includes 3 large towels, 2 medium towels and 2 small towels, with a 10 cm space between the towels, and 20 cm at each end of the line. What is the total length of the line?

Part 3

Read this extract about towel quality and answer the questions.

Towel manufacturers use a measure called 'gsm' to describe the weight of towels. This stands for 'grams per square metre'. The heavier the gram weight per square metre, the thicker and more absorbent the towels are. Very thin towels weigh between 300 and 400 grams per square metre, whereas very thick towels can be as heavy as 900 grams per square metre. This graph shows the relationship between grams per square metre and the quality of the towels.

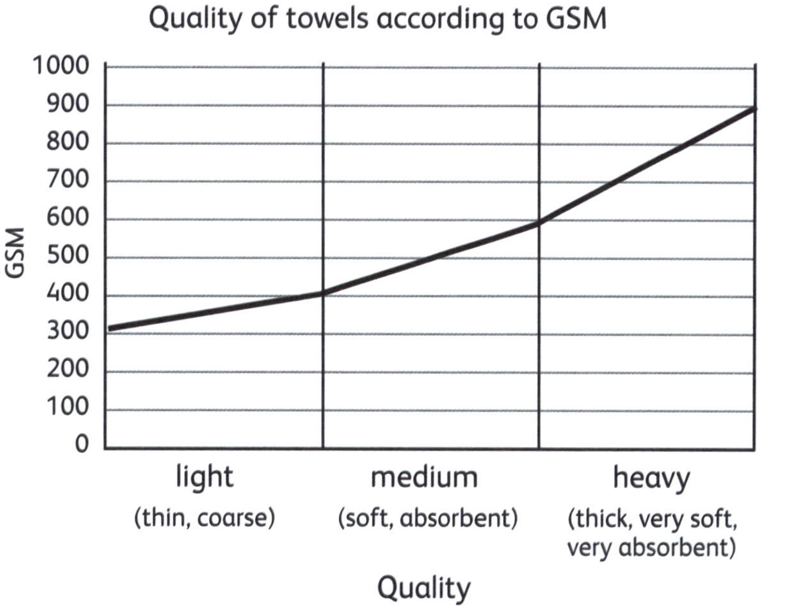

1 Complete these statements about towel qualities:

a A medium towel weighs between _____ and _____ grams per square metre.

b A heavy towel weighs between _____ and _____ grams per square metre.

2 If Joseph uses a towel fabric of 385 gsm, calculate the weight of:

a a medium towel **b** a large towel.

Round off your answers to the nearest gram.

3 Give two reasons why Joseph chose a sensible gsm for his beach towels.

Part 4

Read the information in each question and use it work out the answers.

1 Joseph has to put in an order for some new towels. Half of the towels have typical Jamaican designs, such as pictures of tropical trees and birds. One quarter are striped. The rest have other patterns. What information is missing in order to work out the number of towels he ordered of each type?

2 When the order arrives, $\frac{1}{12}$ of the towels are damaged and $\frac{1}{4}$ are the incorrect designs. He returns all the incorrect and damaged towels and keeps the rest. What fraction of the order does he keep?

3 A customer asks Joseph to make up 12 gift packs with 4 pink towels per pack. He has enough towels for $5\frac{1}{2}$ packs.

a How many towels does he have?

b How many more towels does he need to make up the order?

4 Joseph's daughter Jasmine is learning about flips, slides and turns at school. She uses what she has learned to design a towel for her dad's store. Use grid paper to draw three rectangles, each 6 × 12 cm. Use your three rectangles to show a towel design that uses each type of transformation.

5 Jasmine sketches a pattern she wants to use in a new design. These are the first three shapes in the pattern:

a Describe the rule she has used to make the pattern.

b Use the rule to work out how many blocks would be in the 10th term of the pattern.

c Write it as a number pattern.

d Jasmine draws *y* number of shapes. Express the height of the shape using the letter *y* in your expression.

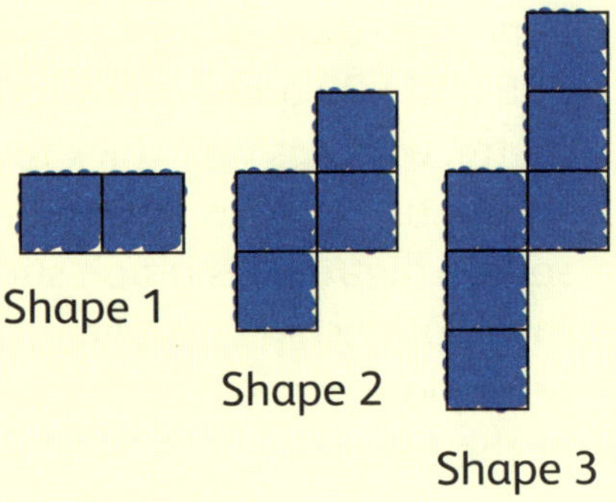

Part 5

Joseph and Jasmine are involved in a conservation project to help protect endangered turtles on nearby Gibraltar Beach, Orechabessa. Use the graph to answer the questions that follow.

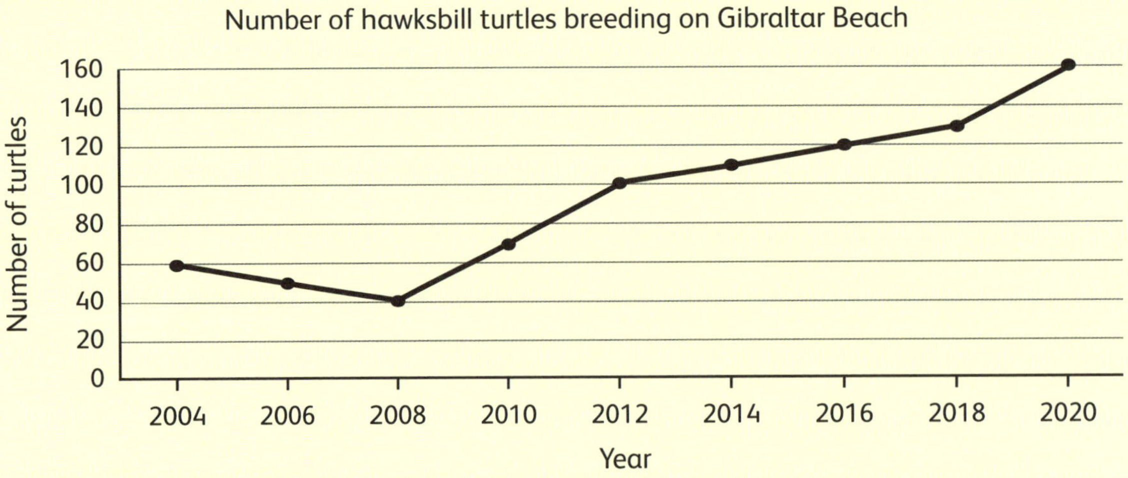

1 **a** Look at the graph on previous page. What does it show?

b Look at the trend of the graph. When do you think conservation initiatives began? Explain your reasoning.

2 Can you think of a reason for the bigger increase between 2018 and 2020?

3 Use the line graph to read or estimate the number of turtles in each year. Copy and complete the table.

Year	Number of turtles
2004	
2005	
2006	
2007	
...	

4 Use the trend of the graph to make a prediction for the year 2021.

5 Predict the number of turtles for the year you are in now.

The line graph below shows information about the world hawksbill turtle population over the past 100 years, since 1920.

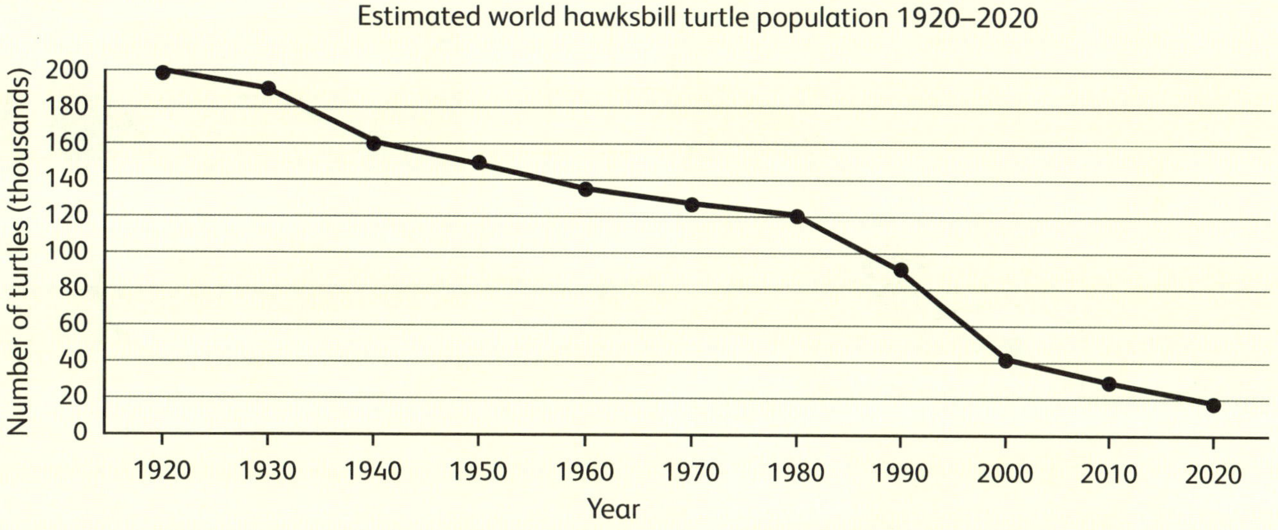

6 In what year was the turtle population first estimated?

a 2000 **b** 1970 **c** 1920 **d** 2020

7 What was the estimated turtle population in 1950?

a 100 000 **b** 90 000 **c** 150 000 **d** 65 000

8 In which year was the turtle population 40 000?

a 1940 **b** 1960 **c** 2000 **d** 2010

9 Which of the following statements is true?

a There has been little change in the turtle population.

b The turtle population has decreased over time.

c The turtle population has increased over time.

d The turtle population was greatest in 1980.

10 Write a description in words of the change in the turtle population shown in the line graph over 100 years. You should write at least two sentences.

11 Explain in your own words why it is better to use a line graph for this data than another graph such as a pictograph or a pie chart for example.

TERM 3

UNIT 1: NUMBER

p225

Multiplication and division

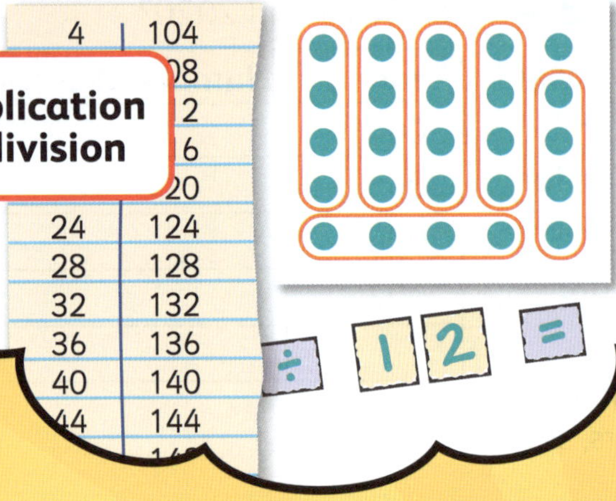

UNIT 2: GEOMETRY

p240

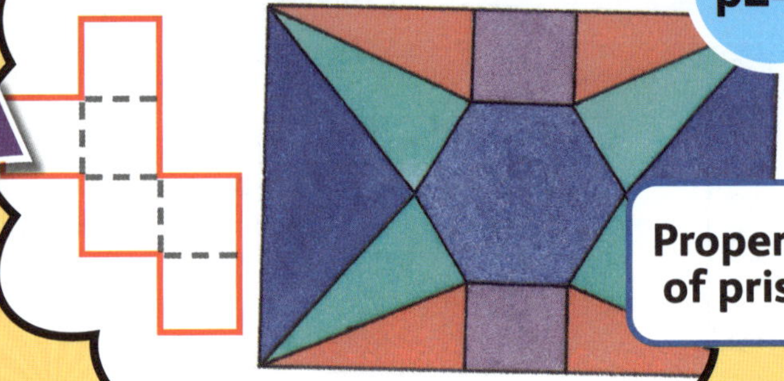

Properties of prisms

UNIT 3: ALGEBRA

p252

Using variables

UNIT 4: STATISTICS AND PROBABILITY

p269

Probability

Unit 1: Number

Chapter 15 Multiplication and division

In this chapter you will explore this focus question:

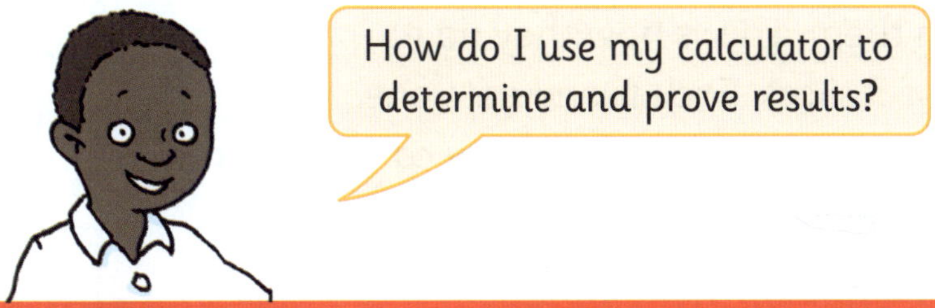

To answer this question, you will:

- define and use terms such as dividend, quotient, divisor and remainder
- recall multiplication and division facts
- solve problems involving division with and without remainders
- work with a calculator.

Starting point

A pet shop owner has a tank of goldfish.

1 Estimate the number of goldfish in the picture. Write your estimate. Then count. Check your answer with a partner.

2 The owner wants to put half of the goldfish into a different tank. Together with your partner, suggest two different ways she could work out how many to put into each tank.

3 Write a fact family of division and multiplication facts about this set of goldfish. Use the number 7 in your fact family.

4 Write two of your own story problems about the goldfish. Exchange problems with your partner and solve. Explain how you knew which operation to use.

Division is equal sharing

Key maths idea

Division is the operation that we use when we want to share an amount equally.

Think about this example:

A loaf of bread has 18 slices. Three friends decide to share a loaf equally between them. This is how they do it:

They take out a set of 3 slices. Each friend gets one.

They repeat this until there are none left. They can repeat the subtraction six times.

At the end, each friend gets 6 slices. There are no **leftovers**.

3 is the **divisor**. We want to share the dividend into 3 equal parts or groups.

dividend ÷ divisor = quotient

$$18 \div 3 = 6$$

18 is the **dividend**. It is the amount we want to share out equally.

6 is the **quotient**. It is the amount in each equal group or part.

dividend ÷ divisor = quotient

Sometimes, we cannot divide the dividend exactly. The leftover amount is called the **remainder**. Imagine sharing the loaf of bread among 5 friends.

$18 \div 5 = 3$ remainder 1

Each person gets 3 slices and there is 1 left over.

Key words

leftovers
divisor
dividend
quotient
remainder
relationship
inverse

We can think of division as repeated subtraction, just like multiplication is repeated addition.

You can use the **relationship** between multiplication and division to help you. Division is the **inverse** of multiplication. You know that $6 \times 3 = 18$. So $18 \div 3 = 6$.

1 Work in pairs. Use blocks, coins or counters to demonstrate that:
   a multiplication is the same as repeated addition
   b division is the same as repeated subtraction.

2 Make cards with the definitions of these terms. On each card, make a diagram to show what the term means. Explain your diagram to a partner.
   a dividend
   b quotient
   c divisor
   d remainder

3 Revise your division and multiplication facts using this 'round robin' game. One student states a division or multiplication fact, for example, 25 ÷ 5 = 5. The next student has to start with the quotient and give a different multiplication or division fact, from another family (you cannot use the same three numbers). In this example, 5 × 6 = 30 or 5 ÷ 1 = 5, are both possible correct answers because they begin with 5 and use a different set of numbers.

4 Work in groups. Use index cards and markers to make number cards with the digits 0 to 9. You will need at least 10 of each digit, as well as cards with the ÷ and = signs. Use the digit cards to create division sentences like the one in the picture below. Follow these instructions.

Each group creates a set of six division sentences (without the answers) and then each group moves to the next group's table to solve their questions. Then make your next set of questions at that table and move to the next one, and so on.

- Start with six division problems that have 3-digit dividends and 1-digit divisors.
- Next create six more division problems, this time using 4-digit dividends and 1-digit divisors.
- Do six more, with 4-digit dividends and 2-digit divisors.
- For the last set of six questions, use 5-digit dividends and 2-digit divisors.

> Use multiplication facts to help you make up your questions. Then you can work out the division sentences using the relationship between multiplication and division.

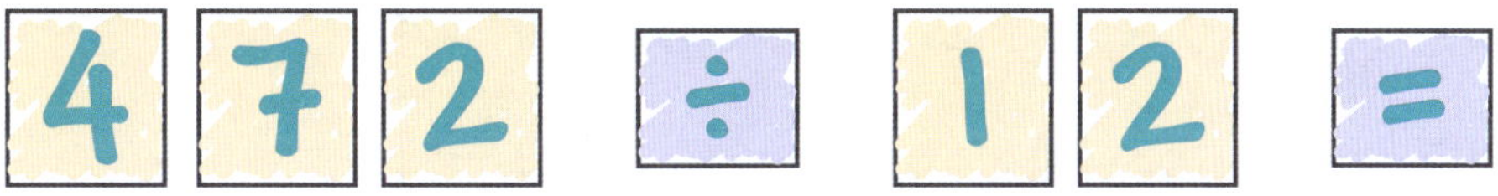

5 Work out the quotient and the remainder for each problem:

a 4 578 g sugar divided into 100 g packets

b 395 blocks divided into towers of 10

c 2 481 ml honey divided into 200 ml jars.

Using a calculator

Key maths idea

A pocket **calculator** can perform many calculations for you. Different types of calculators have different features. Here is a simple calculator.

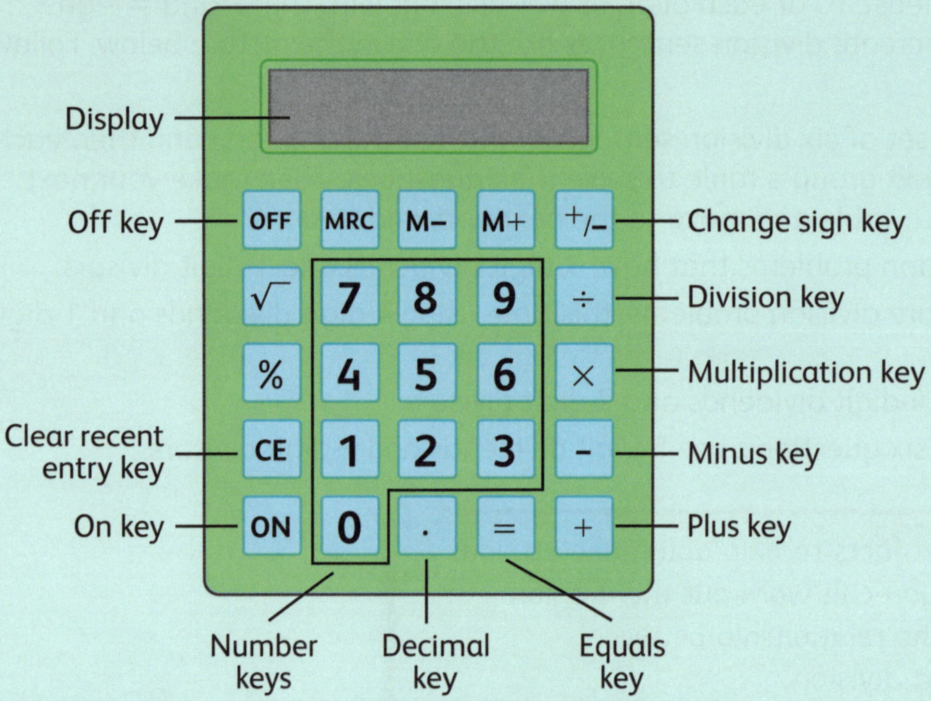

The buttons on the calculator are called keys.

The number keys are numbered from 0 to 9.

The operation keys are:

– + ÷ × =

Most calculators have two 'clear' keys.

The CE key clears only the last number you entered.

The ON/C or ON/AC key clears everything you have entered.

Key word

calculator

1 Work in small groups.

- Decide on any number and write it down. This number will be the answer that you will try and get closest to.
- Next, enter any number into your calculator and then press the multiplication key.
- Then, in less than 5 seconds enter another number that you think will get you to the answer number you wrote down.
- Compare the answer on the display screen to the answer you wrote down and see how close they are.
- Play about ten more rounds using the other operation keys too.

2 Enter these numbers and then turn your calculator upside down to read the word. Write the words in your notebook.

a 663 **b** 7 738 **c** 55 178 **d** 37 818

3 Look carefully at the words you made in question 2. Use them to help you make these words on the calculator screen. Write down the numbers you used.

a BILL **b** SHELL **c** HISS **d** BOSS

Try and make more of your own words.

Maths detective: Calculator games

Case 1

Follow the instructions to find the answers.

Number clues	Word clues	Number on display	Word on display (turn calculator upside down)
a 300 + 214	Not hers		
b 1 852 × 2	A _____ in a sock		
c 985 – 214	Not well		
d 393 + 212	Another name for a pig		
e 5 000 – 1 493	Not win		
f 676 ÷ 2	It makes honey		
g • multiply 17 by 2 • multiply by 1 000 • divide by 6 • take away the numbers to the right of the decimal • subtract three	A nice breakfast		
h • enter 6.2 • multiply this by itself • add 0.23 • divide by 50	A friendly greeting		

Case 2

Use a calculator to investigate the pattern.

1	9 – 1	98 – 21	987 – 321	9 876 – 4 321
2	4 ÷ 4	40 ÷ 4	400 ÷ 4	4 000 ÷ 4
3	8 – 5	88 – 15	888 – 555	8 888 – 5 555

Case 3

Guess what these answers will be and then check on your calculator.

1 98 765 – 54 321

2 987 654 – 654 321

Case 4

159 children are going on a school trip to a museum. Answer the questions and then use your calculator to see if your answers are correct.

1 The teachers put the children into three groups. How many children are in each group?

2 The buses to transport the children can each carry 28 children. Will five buses be enough?

3 The children each paid $400 for the trip. How much money is this altogether?

4 At the museum, the children enter in groups of eight. How many groups will there be?

Case 5

An elevator has this safety notice.

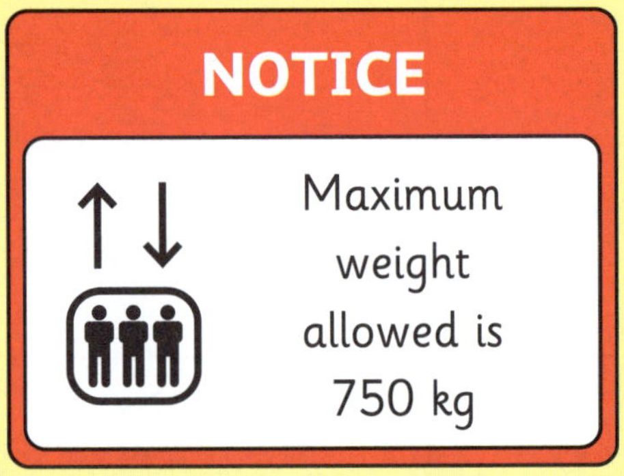

10 people are waiting for the elevator. The mean weight per person is 80 kg. How many of them should get into the elevator? How many should wait for the next elevator? Work it out. Then check your answer with your calculator.

Explore divisibility rules

Key maths idea

Divisibility means 'the ability to share equally without leaving a remainder'.

For example: 30 ÷ 3 = 10

30 is **divisible** by 3. That means 3 goes into 30 without leaving a remainder.

3 and 10 are both **factors** of 30.

30 is a **multiple** of 3. It is also a multiple of 10.

30 has some other factors too.

The factors of 30 are: 1, 2, 3, 5, 6, 10, 15 and 30.

Factors and multiples make some interesting **patterns**. When you learn to see these patterns, you can easily remember your division and multiplication facts. You can also use these patterns to work out divisibility rules. A divisibility rule lets you work out if one number is divisible by another number, just by looking at it.

Key words

divisibility

divisible

factor

multiple

patterns

Work with a partner or group. The questions that follow will help you to explore multiplication and division patterns, and divisibility rules.

1 Even numbers are always divisible by 2 and odd numbers are not.

- **a** Explain what you understand by even and odd.
- **b** How can you tell if a large number is divisible by 2, for example 48 974?
- **c** Write a divisibility rule for numbers that are divisible by 2.

2 Some students started writing down the first 50 multiples of 4. They wrote the first 25 multiples of 4 in one column and then started writing the next 25 in the next column.

- **a** What pattern do you notice in the ones place of the multiples?
- **b** The 25th multiple of 4 is 100. What multiplication sentence can you make from that fact?
- **c** What pattern do you notice in the second column?
- **d** Write the next four multiples of 4 that are missing from the second column.
- **e** Predict what the 50th multiple of 4 will be.
- **f** Cal says: 'This shows us that multiples of 100 are always divisible by 4'. Is she correct? Why?
- **g** The divisibility rule for 4 is that if the last 2 digits of a number are divisible by 4, then the whole number is divisible by 4. For example in the number 8 948, the last two digits make 48. 48 ÷ 4 = 12, so 8 948 is divisible by 4. Why does this rule work?

4	104
8	108
12	112
16	116
20	120
24	124
28	128
32	132
36	136
40	140
44	144
48	148
52	152
56	156
60	
64	
68	
72	
76	
80	
84	
88	
92	
96	
100	

3 Kim says:

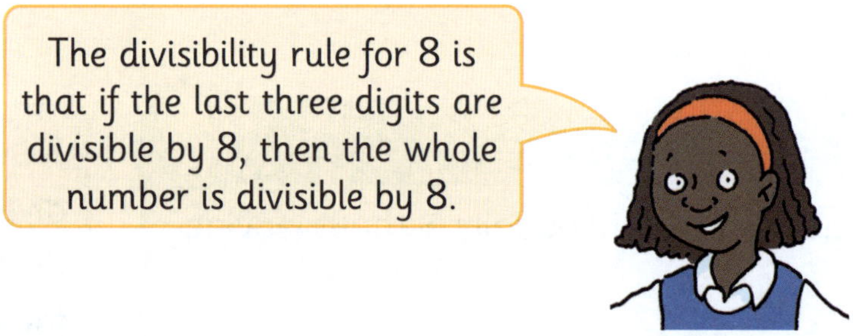

- **a** Predict whether you think Kim's rule will work. Give reasons for your answer.
- **b** Find ways to check Kim's rule. You can use your calculator to help you.

4
- **a** Write the first ten multiples of 10. What makes it easy to work out whether a number is divisible by 10?
- **b** Write the first ten multiples of 5. What relationship do you notice between the multiples of 5 and the multiples of 10?
- **c** Write a rule for working out whether a number is divisible by 5.
- **d** One hundred = 100. Two hundred = 200.
  15 hundred = 1 500.
  What can we use to work out whether a number is divisible by 100?

> Look at the number in the ones place.

5
- **a** Use your calculator to come up with ten different 4-digit multiples of 3, such as 3 735, 2 988, and some others.
- **b** For each number, add together the digits of the number.
  For example, for 3 735, add together $3 + 7 + 3 + 5 = 18$. Then add the digits of the sum again until you reach a single-digit number. $1 + 8 = 9$.

Do this with several examples. What pattern do you notice? Use the pattern to help you develop a divisibility rule for 3.

6 Repeat the exercise from Exercise 4, but this time use your calculator to generate multiples of 6. Use your calculator to add together the digits.
- **a** What rule can you work out for divisibility by 6?
- **b** Use this rule to predict the rule for divisibility by 9. Use your calculator to check.

Division with remainders

Key maths idea

A **remainder** is the amount left over when the dividend is not divisible by the divisor.

For example: $22 \div 4 = 5$ remainder 2

Sometimes we need to use remainders to help us work out longer division problems. We can divide using a 'division house' over the dividend. You divide the digits in each place from left to right. If there are remainders, you carry them to the next digit. Look at these examples.

Example 1

$8\,920 \div 4$	
$4 \overline{)8920}$ with 2 above	Look at the first digit: $8 \div 4 = 2$ Write it above the first digit.
$4 \overline{)89^{1}20}$ with 22 above	Look at the next digit. $9 \div 4 = 2$ r 1
$4 \overline{)89^{1}20}$ with 223 above	Carry to the next place. $12 \div 4 = 3$
$4 \overline{)89^{1}20}$	$0 \div 4 = 0$

If the first digit in the dividend is too small to divide, you look at the first two digits together.

Example 2

$39\,078 \div 5$	
$5 \overline{)39^{4}078}$ with 7 above	5 does not go into 3, so look at the first two digits together. $39 \div 5 = 7$ r 4
$5 \overline{)39^{4}078}$ with 7 8 above	Carry the 4 to the next place. $40 \div 5 = 8$
$5 \overline{)39^{4}07^{2}8}$ with 7 81 above	Nothing to carry. $7 \div 5 = 1$ r 2
$5 \overline{)39^{4}07^{2}8}$ with 7 81 5r3 above	Carry the 2 to the next place. $28 \div 5 = 5$ r 3

Key word

remainder

1 Divide using the method you have learned here. Show your working.

a 248 ÷ 2
b 5 875 ÷ 5
c 9 418 ÷ 3
d 7 761 ÷ 4
e 9 974 ÷ 5
f 8 023 ÷ 7
g 3 804 ÷ 6
h 3 284 ÷ 5

2 Use division to help you work out these fractions of whole numbers.

a one third of 6 396
b one fifth of 2 240
c one tenth of 8 850

3 Use what you know about division and multiplication to help you work out the missing terms.

a divisor = 8; quotient = 7; dividend = _____
b dividend = 100; quotient = 50; divisor = _____
c divisor = 2; dividend = 80; quotient = _____

4 a Danny exercised every day for a week. His total exercise time was 455 minutes. How long did he exercise each day?

b Caryn walked up 9 flights of stairs. She counted 162 steps in total. If there is an equal number of steps on each flight, how many steps are on each flight?

5 147 students go on a field trip.

a The students are from three classes, each with an equal number of students. How many students are there per class?
b 25 students can fit on each bus. How many buses should the school book for the trip?
c Each student collects 4 soil samples on the field trip. How many soil samples do they collect altogether?

6 Work with a partner. Explain how to work out the answers to these questions using number facts you know and the relationship between multiplication and division.

a 2 000 ÷ 20
b 25 500 ÷ 25
c 30 000 ÷ 15
d 42 ÷ 21
e 44 440 ÷ 22
f 24 ÷ 12

7 Use multiplication by 2-digit numbers to generate 5-digit numbers.
Use the fact families you used to create division facts that divide a 5-digit number by a 2-digit number.

The relationship between division and fractions

Key maths idea

$25 \div 4 = 6$ remainder 1

Another way you can think about this is by thinking about **fractions**.

How many whole groups of 4 are there in 25?

Each four is one whole group.

($\frac{4}{4} = 1$ whole)

$25 \div 4 = \frac{25}{4}$

$\frac{25}{4} = 6\frac{1}{4}$

Write your final answer as a **mixed number.**

Can you see that you can write a division sentence as a fraction? The dividend becomes the numerator and the divisor is the denominator.

Another way to do this is by using decimals.

We can write $6\frac{1}{4}$ as a decimal: 6.25.

If you don't know the answer, you can work it out using a division house:

$$\begin{array}{r} 6.25 \\ 4\overline{)25.{}^{1}0{}^{2}0} \end{array}$$

Don't forget the decimal point in the same position in the quotient.

Write a decimal point after the whole number. Write as many zeroes as you need to work out the **decimal fraction**.

Now look at this example.

I have 3 candy bars. I want to divide them fairly among four people.

$3 \div 4 = 0$ remainder 3

Another way is to think of this question:

How many whole sets of four are there in 3?

Remember we can write a division sentence as a fraction.

$3 \div 4 = \frac{3}{4}$

Key words

fraction

mixed number

decimal fraction

1 Write each improper fraction as a division sentence. Then simplify the improper fraction. What do you think they have in common?

a $\frac{10}{5}$ **b** $\frac{18}{6}$ **c** $\frac{12}{3}$ **d** $\frac{20}{5}$ **e** $\frac{28}{7}$

2 Simplify these improper fractions.

a $\frac{3}{2}$ **b** $\frac{5}{2}$ **c** $\frac{7}{2}$ **d** $\frac{9}{2}$

Use your answers to suggest a rule about dividing odd numbers by 2.

3 Calculate. First write the quotient as a whole number with a remainder. Then write it as a mixed number.

a 32 ÷ 5 **b** 23 ÷ 3 **c** 55 ÷ 9 **d** 32 ÷ 10

e 40 ÷ 6 **f** 62 ÷ 3 **g** 85 ÷ 4 **h** 97 ÷ 3

4 Copy these division problems and fill in the missing numbers. Then write the answers as mixed numbers.

a 39 ÷ 7 = _____ r 4 **b** 61 ÷ 8 = _____ r 5

c 77 ÷ 9 = _____ r 5 **d** 18 ÷ 5 = _____ r _____

e 19 ÷ 2 = _____ r _____ **f** 299 ÷ 3 = _____ r _____

5 Divide. Show your working.

a 232 ÷ 2 **b** 487 ÷ 5 **c** 423 ÷ 3

d 904 ÷ 8 **e** 640 ÷ 5 **f** 228 ÷ 2

6 Divide using a 'division house'. Show your working.

a $4\overline{)500}$ **b** $8\overline{)104}$ **c** $9\overline{)288}$

d $3\overline{)172}$ **e** $4\overline{)156}$ **f** $6\overline{)636}$

7 Divide. Show your working out. If the quotient has a remainder, write it with a remainder and also write it as a mixed number.

a 425 ÷ 25 **b** 435 ÷ 25 **c** 511 ÷ 30

d 444 ÷ 18 **e** 1 711 ÷ 7 **f** 2 907 ÷ 16

g 5 013 ÷ 3 **h** 7 654 ÷ 6 **i** 8 597 ÷ 12

j 99 526 ÷ 3 **k** 41 174 ÷ 17 **l** 56 091 ÷ 21

8 Rhona's mom bakes 150 cupcakes. She sells them in boxes of nine.

a How many boxes can she make up?

b How many are left?

Real-world maths: Catering careers

Chefs, caterers and bakers use division every day when they work out how much they need of each ingredient, serving sizes of meals and how to arrange their products on trays or in boxes.

Fried johnnycakes

450 g flour
20 g sugar
10 g baking powder
5 g salt
30 g butter
1 cup of milk
4 cups oil for the pot

Method:

1 Mix together the dry ingredients.
2 Rub the butter into the dry ingredients until the mixture looks like fine bread crumbs.
3 Mix in the milk. The dough should be stiff and a bit dry. Let it rest for 10 minutes.
4 Use your hands to push the dough into a ball. Divide it into 16 small balls.
5 Heat the oil and fry the balls of dough.
6 Serve with ackee and saltfish or any dish that has gravy!

1 A chef has a 12.5 kilogram bag of flour. How many full batches of johnnycakes can she make? Check with a partner to make sure you got the same answer.

2 Work out how much of each of the following she will need:

- **a** sugar
- **b** baking powder
- **c** salt

3 She has a 500 g block of butter. How much more butter will she need?

4 1 litre = 4 cups of milk. Work out how many bottles of milk she needs if she buys:

- **a** 1 litre bottles
- **b** 2.5 litre bottles.

5 **a** If each batch makes 16 johnnycakes, how many johnnycakes does the chef make altogether?

b She arranges 30 johnnycakes on each tray. How many full trays can she make?

6 Choose a simple recipe that you like to make at home. Write the instructions or bring a copy of your recipe to school. Write three division questions about your recipe and exchange them with a partner.

What I have learned

1 Use this division sentence: 48 ÷ 6 = 8.
Label the division sentence to show the dividend, divisor and quotient.

2 Write three mixed numbers and then express them as improper fractions.

3 Write three whole numbers and then express them as improper fractions.

4 10 ÷ 3 = 3 remainder 1
How can you write this as a fraction?

5 True or false? Give a reason for each answer.

- a Multiplication and division are inverse operations.
- b When the divisor is a factor of the dividend, the answer always has a remainder.
- c When you divide an odd number by an even number, the number always has a remainder.
- d A mixed number can never be equivalent to an improper fraction.

6 Use these division sentences to explain your answers to the question below.

48 ÷ 8 = 6 48 ÷ 7 = 6 remainder 6

Nelly says: 'It is impossible that both 7 and 8 go into 48 six times. One of the statements must be wrong.' What is Nelly's mistake?

Practice questions

1 Match the division with its inverse multiplication.

a	20 ÷ 10 = 2	11 × 3 = 33
b	52 ÷ 4 = 13	4 × 12 = 48
c	55 ÷ 11 = 5	20 = 2 × 10
d	48 ÷ 12 = 4	52 = 13 × 4
e	33 ÷ 3 = 11	5 × 11 = 55
f	45 ÷ 3 = 15	45 = 3 × 15
g	132 ÷ 4 = 33	411 × 10 = 4 110
h	4 110 ÷ 10 = 411	4 × 33 = 132

2 Find the quotient and show your working out. Give your answers with remainders and as mixed numbers.

a 49 ÷ 3 b 56 ÷ 3 c 543 ÷ 6 d 423 ÷ 8
e 7 450 ÷ 12 f 2 306 ÷ 10 g 16 544 ÷ 4 h 32 191 ÷ 11

3 Write whether these numbers are divisible by 2, 3 or 4. Some numbers may be divisible by more than one number.

a 54 b 723 c 1 368 d 270 e 2 870
f 75 g 1 320 h 907 626 i 672

4 What is the quotient when 75 is divided by 3?

5 Share 123 pencils equally among eight children. How many are left over?

6 Tom washes his car every 5 days. How many times will he wash it in 8 weeks?

7 a Use a calculator and find these answers.

4×9 4×99 4×999 $4 \times 9\,999$ $4 \times 99\,999$

b What do you notice about the pattern?

8 Enter these numbers and then turn your calculator upside down to read the word. Write the words in your notebook.

a 5 607 (Clue: What do frogs sit on?)

b 0.08 (Clue: What does a ghost say?)

c 577 345 (Clue: What do you find at the beach?)

d $7\,100 \div 10$ (Clue: A black liquid.)

e $1\,821 \div 3$ (Clue: A part of a tree.)

15 Self check

Check you can do this:	If you have difficulty, try this:
1 Explain what the terms dividend, quotient, divisor and remainder mean.	Revise the work on pages 226 and 227.
2 Divide 5-digit numbers by one or 2-digit numbers.	Revise the work on page 227.
3 Test numbers for divisibility by 2, 3 and 4.	Revise the work on pages 231 to 232.
4 Express as a mixed number the answer to a division problem with a remainder.	Revise the work on pages 235 to 236.
5 Convert answers with remainders to mixed numbers.	Revise the work on pages 235 to 236.
6 Recall division facts up to $100 \div 10 = 10$.	Revise your multiplication and division facts and refer to the divisibility rules on pages 231 to 232.
7 Solve problems using multiplication and division.	Revise the work you did on page 234 and any other problem solving work from this chapter, including the real-world maths on page 237.
8 Use a calculator to investigate number patterns and use a calculator to check answers.	Revise the work on pages 228 to 229.

Unit 2: Geometry

Chapter 16 Properties of prisms

In this chapter you will explore this focus question:

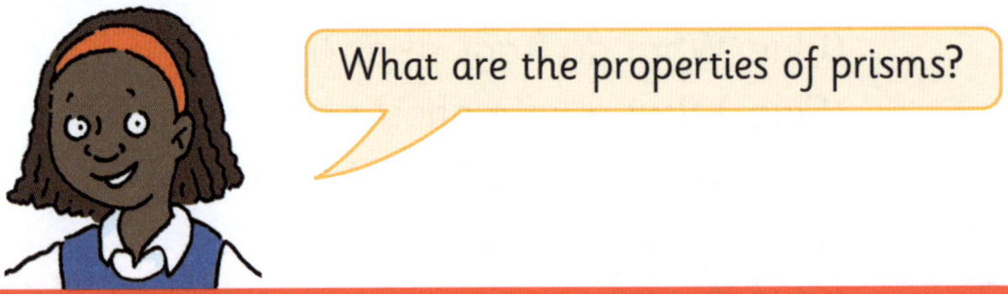

To answer this question, you will:

- develop models of solid shapes using a variety of tools
- work with nets of different prisms, specifically cubes and cuboids
- identify prisms from their nets.

Starting point

Elroy sketched these solids:

Discuss each question with a partner or as a class.

1 For each of these plane shapes, identify one or more solids in the sketch that has a face to match:

a circle

b square

c rectangle

d hexagon.

2 How would you sort and group the solids in the picture? Explain some different ways you could sort them.

3 Elroy says one solid is the odd one out. Which one do you think he means? Give a reason for your choice.

4 Revise what you remember about plane shapes by identifying the odd one out in each set. If you can't remember any of the terms, use a dictionary to look them up or look up examples online. Some sets have more than one possible answer. Remember to give reasons for how you identified the odd one out.

a circle, oval, rhombus

b parallelogram, triangle, quadrilateral, rhombus

c trapezium, square, rectangle, right-angled triangle

Revisit what you know about plane shapes

Key maths idea

You have looked before at open and closed figures. A closed figure is also called a **plane shape** or a **2D** shape. 2D stands for **2-dimensional**, because we can measure two **dimensions** of these shapes – length and width. 2D shapes do not have thickness or depth.

Polygons are 2D shapes that have **straight** sides which meet at **corners** (or **vertices**). Shapes with curved sides, such as circles or ovals, are not polygons.

Key words

plane shape
2D
2-dimensional
dimension
straight
corner
vertices

1 Look at these carpet designs. For each one, describe the shapes that the designer used to make the pattern.

a

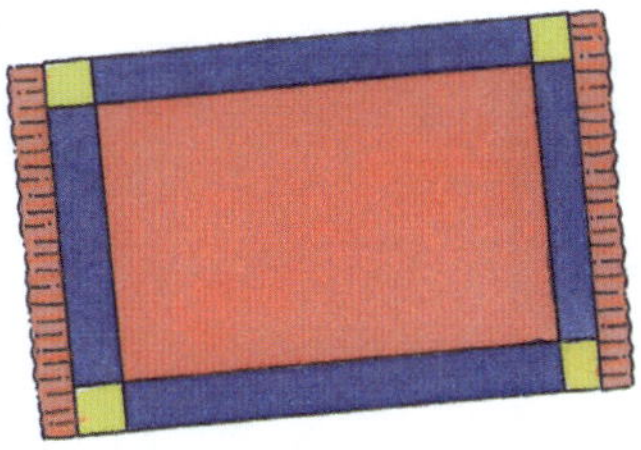

b

c

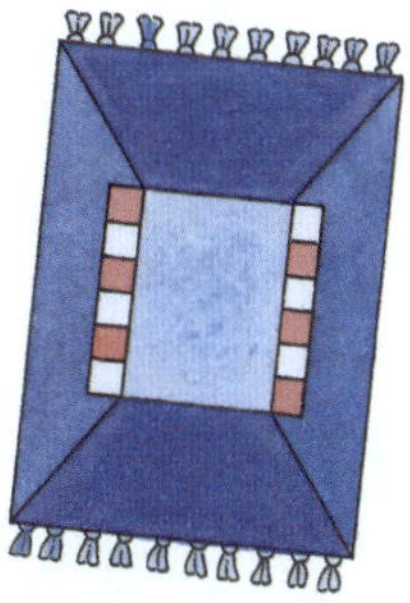

d

e

f

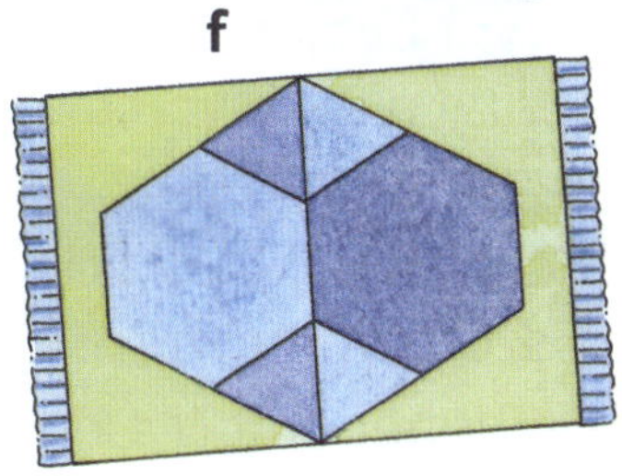

2 Use a ruler. If you have a set-square, you can use it to help you make the right-angled corners. If not, trace around the corner of a book or make a right angle by folding a piece of paper over twice. Draw these shapes:

- **a** a quadrilateral with all its sides equal
- **b** a right-angled triangle
- **c** a quadrilateral with two pairs of parallel sides but no right angles
- **d** an irregular pentagon with one right angle.

3 Look around your classroom. Find real-life examples of:

- **a** triangles
- **b** rectangles
- **c** circles.

3D solids

Key maths idea

Objects that take up space have three dimensions: length, width and height. They are called **solids** or **3D** shapes. Here are some examples of 3D solids:

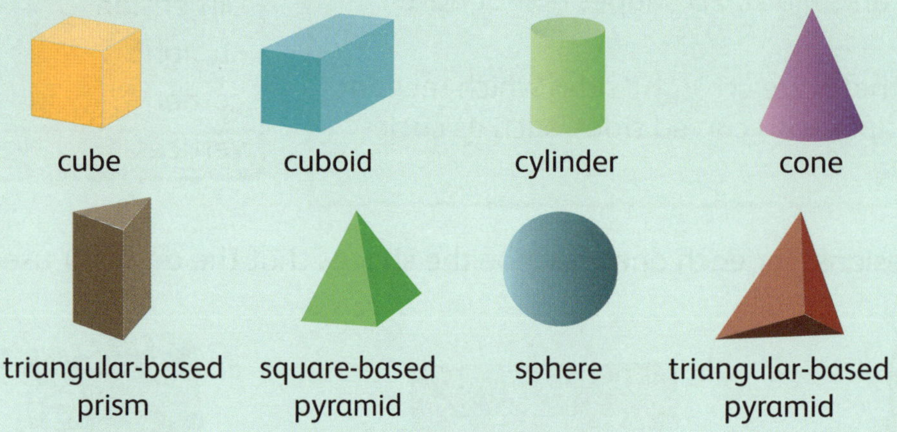

The flat side of a solid is called a **face**. **Cubes** and **cuboids** have six faces. A cylinder has two circular faces and a curved surface. Spheres have no flat faces, only a curved surface.

An **edge** is the line where two faces meet.

The point where three or more faces meet is called a **vertex** (plural: **vertices**).

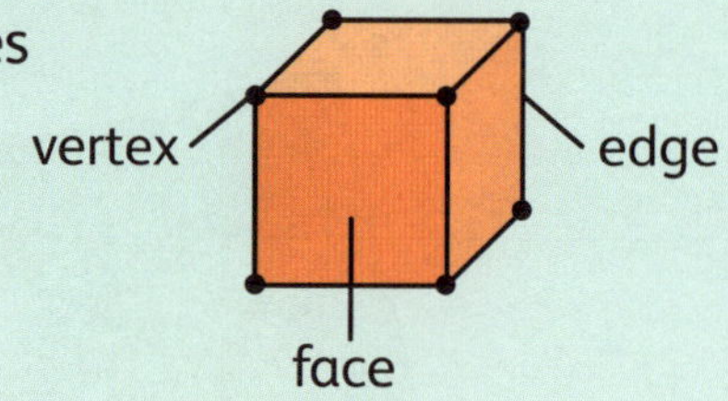

We can classify solids in different ways, according to:

- number of faces, edges or vertices
- types of faces or surfaces.

Key words

solid
3D
face
cube
cuboid
edge
vertex
vertices

1 Look at the solids shown in the diagram above. Identify real-life objects that have each shape.

2 Compare a cube and a cuboid. Copy and complete this table showing their differences and similarities.

	Cube 	**Cuboid** 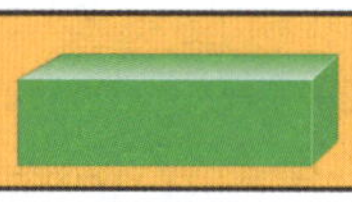
Number of edges		
Number of vertices		
Number of faces		
Shape of faces		

Maths detective: Creating 3D solids

For this activity you will use craft materials to make a variety of 3D solids.

You will need some or all of these materials:
- pipe cleaners or wire that you can bend into shapes
- fudge sticks joined with tape or glue
- toothpicks or cotton earbuds
- small balls of adhesive putty to join the edges at points
- beads to mark the vertices.

Case 1

Work in groups. Collect your materials and use them to make:
- two cubes of different sizes
- a cylinder
- a triangle-based pyramid
- two different cuboids
- a square-based pyramid
- a triangular prism.

Colourful pipe cleaners are useful for making 3D solids, but there are also many other materials you can use if these are not available.

Case 2

In Case 1, you had to make solid shapes by constructing their edges. How would you construct the same shapes from cardboard? You will need to cut out the shape of each face. For each shape you constructed in Case 1, draw a sketch of the faces you would need to stick together to make the same shape.

Nets

Key maths idea

In Case 2 on page 243 you had to think about the flat shapes you would need to make a model of a 3D solid. A **net** is a 2D shape or pattern that you can fold to make a 3D solid. Usually, when you draw a net, you draw the exact shapes of the faces. The nets we use in real life sometimes also have tabs that you can use to stick the faces together.

Key word
net

1 Trace the nets on this page onto paper. Cut them out and then fold them to make solids. The solid lines show you which lines to trace and cut. The dotted lines are fold lines.

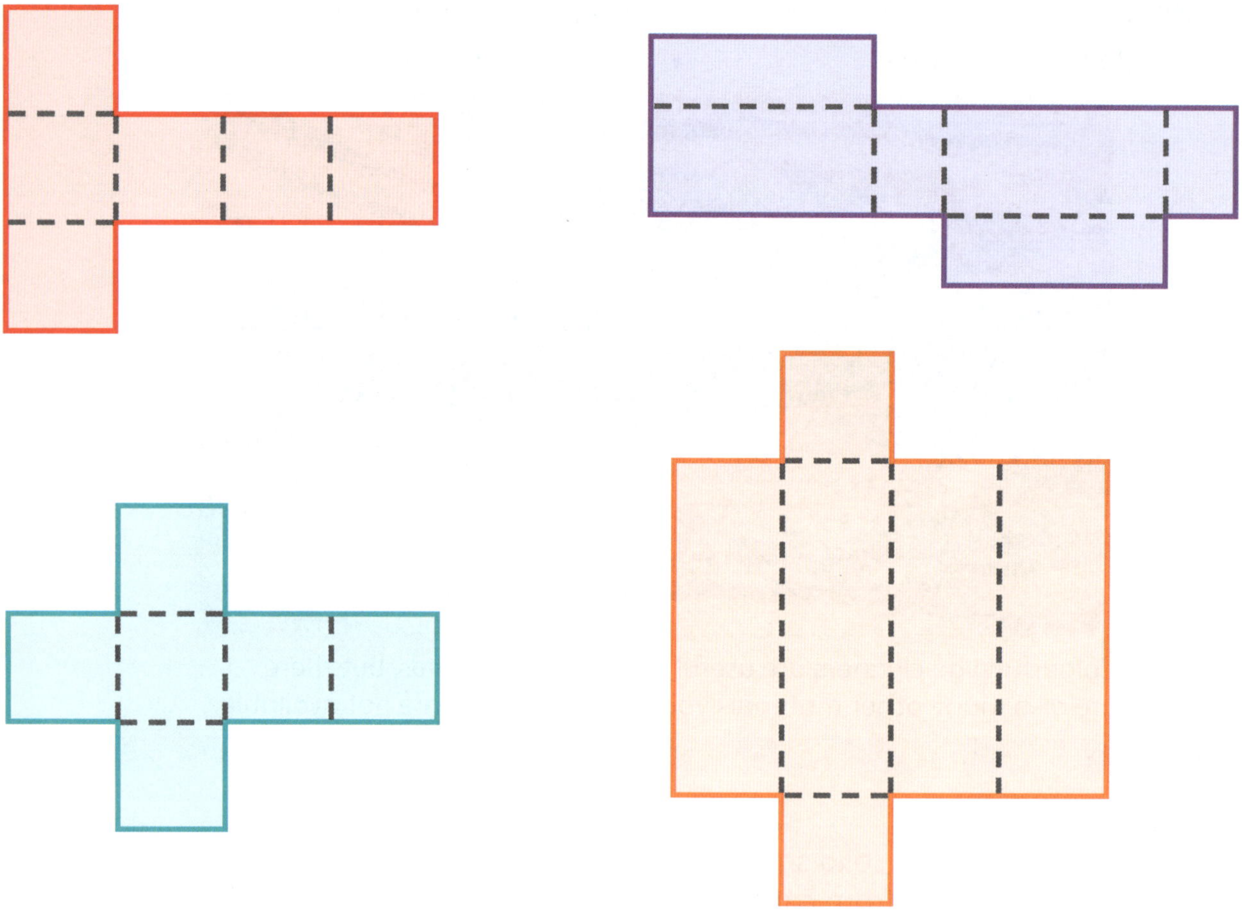

2 Work with a partner. Sketch two other net designs that you could use to make:

a a cube

b a cuboid.

3 Identify which of these nets can be folded into a cube and which cannot. Explain your answers to a partner.

a

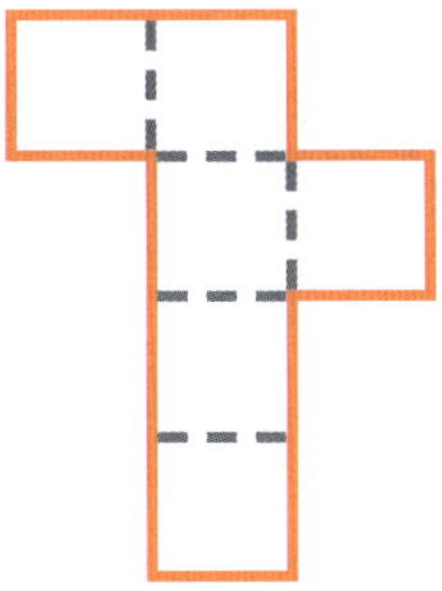

b

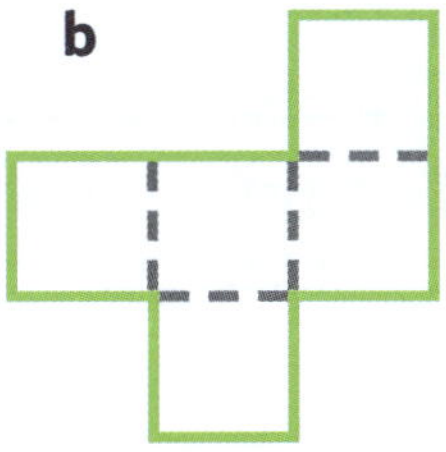

c

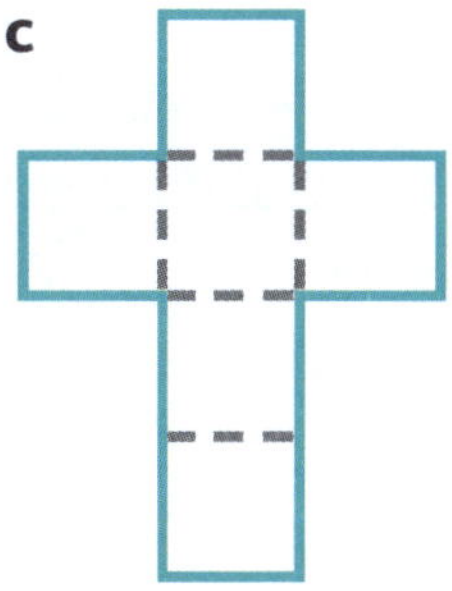

d

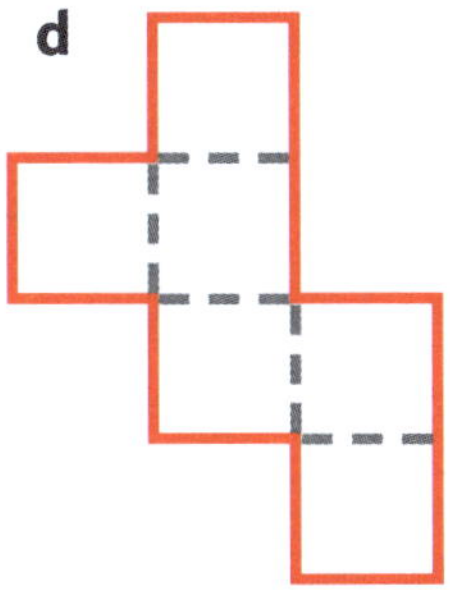

e

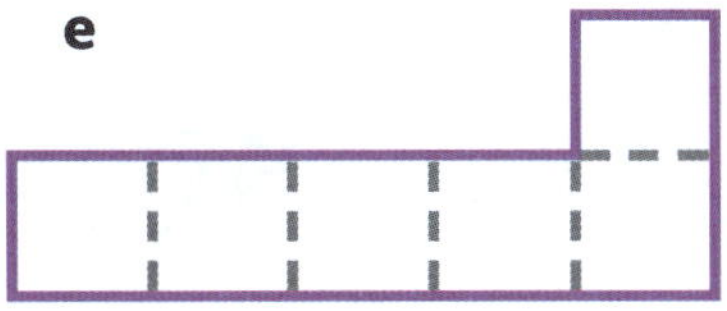

4 Eleven different nets are possible for a cube. How many more can you find, besides those shown on this page? Use this technique: roll a cube on a sheet of paper, tracing its outline each time you roll it over. Number the blocks from 1 to 6 to track your rolls.

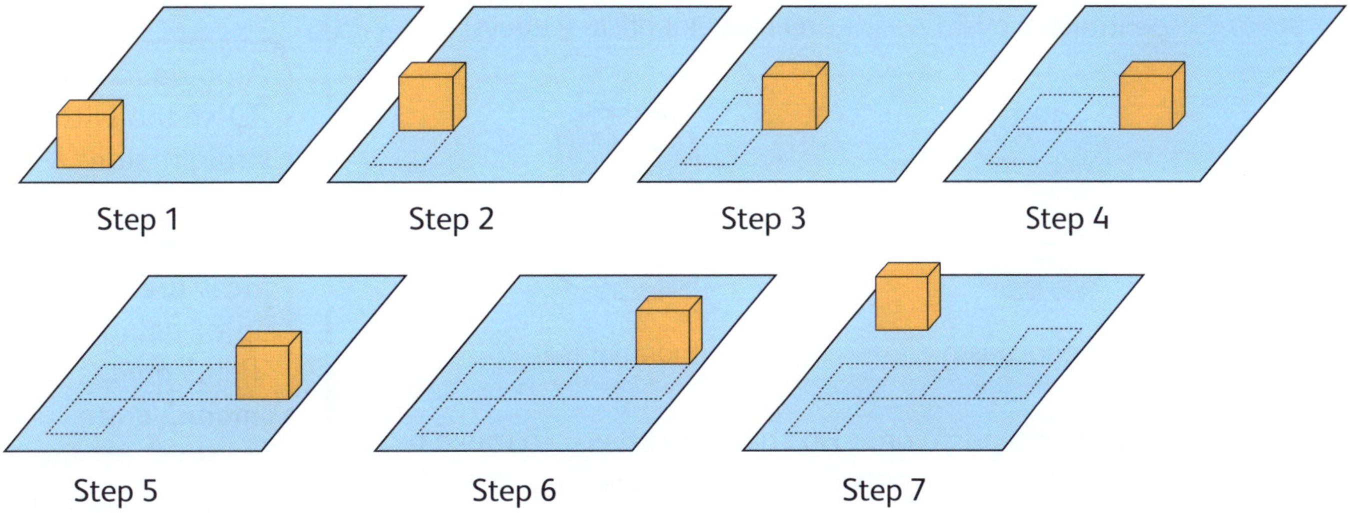

5 What shape does each of these nets make?

a

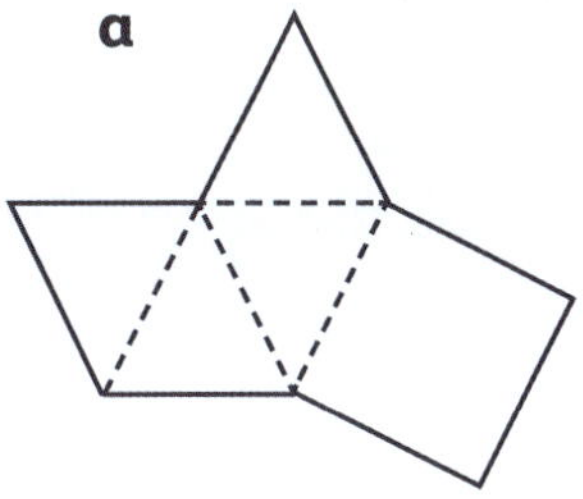

b

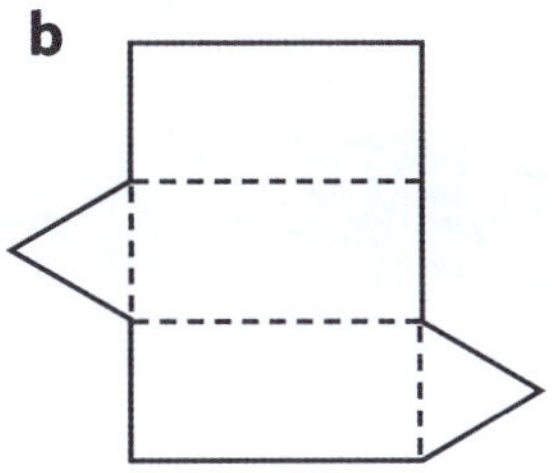

c

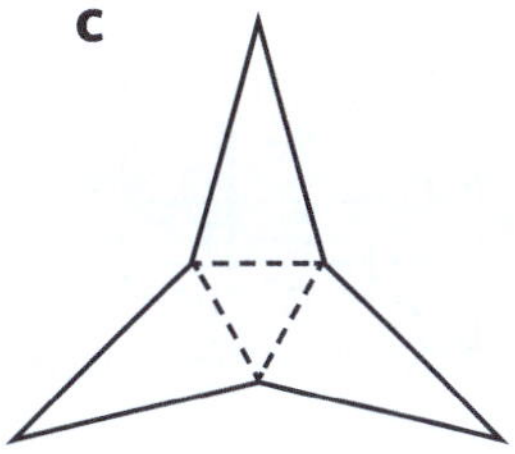

d

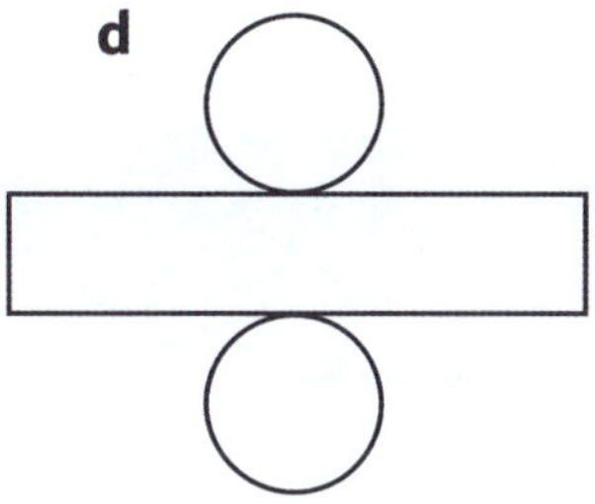

Prisms

Key maths idea

A **prism** is a polyhedron that has identical polygons as both its end faces. These end faces are joined by rectangular faces. Here are some more examples of prisms:

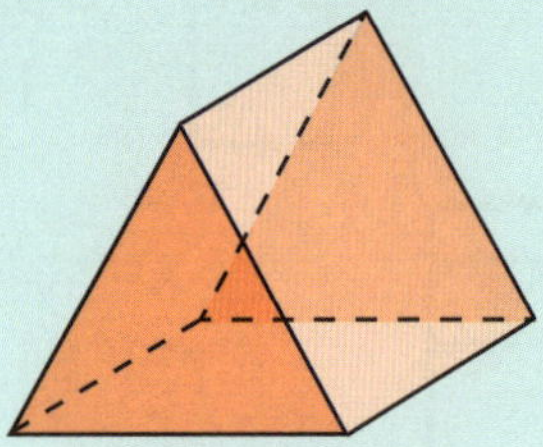
triangular prism

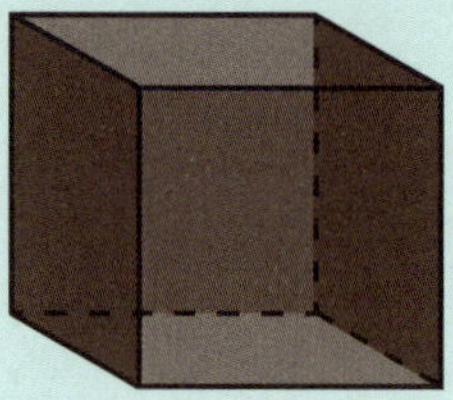
square prism (cube)

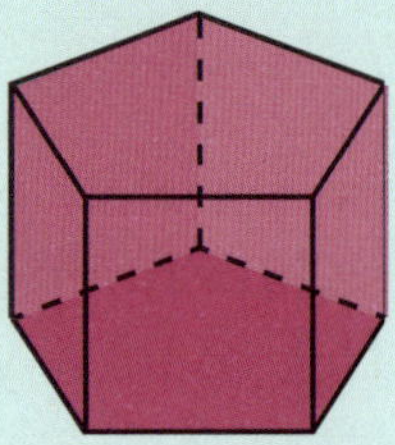
pentagonal prism

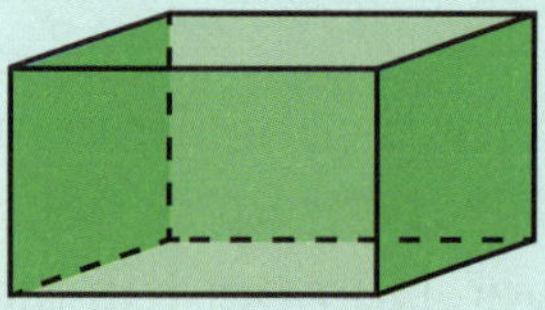
rectangular prism (cuboid)

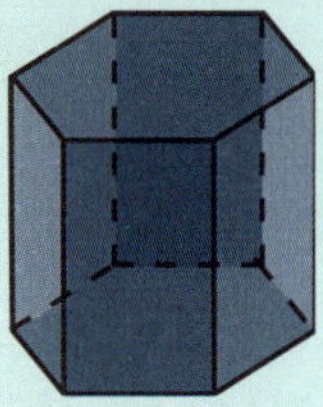
hexagonal prism

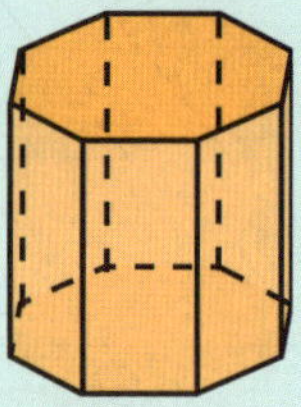
octagonal prism

The end faces of a prism are also called its **bases**. To make a **cross-section** of a prism, you slice across the solid, parallel to the base. This produces an identical shape to the base shape.

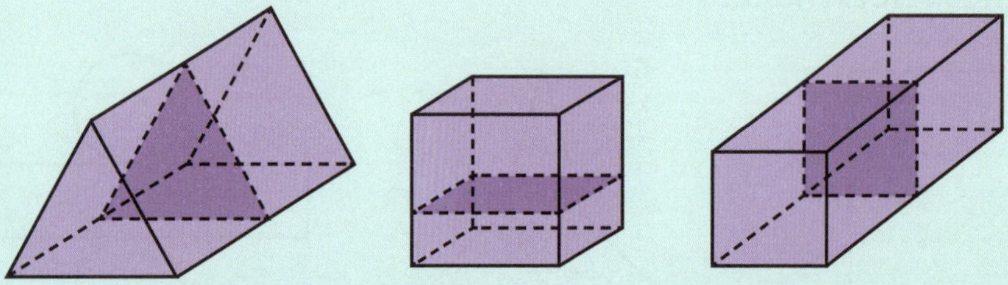
Cross-sections of some prisms

Key words

prism
base
cross-section

A polygon is a 2D shape with straight sides. A polyhedron is a 3D solid whose faces are made up of polygons. 'Poly' means 'many' and 'hedron' means 'surface'. Why is a cylinder not a prism?

You will need:
- some colourful plasticine or clay
- a wooden board or other flat surface to work on
- a ruler or old plastic card
- dental floss or thread (for cutting cross-sections).

1 Use the clay to make an example of each type of prism:

a cuboid **b** cube **c** triangular prism **d** hexagonal prism.

To make each shape, first roll your clay into a ball. Then press it against the board to make the sides flat. You can also use a ruler or plastic card to help flatten the sides and neaten the edges.

2 Use a knife or a piece of thread to slice through each shape. What do you notice about the cross-sections?

Maths detective: Games with cuboids

Have you ever played this game?

It starts with a tower of wooden blocks arranged in a special way. Each player takes a turn to remove a block without toppling over the whole tower.

1 Use the picture to help you to work out the answer to these questions:

a How many cuboids make up each layer?

b If there are 18 layers at the start of the game, how many cuboids are there?

c What do you notice about the relationship between the width and the length of each cuboid?

2 Imagine you wanted to make your own game set. Your aunt is good at woodworking and she asks you to give her the dimensions (length, width and thickness) of the blocks you need. What would you tell her?

3 With each turn, it becomes increasingly difficult to keep the tower standing. What do you think is the highest number of turns the tower could remain standing? Give a reason for your answer.

Why isn't it a cube or cuboid?

Key maths idea

Sometimes a good way to understand a rule is to think about situations that do not satisfy the rule.

None of these solids are cubes or cuboids. Look carefully at each solid. Discuss why it is not a cube or cuboid. Can you find any properties it has in common with cubes or cuboids? Explain why someone might mistakenly categorise it as one of those solids.

a

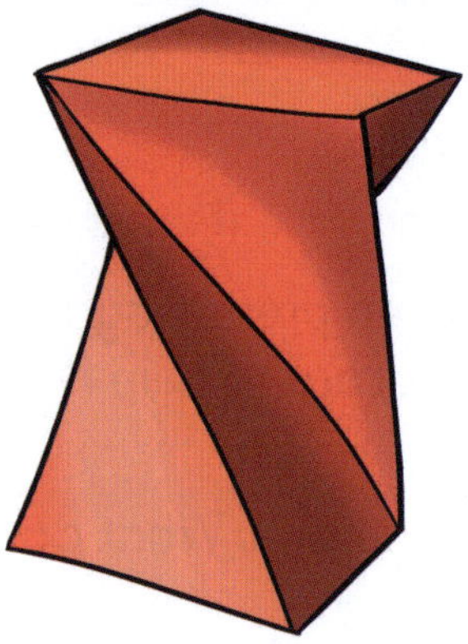

b

c

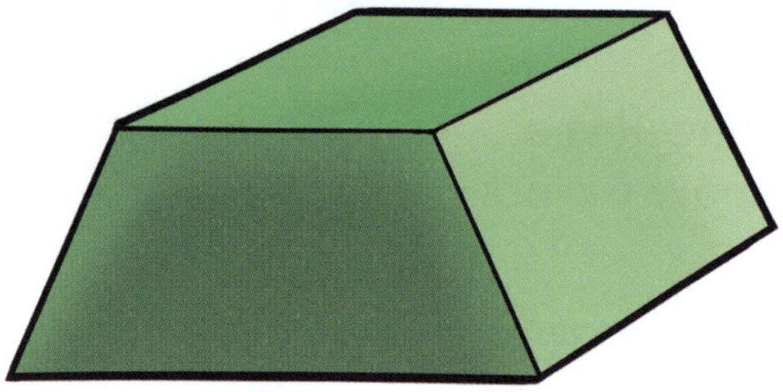

d

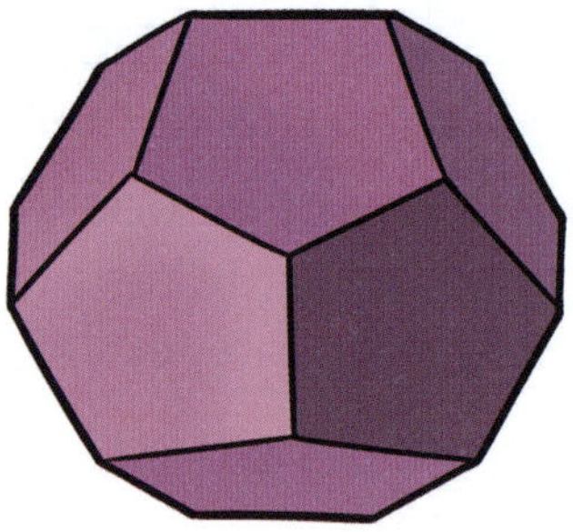

e

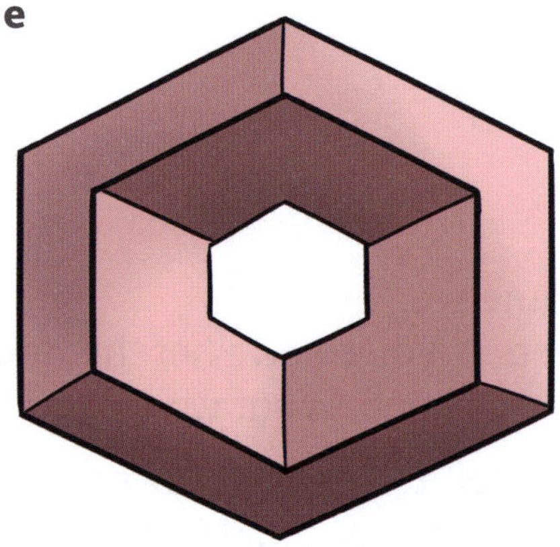

f

Real-world maths: Building construction for a sustainable future

You will need:

- a camera or smartphone, or some photographs or postcards of a city
- paper and card, thin wooden sheeting or similar materials
- pencils
- a ruler
- scissors
- tape
- glue.

1 **a** Identify five 2D shapes and five 3D solids in this picture.

b Which objects are cubes or cuboids?

2 **a** Look around at school and at home and identify structures and objects that match the shapes of cubes and cuboids.

b Use a digital camera or a camera on a phone to take some photographs of buildings in your community. You can also trace photographs from books or postcards.

c Sketch some different buildings created from geometric solids (cubes and cuboids).

3 Work in groups. Research how architects are designing buildings in new ways to use resources sustainably. Find out about any of these features:

a renewable sources of energy (for example, solar power)

b energy-efficient ways of cooling a building in a hot climate

c safety features that can help to protect buildings in the case of tropical cyclones

d features that can assist with recycling.

Draw sketches to show how building designs can include these features.

Geometry

4 Work in groups to create a 3D model of a school for the future. Think about features that you would like to see in schools, for example:
- technology labs
- a school garden
- sports facilities
- energy-efficient features such as those you researched in Exercise 3.

5 Create your model. It should show the geometric solids you have learned about in this unit. You can make your model from materials such as card or paper, or you can present it using computer software (for example, SketchUp or other drawing software). The picture shows some examples of models made from cardboard.

6 Prepare a class presentation. Imagine you are presenting a plan to the Ministry of Education to help design the next generation of Jamaican schools.

7 Write a journal entry based on your experience doing this project.

What I have learned

Create cards with illustrations and simple definitions of the following terms from this chapter.

prism | cube | cuboid | edge | net | cross-section | face | vertex

Practice questions

1 Draw any 3D solid and label a face, an edge and a vertex.

2 Explain which of these is not a prism. Give reasons for your answer.

3 This model was built from wooden cubes. How can you work out whether it is a cube or a cuboid?

4 Draw a solid like this one in your notebook.

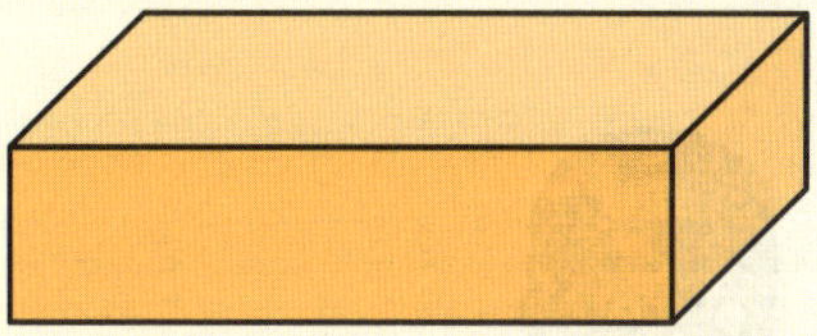

- a Write the name of the solid.
- b Label it to show an edge, a face and a vertex.
- c How many edges does it have in all?
- d What is the sum of the faces and the vertices?
- e Name four everyday objects that have this shape.

5 Draw a net of a cube and a net of a cuboid.

- a Write three things that are the same in a cube and a cuboid.
- b Write two things that are different between a cube and a cuboid.

16 Self check

Check you can do this:	If you have difficulty, try this:
1 Identify cubes, cuboids and other 3D solids correctly.	Revise the 3D solids on pages 242 to 243.
2 Give the number of faces, edges and vertices of a prism.	Revise the work on prisms on pages 246 to 247 and explore models of the prisms you constructed in this chapter.
3 Construct cubes and cuboids from nets and from other materials.	Revise work on nets on pages 244 to 245.
4 Sketch and construct model buildings.	Go over the work you did on pages 249 to 250.

Unit 3: Algebra

Chapter 17 Using variables

In this chapter you will explore this focus question:

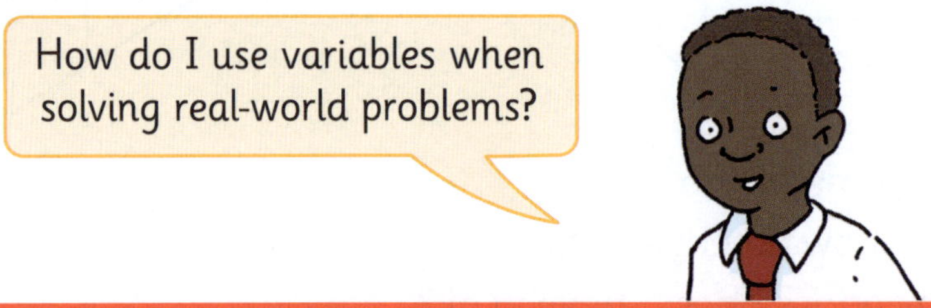

To answer this question, you will:

- identify operations to use to solve a problem
- solve word problems using algebraic expressions
- use substitution in simple formulae.

Starting point

1 Martine has been making mosaic pictures using coloured stones. She uses the same number of stones for each picture.

a If you know how many pictures she will make, how can you work out how many beads she will use? Express this using *p* for the number of pictures and *s* for the number of stones.

b If she uses 8 stones per picture and she makes 5 pictures, how many stones does she need?

c If she uses 10 stones per picture and she makes 7 pictures, how many stones does she need?

Revisit what you know about solving problems

Key maths idea

In maths we use four basic **operations** to solve many different kinds of problems involving numbers.

In Chapter 13, you started working with algebra. **Algebra** is a branch of mathematics where we use **variables**. These are letters that represent unknown quantities or amounts.

add	+
subtract	−
multiply	×
divide	÷

An **algebraic expression** can use a combination of variables and numbers, for example:

'$2 + f$' means '2 more than f'

'$2q$' means '2 times q' or 'double q'

'$2q + f$' means 'multiply q by 2, then add f'.

Key words

operations
algebra
variables
algebraic expression

1 Copy the table. Categorise the mathematical words according to the operation that they relate to.

plus	less than	quotient	more than
minus	division	sum	subtraction
addition	times	decrease by	double
product	increase by	multiplication	share

+	−	×	÷

2 Insert the correct operations sign into each number sentence to make it true.

a 100 ☐ 2 = 50 **b** 5 ☐ 5 = 25 **c** 8 ☐ 2 = 6

d 336 ☐ 3 = 112 **e** 34 ☐ 18 = 52 **f** 3 ☐ 9 = 27

g 123 ☐ 37 = 160 **h** 500 ☐ 275 = 225 **i** 54 ☐ 6 = 9

3 Write these number sentences in words. The first one has been done for you.

a $3f + 23$ Multiply f by three, then add twenty-three.

b $p - 10$ **c** $190 \div 5$ **d** $8 \times w$ **e** $\frac{1}{4}$ of m

Algebraic expressions

Key maths idea

An expression can be a number or a combination of numbers and operations signs. Algebraic expressions use letters as well as numbers and operators. They can contain more than one variable.

Example

$m + n + p$ means 'the sum of m, n and p'.

Algebraic expressions can represent one-step, two-step or multi-step situations.

Example

$5y + 2x + w$ is an algebraic expression that means 'the product of 5 and y, plus the product of 2 and x, plus w'.

Remember, a **variable** is a letter used to replace a number in an algebraic expression or equation. The number that appears in front of a variable, like the 7 in $7y$ is called the **coefficient** or **constant**. It tells us the number multiplied by the variable.

Key words

variable
coefficient
constant

1 Change these word phrases to algebraic expressions:

- **a** the sum of a, b and c
- **b** seventeen less than q
- **c** the product of j and k
- **d** triple x
- **e** the product of 3 and d minus the product of 2 and e
- **f** the product of 2 and j divided by v.

2 Now describe these algebraic expressions in words:

- **a** $z - b$
- **b** $7y + r$
- **c** $5n - 10$
- **d** $x + y$

3 If $k = 10$, work out the value of these expressions:

a $3k$ b $k + 1$ c $k - 1$ d $k \div 5$ e $2k + 5$

4 Write two algebraic expressions of your own. Swap with a partner and challenge each other to write the expressions as word phrases.

Using algebraic expressions to describe patterns

Key maths idea

Pattern blocks come in sets with shapes of specific sizes that you can fit together to make patterns. If you have a set of pattern blocks in your classroom, you can use them to make patterns. If you do not have a set, your parent or teacher may be able to print and cut out the pattern blocks for you to use. You can also find websites online that let you work with pattern blocks on a computer or tablet.

You can use algebraic expressions to help you describe a **pattern**. This can help you develop a **formula** so that you can **compute** the value of **terms** in the pattern.

Key words

pattern
formula
compute
terms

Algebra

1 This pattern is made from triangles. Each side of each triangle is 2 cm long.

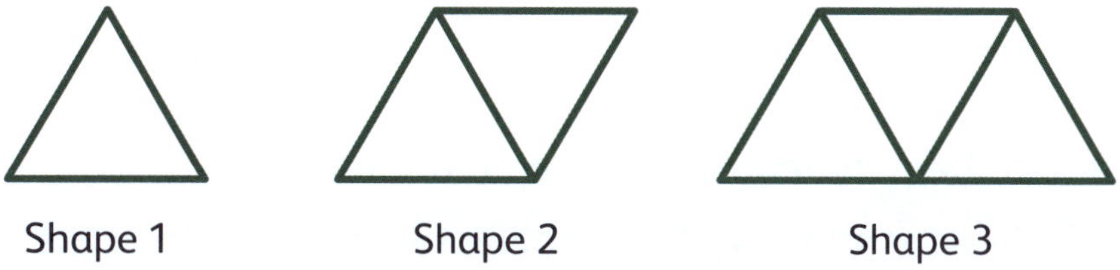

Remember: the perimeter of a shape is the distance around its border.

a Work out the perimeter of each shape in the pattern.

b What number pattern do the perimeters of the terms follow?

c How can you work out the perimeter of the 10th shape in the pattern? Do a sketch to check if your answer was correct.

2 This pattern is made from hexagons.

The perimeter of Term 1 = 6 side lengths.

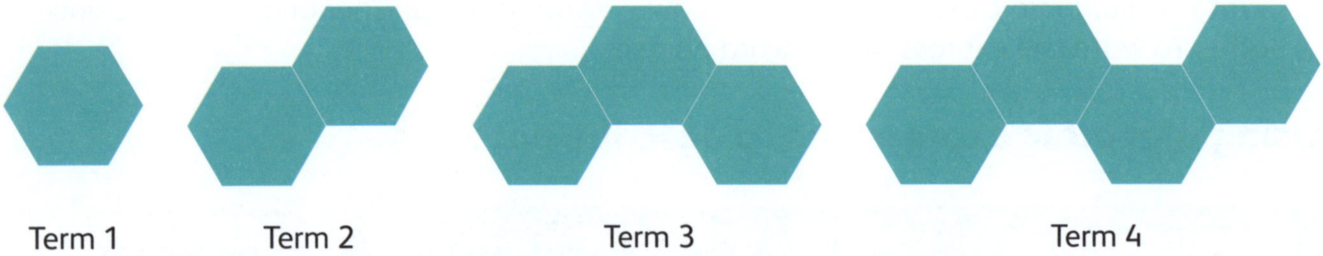

a Helen says: 'To work out the perimeter of each shape, just multiply the number of hexagons by the term number.' Why is Helen's statement incorrect?

b Complete the table to show how the pattern grows.

Term	Number of shapes	Number of side lengths in perimeter
1	1	6
2	2	**A**
3	3	**B**
4	4	**C**
5	5	**D**

c Use the table to help you to work out the perimeter of Term 10.

d Use the table to help you work out the perimeter of Term *n*.

3 This pattern is made from squares and hexagons. The side lengths of the squares are equal to the side lengths of the hexagons.

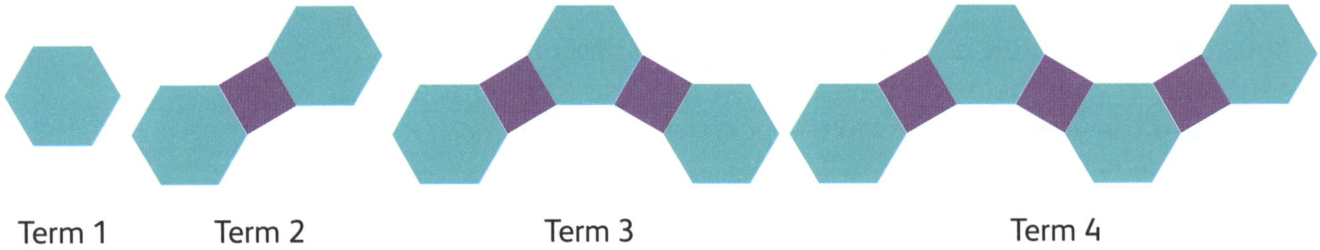

Work with a partner.

a Develop a table like the one in Exercise 2 to show how the perimeter increases with each term.

b Write a rule for expressing the perimeter of Term *n* in the sequence.

4 If you have access to a computer or a tablet, do an online search for 'interactive maths pattern blocks'. Use one of the online tools you find to make your own pattern with two or more shapes. Use your pattern to create a number rule that describes the perimeter of each term in the pattern.

5 A carpenter uses cubes and cuboids to design these chairs and tables for a primary school. There are four cube-shaped seats around each table.

a Complete the table to explore how many students can be seated for each number of tables.

Number of tables	1	2	3	4	5	*n*
How many students can be seated	4	**A**	**B**	**C**	**D**	**E**

b Write a sentence to describe the relationship between the number of seats and the number of tables.

c This sketch shows how many seats fit if two tables are pushed together:

Complete the table to show how many seats fit around the tables if the teacher pushes them together in a longer row.

Number of tables pushed together in a row	1	2	3	4	5	*n*
How many students can be seated	4	6	8	**A**	**B**	**C**

d Write another sentence to describe the relationship between the number of seats and the number of tables (pushed together in a row).

6 Work with a partner. Explore different shapes that you could make with the tables, and how the different shapes affect how many students could sit in each arrangement.

Algebra

Patterns and rules

Key maths idea

Think about a balance scale. When the pans are at the same level on both sides, this tells us that both sides have the same mass. They are **equal**. You can think about an **equation** like a balanced scale. The expressions on each side of the equal sign have the same value.

Example 1

We usually say 'like chalk and cheese' to describe things that are very different. But in the picture, 6 sticks of chalk weigh the same as the wedge of cheese.

Imagine the mass of one stick of chalk = x and the mass of one wedge of cheese = y. We can write this equation:

$6x = y$.

We use algebra to represent stories that we read in word problems. We can then use algebra to solve the problem. The final **solution** to an algebra problem is when we state the value of the variable.

To solve the equation $6x = y$, we would need information about x or about y. For example, if we know that each stick of chalk weighs 30 g, we can work out the mass of the cheese:

$6 \times 30 = 180$

Whatever the mass of the stick of chalk, the cheese is six times that.

When an equation only has one variable, you can use the equation to help you work out the value of the variable.

Example 2

Anna blows up k number of balloons. Then she blows up 28 more. The total is 50. What is the value of k?

$$k + 28 = 50$$
$$k + 28 - 28 = 50 - 28$$
$$k = 22$$

Key words

equal
equation
solution

Example 3

Three sisters each write n number of party invitations. They give out all eighteen invitations at school. How many invitations did each sister have?

The problem tells us that there are three lots of n number of invitations, so we write:

$3n = 18$

Then we solve the equation:

$n = 18 \div 3$

$n = 6$

Solve each problem using algebra. Don't forget to clearly show each step of your working out.

1 A florist has four buckets, each containing m number of flowers. Altogether there are 112 flowers. How many flowers are in each bucket?

2 Kiesha and her brother each collected x cans of litter at a beach clean. Beth collected 30 cans. Altogether, they collected 88 cans. How many cans of litter did Kiesha's brother collect?

3 A beach vendor sells b number of bottled drinks in a day. This is made up of 25 bottles of water, 18 bottles of soda and 27 bottles of juice. What is the value of b?

4 A chef makes y samples of a new dish. She gives out 9 samples in the morning, 18 in the afternoon and 25 in the evening. How many samples did she make?

Algebraic equations with multiple steps

Key maths idea

Sometimes equations contain multiple steps.

$4y + 6 = 22$

First take away six from both sides:

$4y + 6 - 6 = 22 - 6$

$4y = 16$

But this is not the solution. To solve the equation, we need to find the value of y. We need to take another step. We know what $4y$ is worth, so we divide to find out the value of y.

$y = 16 \div 4$

$y = 4$

Solve these equations.

1 $2x + 3 = 5$

2 $7 + y = 15$

3 $6x \times 3 = 72$

4 $4y \div 2 = 20$

5 $5x - 6 = 9$

6 $7y + 13 = 90$

Substitution

Key maths idea

Look at these number sentences:

7 +  = 20

100 −  = 40

 × 3 = 27

24 ÷  = 8

What is the value of each shape?

How can you work it out?

How can you check whether your answer is correct?

A **substitute** is a replacement. You may have had a substitute teacher before or seen a substitute player in a sports match. In algebra, if you know the value of a variable, you can use **substitution** – you replace the letter with the known number or value. This allows you to work out the value of the expression or to check if an equation works.

Example 1

Take the expression $a + b + c$

If we know that $a = 5$, $b = 12$ and $c = 30$, then we can work out the value of the expression.

$a + b + c$

$= 5 + 12 + 30$

$= 47$

So the value of $a + b + c$ is 47.

What if the expression was $2a + b + c$? What do we need to do differently?

There are now two lots of a, so we add 2 × 5 instead of 5.

$2a + b + c$

$= 10 + 12 + 30$

$= 52$

The value of $2a + b + c$ is 52.

Key words

substitute

substitution

Example 2

Sometimes, the variable can change.

For example, Deon is 5 years older than Nico. Let Nico's age = n and Deon's age = d.

$d = n + 5$

When Nico is 10 years old, $d = 10 + 5$, so Deon's age is 15.

When Nico is 25 years old, $d = 25 + 5$, so Deon's age is 30.

1 Work out the value of each shape. Use substitution to check your answers.

a ✱ × 10 = 90 **b** $\frac{1}{2}$ of ♥ = 100 **c** ■ − 22 = 45

d 15 ÷ ▼ = 3 **e** ● + 77 = 100 **f** ▲ + 40 = 140

2 Make your own number sentences using shapes for a partner to solve.

3 Calculate the value of each expression using the given values.

a $2x + 5$, when $x = 15$

b $3p - 1$, when $p = 20$

c $150 - 2h$, when $h = 17$

d $3w + 2n$, when $w = 8$ and $n = 9$

e $f + g + 3i$, when $f = 1\,000$, $g = 100\,000$ and $i = 200$

f $25m \div 2$, when $m = 10$

4 Work out the value of each expression if $a = 8$, $b = 20$, $c = 3$, $x = 11$, $y = 15$ and $n = 4$.

a $x + y$ **b** $c \times a$ **c** $b \div n$

d $2x - b$ **e** $6b \div c$ **f** $5a \div b$

5 Use these values to work out whether each equation below is true or false.

$r = 18$, $p = 24$, $t = 30$, $m = 50$, $w = 6$

a $r = 3w$ **b** $p = m - t$ **c** $p = r + w$

d $m = 2t$ **e** $t - w = p$ **f** $p - r = w$

6 The values of some variables are given below. Use these to write two statements of your own that are true and two that are false. Then swap with a partner and challenge your partner to decide which are true and which are false.

$a = 4$, $b = 20$, $c = 5$, $d = 40$, $e = 16$, $f = 8$

Maths detective: A substitution challenge

In this substitution challenge, each letter of the alphabet has a value as shown in the table below. You will use the table to create algebraic expressions which will lead you to find letters that spell words.

Example

Write an algebraic expression for the word 'maths'.

Step 1: Write down the values of each of the letters in the word 'maths'.

Step 2: For each letter, work out an algebraic expression using the values of other letters.

Let's start with *m*, which has a value of 30. Look for other numbers in that table that you can use to reach 30. You can see that $b = 15$ and $z = 2$, so an algebraic expression for the letter *m* could be $b \times z$.

If you were solving the puzzle, you would work out the value of the expression is 30 and see that 30 gives you the letter *m*.

Do the same thing for every letter until you have a set of expressions to spell out the word. In this case, maths: $b \times z$, $h - q$, $j \times v$, $q + a$, $v \times f$.

a	*b*	*c*	*d*	*e*	*f*	*g*	*h*	*i*	*j*	*k*	*l*	*m*
4	15	48	33	5	20	8	16	9	7	42	10	30
n	*o*	*p*	*q*	*r*	*s*	*t*	*u*	*v*	*w*	*x*	*y*	*z*
6	100	63	12	40	60	21	24	3	1	27	11	2

Case 1

What words do the following algebraic expressions spell?

1 $t - n$, $b - y$, $b + 2v$

2 $a \times b$, $s \div z$, $q - v$, $s \div n$, $l \div z$

3 $7n$, $b - n$, $s \div l$, $a \times z$

4 $4b$, $10w$, $5f$, $y - l$

Case 2

Write an algebraic expression for each of these words.

1 wink 2 drop 3 fly 4 cub

Case 3

Use the values in the table to write expressions for two words of your own. Then challenge a partner to work out what they are. Your words should contain only 3 or 4 letters.

Look for relationships between the values to help you. Think about times tables!

Case 4

Write algebraic expressions to spell out your name. Give it a try!

Constructing one-step and two-step problems

Key maths idea

You are going to write problems of your own for a classmate to solve. You need to think carefully about the situations you will describe and the numbers you will use. You may use any letter of the alphabet as your variable. Before you share your problems, test the problem and make sure you can solve it for yourself.

I won't use the letter o because it can be easily confused with zero!

Example 1

This is a one-step problem:
Michelle has x number of counters. She gave 24 to Mia and now has 30 left. How many counters did Michelle have to begin with?

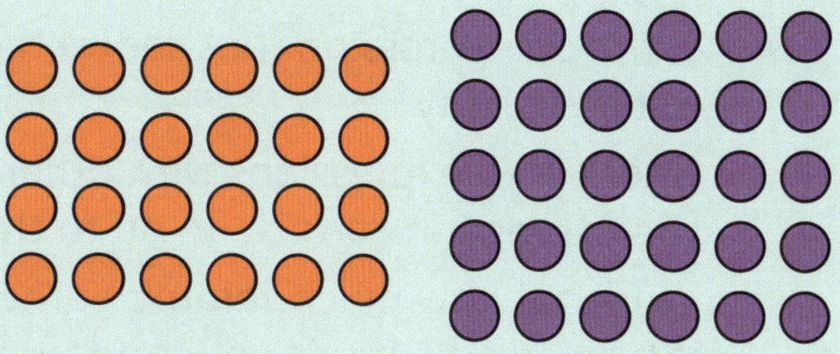

- The situation is giving out counters.
- The numbers are 24 and 30.
- The variable is x.

Solve the equation to check:

$x = 24 + 30$

$x = 54$

Example 2

Here is a two-step problem:

Michelle has x number of counters. She gave 24 to Mia and then she gave 19 to Ruby. Michelle now has 27 counters left. How many counters did Michelle have to begin with?

- The situation is still giving out counters.
- The numbers are 24, 19 and 27.
- The variable is x.

Solve the equation to check:

$x = 24 + 19 + 27$

$x = 70$

Here are some possible situations. Choose two of the situations:

- giving out pencils
- sharing sweets
- collecting litter at a forest clean-up
- selling raffle tickets.

1 Write a one-step problem for each situation you have chosen.

2 Write a two-step problem for each situation you have chosen.

3 **a** Think of a situation of your own. Write a one-step problem and a two-step problem for that situation.

b Work with a partner. Swap and solve each other's problems.

Real-world maths: In a restaurant

1 At a restaurant, the waiting staff get paid according to the following rule:

For each shift, there is a flat fee of \$1 000, plus an hourly rate of \$600 per hour. For Sundays or public holidays, the hourly rate is double the normal rate.

a Let the numbers of hours worked = h. Write an expression that shows the total pay for a shift on a normal workday.

b Write an expression that shows the total pay for a shift on a Sunday or public holiday.

c Check your expressions with others in your class. Once you are sure you have the expression correct, go on to the next question.

d Work out how much Jerome earned if he worked 4 hours on Monday, 6 hours on Tuesday and 8 hours on Wednesday.

e Work out how much Niki earned if she worked 3 hours each day from Monday to Friday, but one of the days was Independence Day.

2 The same restaurant has the following offers:
- four burgers for the price of three
- seven burgers for the price of five.

The cost of two burgers is \$950.

a Work out the normal cost of one burger.

b Work out the price of each burger if you order four.

c Work out the price of each burger if you order seven.

d Which is the better deal?

Maths detective: Algebra games

Look at the table. Then solve the number value cases.

Case 1

1 Assign a number value to each letter of the alphabet, for example $a = 1$, $b = 2$, $c = 3$ and so on. You can copy and complete this table with your values.

a	*b*	*c*	*d*	*e*	*f*	*g*	*h*	*i*	*j*	*k*	*l*	*m*
n	*o*	*p*	*q*	*r*	*s*	*t*	*u*	*v*	*w*	*x*	*y*	*z*

2 Find the sum of p, q and z.

3 Find the product of n and s.

4 Find the difference between y and t.

Case 2

Write your own equations to show sums of letters that are equivalent to other letters. For example, if $a = 1$, $b = 2$ and $c = 3$, $a + b = c$.

1 How many different sums can you come up with that have a total of y?

2 What expressions can you make that are equal to $2p + 5$?

3 Find values that can help you complete these statements:
- Half of _____ is equal to 2 _____ + 3.
- _____ squared is equal to the difference between _____ and _____ .
- The product of _____ and _____ is equal to the sum of _____ and _____ .

4 Try to use all four operations in an equation using at least five different variables.

Algebra

Case 3

A card game has three characters. Each character has a different formula for calculating their strength, speed, intelligence and stamina.

1 Work out the strength, speed, intelligence and stamina levels for these values:
- $a = 5, b = 8, c = 7$
- $a = 20, b = 50, c = 15$

2 Write some sentences about each character's strongest and weakest points.

3 Look at the character cards. Work in groups. Create your own characters with expressions for strength, speed, intelligence and stamina. Use substitution to work out which character has the greatest strength, which character has the highest stamina and which has the best overall score.

CHARACTER A

Strength:	$2c - a$
Speed:	$5a - 2b + 20$
Intelligence:	$50 + a$
Stamina:	$4c + a$

CHARACTER B

Strength:	$6a - b$
Speed:	$8a - 3b$
Intelligence:	$7c + 10$
Stamina:	$2a + 4b$

CHARACTER C

Strength:	$a + 10$
Speed:	$100 - 3a$
Intelligence:	$10c - b$
Stamina:	$65 - 3c$

What I have learned

Draw your own mind map to illustrate what you learned in this unit. Your mind map should define these terms and give examples:

- variable
- four operations
- algebraic expression
- algebraic equation
- substitution
- two-step problem.

Practice questions

1 Which is an algebraic expression? Give a reason for your answer.

a $x + 7 = 20$ b $6y - 5 = 100$ c $4x + 6 - 2$

2 Which is an equation? Give a reason for your answer.

a $4x + 2y + 7$ b $8y \times 5 = 400$ c $x + y + z$

3 Insert the correct sign into each number sentence to make it true.

a 15 △ 3 = 5 b 12 □ 6 = 72 c 57 ○ 26 = 83

d 2 000 ◇ 1 250 = 750 e 500 ☆ 7 = 3 500 f 4 000 ▽ 250 = 16

4 Solve these equations.

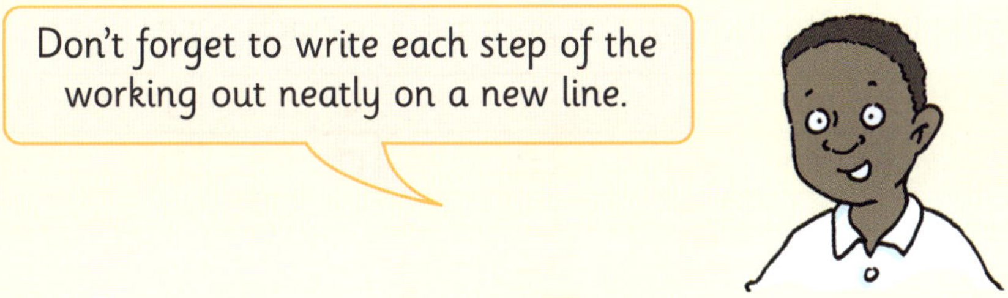

a $9y + 1 = 100$ b $8x = 160$ c $25y - 10 = 40$

d $4x \times 5 = 100$ e $7y \times 6 = 42$ f $3x - 9 = 30$

5 Solve these problems using algebra.

a Five children are each carrying *n* number of books from one classroom to another. If 55 books are moved, how many did each child carry?

b A postal worker has *n* number of parcels to deliver. She delivers 3 parcels to a school, 14 parcels to an apartment block, 7 to a hotel, and she has 18 left in the van. How many parcels did she have to deliver?

c A choir is given *n* pieces of fruit after a concert. If there are 22 people in the choir and each person ate three pieces of fruit, how many pieces of fruit did the choir receive?

6 The values of some variables are given below. Use them to answer the questions.

$a = 10, b = 16, c = 7, x = 9, y = 30, n = 1$

Work out the value of each expression.

a $44n + b$ b $3y \div x$ c $y - 4c$

d $8a - y$ e $2x \times 5n$

Algebra

7 Use the same values as in Exercise 6. Choose the equation that is true.

a $10a = 1\,000$ b $10n = 10$ c $10x = 100$

8 Use the same values again. Choose the equation that is false.

a $a = 10n$ b $y = 3a$ c $b = 2c$

9 a Copy these three trains and then draw the fourth train of the pattern.

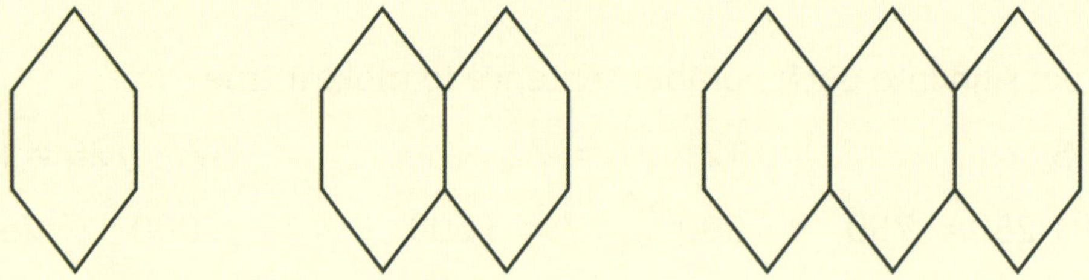

Train 1 Train 2 Train 3

b Complete the table by writing down the perimeter in each case.

Number of hexagons	1	2	3	4	5	10
Perimeter	6	**A**	**B**	**C**	**D**	**E**

17 Self check

Check you can do this:	**If you have difficulty, try this:**
1 Understand algebraic sentences and match algebraic sentences to word sentences.	Revise the explanation of algebraic expressions on page 253 and practice making number sentences using letters to represent unknown amounts.
2 Decide which mathematical operation to use in an algebraic situation.	Revise the activities you did on pages 254 to 255.
3 Solve word problems using algebraic expressions.	Revise the work on pages 258 to 259.
4 Use information to write one-step and two-step problems.	Go back to the work on pages 263 to 264.
5 Solve problems by substituting values for variables.	Revise the work from pages 260 to 261.

Unit 4: Statistics and probability

Chapter 18 Probability

In this chapter you will explore this focus question:

To answer this question, you will:

- identify likely and unlikely outcomes
- make predictions about the outcomes of experiments
- record results of actual experiments.

Starting point

Talia and Dante are planting flowers. They have fifteen flowering plants that will produce beautiful red, orange or pink flowers. They don't know which plants will produce which colour flowers yet, but they think that half of the flowers should be red, at least a quarter of the flowers should be pink, and some should be orange.

In a group discuss these questions. For each question, first ask yourself: Is it possible? Is it probable? Then share your ideas with the class.

- Which colour do you think most flowers will be?
- Do you think there will be more orange flowers or pink flowers?
- Could there be no orange flowers?
- Could all of the flowers be the same colour?
- Could there be no flowers?

Real-world maths: Raffles and lotteries

Work with a partner. Talk about competitions like raffles that you know about. Look at the pictures to help you. Where have you seen raffles, draws or games of chance taking place? Have you ever taken part in a raffle or other game like this? What was it like? What advantages and disadvantages are there of taking part in raffles and games like this? Share your ideas with the class or a small group.

Bingo balls

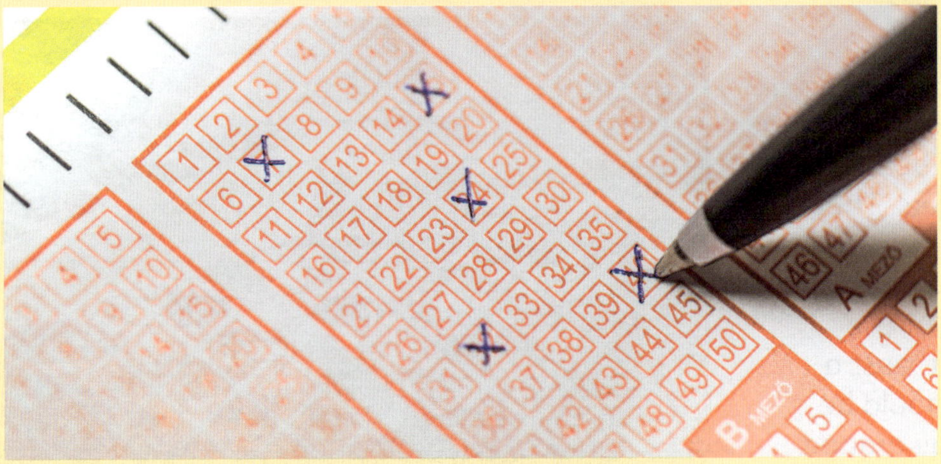

Lottery ticket

Lucky draw

Probability

Key maths idea

Probability is about how likely it is that an **event** will have a specific **outcome**. This is also known as **chance**. For example, when you roll dice, there are six different possible outcomes for each die.

Some events are **certain** to happen. For example, when you roll a pair of dice, it is certain you will get a total between 2 and 12. It is **impossible** to get a total of 20.

Most outcomes are neither certain nor impossible. It is **unlikely** that you will roll double sixes three times in a row, although it is **possible**.

Key words

probability
event
outcome
chance
certain
impossible
unlikely
possible

1 Unscramble the letters to find the words related to probability. For each word, make a sentence that shows what the word means.

- **a** ratnice
- **b** mouteco
- **c** tenve
- **d** misslobpie
- **e** yelilk

2 Name two events that are certain to happen.

3 Name two events that are impossible.

4 There are six oranges and two apples in a bag. Without looking in the bag, you choose a piece of fruit. Describe the probability of picking an apple.

5 100 raffle tickets are sold to win a TV. You bought two tickets.

- **a** Describe in words how likely you are to win the prize.
- **b** Why do you think people choose to buy raffle tickets for draws like this?

The probability scale

Key maths idea

The **probability scale** allows us to express how likely it is that an event will happen by using numbers.

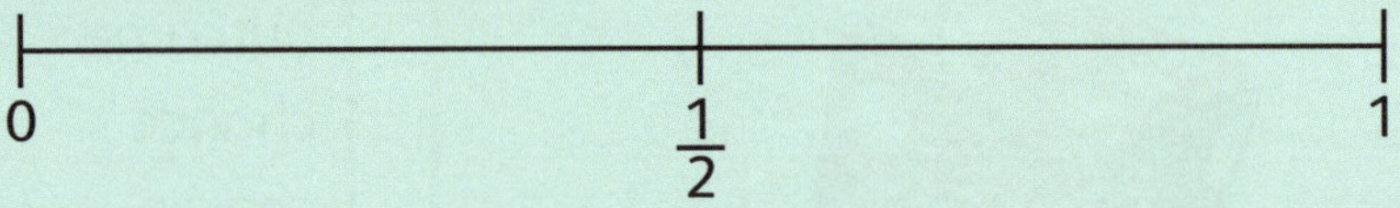

The probability scale goes from 0 to 1. Events that are impossible have a probability of 0 (zero). Events that are certain to happen have a probability of 1. Other events have probabilities between 0 and 1. The closer to 1 a probability is, the more likely it is to happen.

Think about tossing a coin.
There are two possible outcomes: heads or tails. One is as likely as the other. The outcomes are equally likely.

The probability of the coin landing on heads is 0.5 or $\frac{1}{2}$ or 50%.

The probability of the coin landing on tails is also 0.5 or $\frac{1}{2}$ or 50%.

We could also say the outcomes are 'even' or 'fifty-fifty'.

We can use decimals, fractions or percentages to describe probability numerically.

Key word
probability scale

Note that $0.5 + 0.5 = 1$. If you add together the probabilities of all of the outcomes of an event, the total is always 1.

1 Copy the probability scale. Write the words below in the correct place on the scale.

- Certain
- Impossible
- Likely
- Very likely
- Very unlikely
- Unlikely
- Even

2 Copy the probability scale again. Write the letters for each event below in the correct place on the scale. Ask yourself: Is each event possible? Then ask yourself how likely it is.

- **a** Throwing a die and scoring 2
- **b** Staying silent for the rest of the day
- **c** Aliens landing on the school roof
- **d** Sleeping tonight
- **e** Eating hot food today
- **f** Feeding an animal today.

3 Copy the probability scale again. Think of five events of your own. Write them in a numbered list and put the numbers in the correct places on your probability scale.

Sample space

Key maths idea

The **sample space** describes the total number of possible outcomes for any event.

So, for tossing a coin, the sample space is two because there are two possible outcomes, and for throwing a die, the sample space is six because there are six possible outcomes.

We show the sample space by writing all the **possible outcomes** inside a pair of brackets and separating each outcome with a comma.

Sample space for tossing a coin = {heads, tails}

Sample space for throwing a die {1, 2, 3, 4, 5, 6}

Key words

sample space

possible outcomes

1 Look at the following three spinners.

a

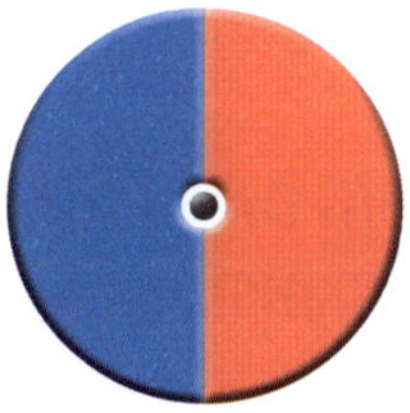

b

c

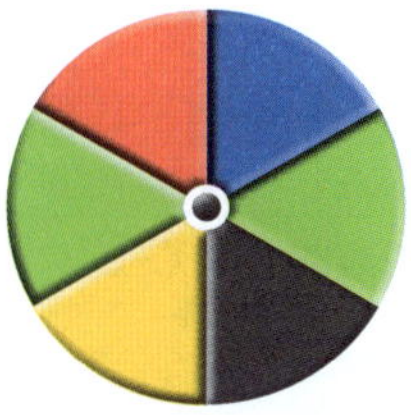

Write the sample space for spinning each spinner.

2 Look at this restaurant menu.
Write the sample space for:

- **a** picking a starter
- **b** picking a main course
- **c** picking a dessert.

Menu

Starters

Soup of the day
Fresh salad

Main courses

Rice and peas
Jerk chicken
Curry goat
Vegetarian pasta
Four cheese pizza

Desserts

Ice-cream sundae
Fruit platter
Key lime pie
Sponge cake

Probability experiments

Key maths idea

You will now carry out a probability **experiment**. First you will make **predictions** about the probable outcomes. Then you will test your predictions against the **actual outcomes**. You can use a tally chart to record each outcome.

Key words
- experiment
- prediction
- actual outcome

Work on your own or with a partner.

You will need:
- some coins
- a die
- a pack of playing cards.

Hint: you can use a random number generator to represent the scores on a die.

Experiment 1: Tossing a coin

You are going to toss a coin twenty times.

1 Make a prediction about how many times the coin will land on heads and how many times it will land on tails.

2 Draw up a tally chart to record your results.

Heads	Tails

3 Carry out the experiment. Record the outcomes in your tally chart.

4 Construct a bar graph to show the results of the experiment.

5 Look back at your predictions. Were you correct? How close or far from the actual outcomes were your predictions? Write some conclusions about the experiment.

Experiment 2: Throwing a die

You are going to throw a die twenty times.

1 Make a prediction about how many times each number from 1 to 6 will be scored.

2 Draw a tally chart to record your results.

1	2	3	4	5	6

3 Carry out the experiment. Record the outcomes in your tally chart.

4 Construct a bar graph to show the results of the experiment.

5 Look back at your predictions. Were you correct? How close or far from the actual outcomes were your predictions? Write some conclusions about the experiment.

6 When you roll a die, you have a 1 in 6 chance of landing on each number. In theory, this means that a pie chart of the outcomes would look like this:

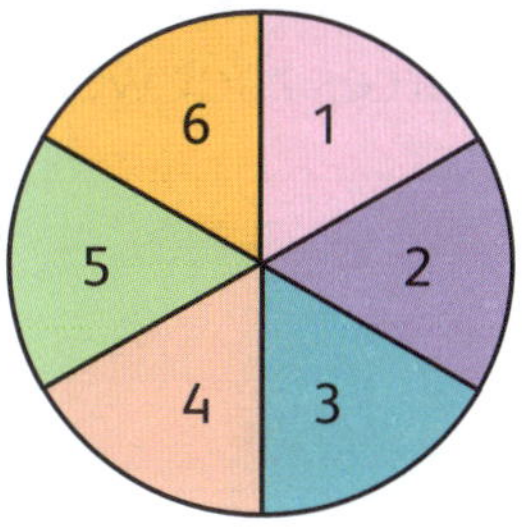

However, when some Grade 4 students rolled a die six times, their outcomes turned out like this:

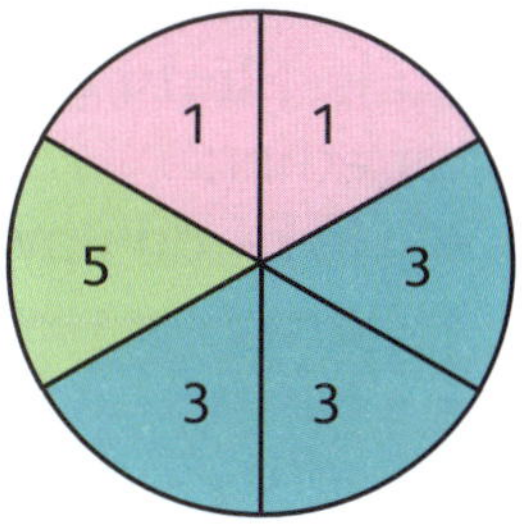

Roll a die 6 times and draw a pie diagram to show your outcomes.

Experiment 3: Choosing a playing card

A standard pack of playing cards has 52 cards (if we exclude the jokers). The cards are split into four groups called suits:

Spades Diamonds Hearts Clubs

You are going to choose a card at random from the pack 32 times.

1 Make a prediction about how many times you think each suit will be picked.

2 Draw a tally chart to record your results.

Hearts	Spades	Diamonds	Clubs

3 Carry out the experiment. Record the outcomes in your tally chart.

4 Construct a bar graph to show the results of the experiment.

5 Look back at your predictions. Were you correct? How close or far from the actual outcomes were your predictions? Write some conclusions about the experiment.

Statistics and probability

Real-world maths: Spinning science!

Work in groups. First read the extract and then do the activity.

Have you ever wondered how it is possible for aeroplanes to fly? In science, you have learned about **forces** – **pushes** and **pulls**. **Flight** is made possible because of four forces that work together:

- **lift** – an upward force
- **drag** (or **resistance**) – a backward force
- **thrust** – a forward force
- **mass** – a downward force.

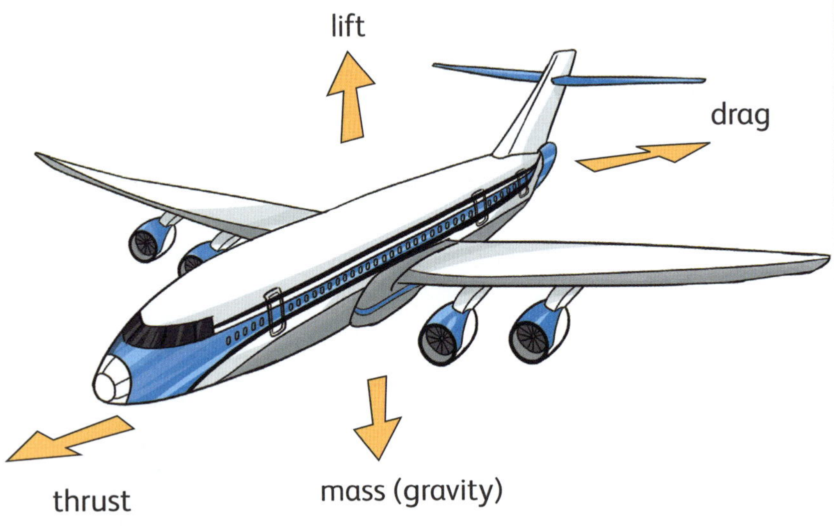

You are going to carry out a probability experiment about the forces of flight using a handmade spinner.

You will need:

- some card
- a ruler
- scissors
- a pencil
- colouring pencils or crayons
- a paper fastener
- a paper clip.

If you don't have a paper fastener and paper clip, you can make an alternative spinner with a pencil in the centre.

Key words

force
pushes
pulls
flight
lift
drag
resistance
thrust
mass

First, make your spinner by following these steps:

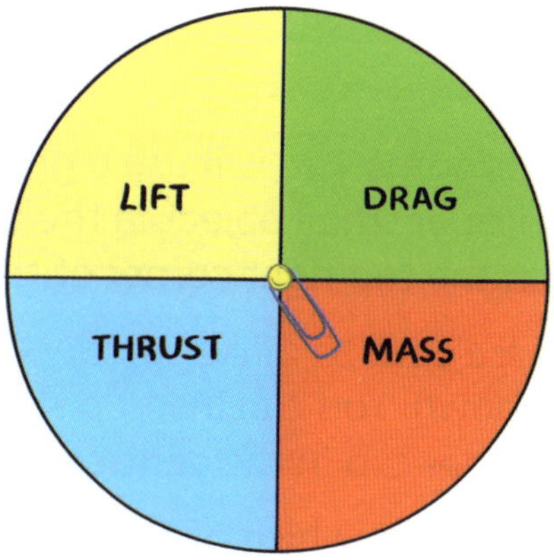

- **Step 1:** Draw a circle on your card. Your circle will be more accurate if you use a pair of compasses to draw it or draw around a circular object such as a mug or cup.
- **Step 2:** Use your ruler to split the circle into quarters.
- **Step 3:** Label each quarter with one of the four forces of flight: lift, drag, thrust and mass.
- **Step 4:** Colour or decorate your spinner as you wish and then cut out the circle.
- **Step 5:** Carefully make a hole in the exact centre of the circle. (You could use the point of a pencil to do this.)
- **Step 6:** Insert the paper fastener into the hole you have made, place the paper clip over the fastener and then fold over the two clips at the back to secure everything in place.

Hint: you could also insert a sharp pencil point through the centre of the spinner until the spinner is approximately a quarter of the way along the length of the pencil. Your spinner is made!

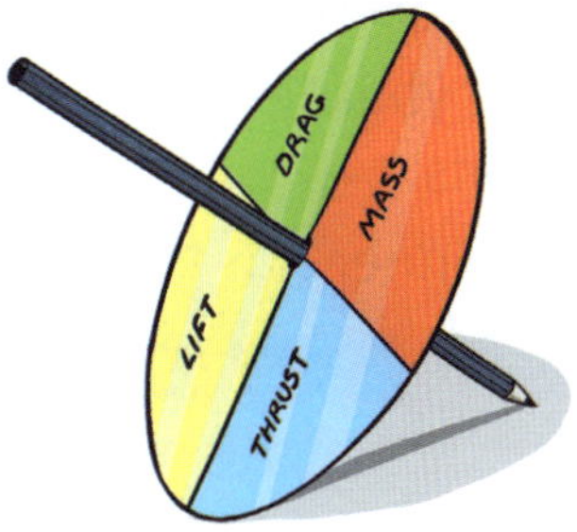

You are now going to spin the spinner 40 times.

1 Make a prediction about how many times the spinner will land on each force.

2 Draw a tally chart to record your results.

Lift	Drag	Thrust	Mass

3 Carry out the experiment. Record the outcomes in your tally chart.

4 Construct a bar graph to show the results of the experiment.

5 Look back at your predictions. Were you correct? How close or far from the actual outcomes were your predictions? What do you think this tells you about the four forces of flight?

6 Write a paragraph to describe what you can conclude from the experiment.

Maths detective: Risky business

Case 1

In this game you will use a grid of squared paper. To begin, both players start with a counter or small object in the bottom left corner. The aim is to be the first player to reach the top right corner of the grid. Work in pairs.

1. Use card to make spinners like the ones shown. You can insert a short pencil or paper clip in the centre as the spinner.
2. Examine which of the spinners is going to provide the best advantage to the player. Discuss your idea.
3. Each player picks a spinner to use. Take turns. Each time you spin, move three blocks in the direction indicated on the spinner.
   - Who finishes first?
   - Compare your results with other pairs.
   - Which spinner has the best probability of helping you win? Why?

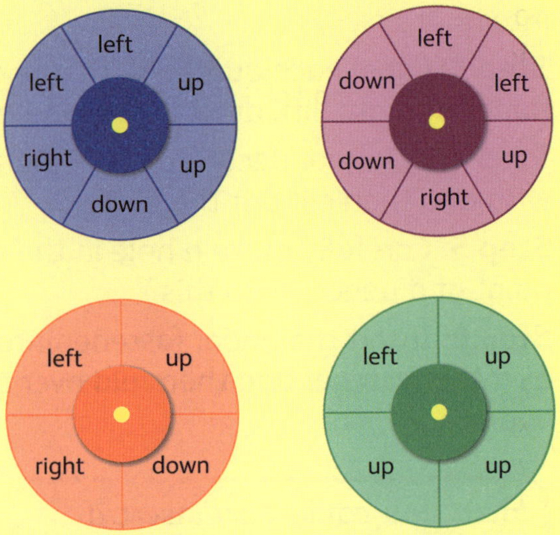

Case 2

What do you understand by the term 'risk'? What risks are there in things people do every day? Which risks are sensible? Which are not? As a class, have a discussion about the advantages and disadvantages of gambling and taking risks. Make a class poster to show your ideas.

What I have learned

In pairs, plan and design a concept map about probability. You can use this concept map to help you or design your own.

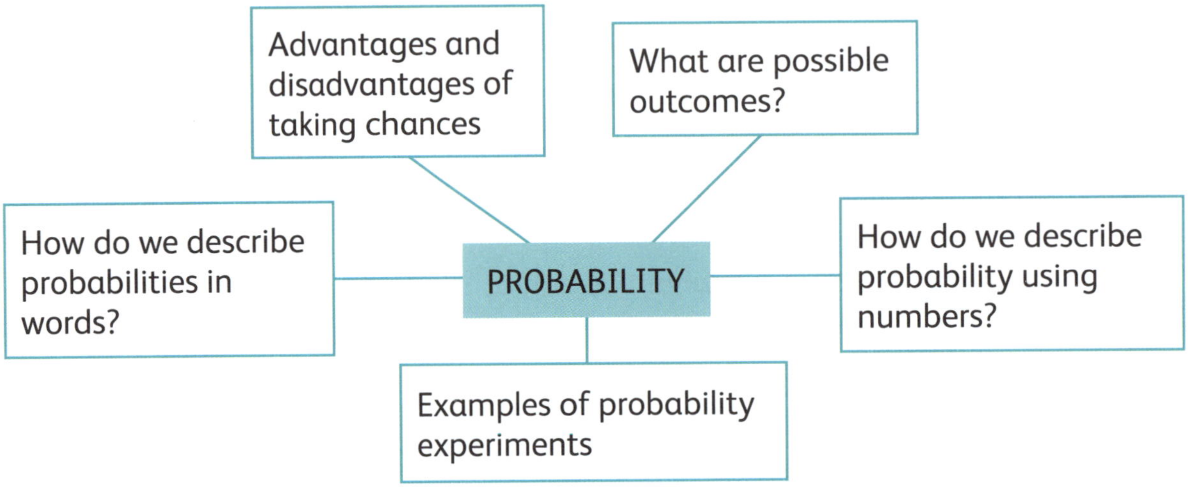

Practice questions

1 Which word describes an event that will definitely happen?

a likely b unlikely c certain d impossible

2 Which word describes two outcomes that are equally likely?

a possible b even c maybe d definite

3 Look at the probability scale. Match the events below to the correct letter on the scale.

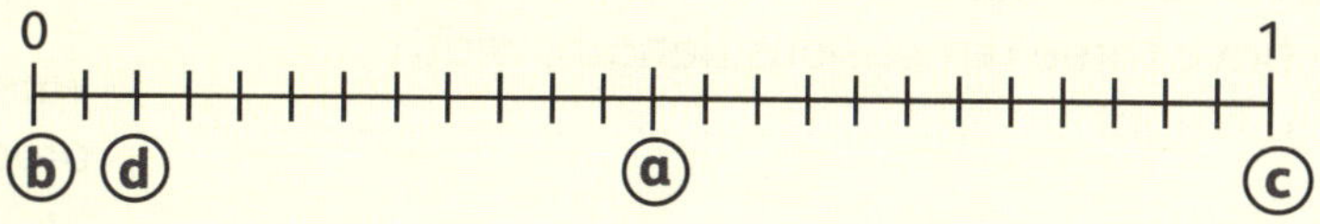

- Chocolate can be found on Mars.
- You will be older tomorrow than today.
- The next person you see will be female.
- The next person you see will have their birthday in August.

4 Another event has been added to the probability scale. Give an example of what you think this event could be.

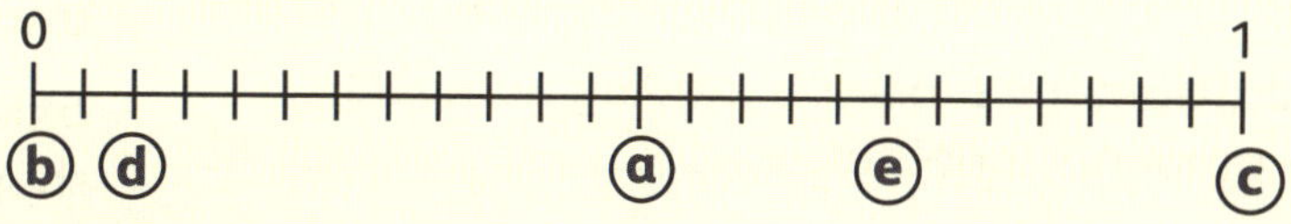

5 Write the sample space for each of these situations.

a Picking a ball from a bag that contains a blue ball, a red ball, a green ball and a black ball.

b Rolling a ten-sided die.

c Turning right or left at a road junction.

d The day it will be tomorrow.

18 Self check

Check you can do this:	If you have difficulty, try this:
1 Describe the probability of an event as likely, unlikely, possible or impossible.	Revise the work on pages 271 to 273.
2 Predict the possible outcomes of an event.	Revise the idea of the sample space on page 273.
3 Conduct simple probability experiments.	Revise the work on pages 274 and 275.
4 Record the actual outcomes of an event.	Revise the work on pages 276 and 277.

Glossary and Index

2D: 2-dimensional, flat, having width and length but not depth 241

3D: 3-dimensional, having or appearing to have three dimensions (length, width and height) 242

A

actual outcome: the result or effect of an action 274

addend: a number that is added to another in forming a sum 203

addition: the process of adding numbers or amounts together 46

algebra: a branch of mathematics where we use letters that represent unknown quantities or amounts 253

algebraic expression: a set of terms with letters and numbers that are combined using addition (+), subtraction (–), multiplication (×) and division (÷) 253

angle: the space between two lines or surfaces at the point at which they touch each other, measured in degrees 93

appropriate: suitable or right for a particular situation or occasion 162

approximate: not completely accurate but close 51, 164

aquarium: a glass container in which fish and other water creatures can be kept 155

architect: a person whose job is to design new buildings and make certain that they are built correctly 168

area: the amount of space a 2D shape takes up on a flat surface 162

arithmagon: kind of a missing number puzzle in the shape of a polygon 195

array: a regular or systematic arrangement 125

ascending: moving upwards 192

associative property: 1. the mathematical principle that says that when we add, it does not matter how we group the numbers 55
2. the mathematical principle that says that when we multiply three or more numbers, the product stays the same no matter how we group the numbers 128

attribute: a quality or characteristic that someone or something has 19

average: the result you get by adding two or more amounts together and dividing the total by the number of amounts 203

axis (pl: axes): a fixed line on a graph used to show the position of a point 109

B

bar graph: a type of graph (a mathematical picture) in which different amounts are represented by thin vertical or horizontal rectangles 109, 209

base: the part on which something stands; for example, the end face of a prism 246

biased: unfairly prejudiced for or against someone or something 112

blueprint: a copy of an architect's drawing with the details and dimensions of the building; floor plan 168

braces: curly brackets used to enclose words or lines to be considered together 19

breath: the air that goes into and out of your lungs 131

breathe: to move air into and out of the lungs 131

breathing: the act or process of taking air into your lungs and releasing it 131

budget: a plan to show how much money a person or organisation will earn and how much they will need or be able to spend 156

C

D

degree: the 360th part of the circumference of a circle, often represented by the sign ° 94

degrees Celsius (°C): a specific temperature on the Celsius scale 82

denominator: the number below the line in a fraction 27

descending: going from a higher to a lower place 192

design: the art of making plans or drawings for something 190

diagonal line of symmetry: a diagonal line that divides an object into two identical halves 175

difference: 1. the way in which two people or things are not the same 18
2. the result of subtracting one number from another 46

digit: any of the numbers from 0 to 9 .... 7, 46

dimension: a measurement of the length, width or height of something ... 160, 241

discrete (data): data that only takes certain values, commonly in the form of whole numbers 203

distributive property: the mathematical principle that says that multiplying the sum of two or more addends by a number produces the same result as when each addend is multiplied individually by the number and the products are added together 128

dividend: the amount we want to share out equally 226

divisibility: the ability to share equally without leaving a remainder 231

divisible: capable of being evenly divided, without remainder 231

divisor: a number by which another number, the dividend, is divided 226

dollar: the standard unit of money used in the US, Canada, Australia, New Zealand and other countries 153

drag: a backward force 276

duration: the length of time that something lasts 71

E

edge: the line of a solid where two faces meet 242

element (of a set): any of the distinct objects that belong to that set 19

endpoint: a point at which a line segment ends 93

entrepreneurship: all the activities of planning, starting and running a business 156

equal: the same in amount, number or size 258

equation: a mathematical statement in which you show that two amounts are equal using mathematical symbols 193, 258

equivalent fractions: the fractions that have different numerators and denominators but are equal to the same value 31, 143

estimate: to make your best guess to get as close to the real answer as you can 30, 52, 164

estimation: a guess or calculation about the cost, size, value, etc. of something 61

event: something that happens 271

exact: strictly accurate or correct 164

expanded form: a way to express a number as the sum of the place values of its digits 9

expanded notation: writing a number to show the value of each digit – it represents a number as the sum of each of its digits multiplied by their place value (ones, tens, hundreds, etc.) 133

experiment: a test done in order to learn something or to discover if something works or is true 274

express: to show 79

expression: a set of terms with letters and numbers that are combined using addition (+), subtraction (–), multiplication (×) and division (÷) 193, 196

F

face: the flat side of a solid 242

face value: the actual value of the digit in a number 7

fact family: a collection of related addition and subtraction facts, or multiplication and division facts, made from the same numbers 128

factor: any whole number that is produced when you divide a larger number by another whole number 126, 231

flight: an occasion when something flies or moves through the air 276

flip: when the source image is reflected in a mirror line; reflection 180

floor plan: a diagram of one room, apartment or entire floor of a building, usually drawn to scale 168

force: an influence on a body or system, producing or tending to produce a change in movement 276

forge: to make an illegal copy of something in order to deceive; for example, fake money 194

formula: a mathematical rule or principle, frequently expressed in algebraic symbols ... 255

fraction: a number usually expressed in the form a/b; a number that results from dividing one whole number by another 142, 235

fraction chart: a visual representation to help students understand and engage with fractions 30

fraction family: it refers to how fractions can be grouped into 'families' of equivalent fractions (e.g.: 1/2, 2/4, 3/6...) 30

fractional number: any number that has parts that are a fraction of a whole (proper fractions, improper fractions, mixed numbers, decimals) 149

full turn: a 360-degree turn 94

functional: helping objects to work properly 90

G

geometric pattern: a set of shapes arranged in a sequence that follows a particular rule 188

goods: things for sale, or the things that you own 156

gram: a unit of mass equal to 0.001 kilograms 76

graph: a way of organising data in a picture so that it is easy to understand 109

greatest: largest; biggest in amount 10

grid: a pattern or structure made from horizontal and vertical lines crossing each other to form squares 160, 178

grid method: a way of multiplying two numbers with more than one digit that involves partitioning numbers into tens and units before they are multiplied 134

group: to form a group or put people or things into a group 18

H

half turn: a 180-degree turn 94

health: the state of being well 132

heptagon: a flat shape with seven straight sides 98

hexagon: a flat shape with six straight sides 98

horizontal (lines): parallel to the ground or to the bottom or top edge of something 96, 178

horizontal line of symmetry: a horizontal line that divides an object into two identical halves 175

hundredth: one of a hundred equal parts of something 147

I

image: picture; a representation of a person or thing 180

impossible: not able to happen 271

improper fraction: a fraction that has a numerator that is greater or equal to the denominator 143

information: data; facts about a situation, person, event, etc. 11, 203

interpret: review, explain and arrive at conclusions 205

intersecting (lines): lines that cross each other at a common point 93

interval: a period of time between two things 71

interview: a conversation for gathering information 114

inverse: opposite in relation to something else 128, 226

inverse operations: pairs of mathematical manipulations in which one operation undoes the action of the other; for example, addition and subtraction, multiplication and division 135

irregular: not regular in shape or form 99

K

key: tell us what the symbols on the graph represent 109

kilo-: one thousand 60, 76

kilogram: a unit of mass equal to 1 000 grams 76

kilolitre: a unit of volume equal to 1 000 litres 65

kilometre: a unit of length, the common measure of distances, equal to 1000 metres 60

kite: it has two pairs of equal sides and the equal sides are next to each other 99

L

largest: biggest in amount 10

lattice method: a different method for solving multiplication by teachers in China.. 137

least: smallest in amount 10

leftovers: anything left or remaining from a larger amount; remainders 226

length: the measurement of something from end to end or along its longest side 160

less: a smaller amount of 188

lift: an upward force 276

like fractions: fractions with the same denominator 32, 143

line graph: a graphical representation of information that changes over a period of time – it is a chart made by joining points using line segments 209

line of symmetry: the line that divides a shape or an object into two equal and symmetrical parts 173

line segment: a section of a line between two points 93

litre: used to measure capacity and volumes – equal to 1 000 cubic centimetres 65

location: position 178

loss: a situation in which a business spends more money than it earns 156

lowest common denominator (LCD): the smallest number that is a common denominator of a certain group of fractions 38

M

mass: 1. the amount of matter in any solid object or in any volume of liquid or gas 76
2. a downward force; gravity 276

mean: the result you get by adding two or more amounts together and dividing the total by the number of amounts 203

median (value): the middle one in a set of values arranged in order of size 203

member (of a set): any of the distinct objects that belong to that set 18, 19

mentally: working out the answers to sums in your head, without using a calculator or working in columns 49

metre: the standard unit for measuring length 60

milli-: one thousandth 60

millilitre: used to measure capacity and volumes – equal to 0.001 litres 65

millimetre: a unit of length that is equal to 0.001 metres 60

minting: the process of making money 194

mirror line: the line that divides a shape or an object into two equal and symmetrical parts 173

mixed number: a combination of a whole number together with a proper fraction 27, 143, 235

mode: the number or value that appears most often in a particular set 203

money: coins or notes that are used to buy things, or an amount of these that a person has 203

more: a larger or extra number or amount .. 188

multiple: a number that can be divided by a smaller number an exact number of times 126, 231

multiplication: the process of multiplying numbers 125

multiplier: a number used to multiply another number 129

N

net: a 2D shape or pattern that you can fold to make a 3D solid 244

nonagon: a flat shape with nine straight sides 98

number: (a sign or symbol representing) a unit that forms part of the system of counting and calculating 7

number line: a straight line on which there is indicated a one-to-one correspondence between points on the line and the set of real numbers 49

number sentence: a combination of numbers and mathematical symbols ... 193, 196

numerator: the number above the line in a fraction 27

O

oblique (lines): lines that run at a slant or an angle 96

observation: an act or instance of watching attentively 114

octagon: a flat shape with nine straight sides 98

open path: a path that starts and ends at different vertices 92

operation: a mathematical process, such as addition or multiplication, in which one set of numbers is produced from another 187, 253

orientation: a geometric notation which says about the position of the object based on the specific direction or with relation to the points of the compass 180

outcome: one of the possible results in a probability experiment 271

overlapping (sets): the sets which have some elements in common 21

P

parallel (lines): lines that stay the same distance apart and never cross 96

parallelogram: a flat shape that has four sides – the two sets of opposite sides are parallel and of equal length to each other 41, 99

pattern: any regularly repeated arrangement 187, 231, 255

pentagon: a shape with five sides and five angles 98

perimeter: the distance around a shape or the length of its boundary 160

perpendicular (lines): lines that meet at a right angle (90 degrees) 96

physical exercise: a particular set of movements intended to improve health or make you stronger 131

pictograph: the pictorial representation of data using images, icons or symbols 209

pie chart: a way of showing information about how a total amount is divided up, consisting of a circle that is divided from its centre into several parts 209

place value: the value of a number decided by its position; for example, whether it is a one, a ten, a hundred, etc. 7

plan: a drawing or diagram of a place or object that shows its dimensions 160

plane shape: a closed figure 241

plot: to make marks to show the position, movement or development of something, usually in the form of lines or curves between a series of points on a map or piece of paper 215

point: a position in space 92

point of intersection: a point at which intersecting lines cross each other 93

polygon: a closed path made from line segments that meet at vertices (corners) 98

population: when you collect data, the population is the group of people or objects you choose to study 11, 111, 205

position: the place where something or someone is, often in relation to other things 188

possible: that might or might not happen 271

possible outcomes: possible results of an experiment or trial 273

power: the number of times that a number is to be multiplied by itself 9

prediction: a statement about what you think will happen in the future 188, 274

prefix: a letter or group of letters added to the beginning of a word to make a new word 85

present: to give, provide or make something known 213

price: the amount of money for which something is sold 156

prism: a polyhedron that has identical polygons as both its end faces – these end faces are joined by rectangular faces 246

probability: a possibility or chance of anything happening; likelihood 271

probability scale: a way to express how likely it is that an event will happen by using numbers 272

product: the result you get when two or more numbers are multiplied together 126

profit: the difference between the cost and the selling price when you sell something for more than it cost to produce 156

proper fraction: a fraction in which the numerator is less than the denominator 27, 143

pull: a force in which an object moves towards the source of the interaction 276

push: a force that moves an object away from something 276

Q

quadrilateral: a flat shape that has four sides and four inside angles 98

quarter turn: a 90-degree turn 94

questionnaire: a list of questions that several people are asked so that information can be collected about something 11, 114

quotient: the result you get when you divide one number by another 226

R

ray: a line that starts from one endpoint and continues in one direction 93

record: to keep information for the future....108

rectangle: a flat shape with four right angles and two pairs of sides that are equal in length 99

reflection: when the source image is reflected in a mirror line; flip 182

regular (polygon): a polygon that has all its sides equal in length and all its inside angles equal in size 98

relationship: the way in which two things are connected........ 226

remainder: the amount that is left when one number cannot be exactly divided by another 226, 233

representative: typical of, or the same as others in a larger group of people or things........ 112

resistance: a backward force; drag........ 276

rhombus: a flat shape that has four equal sides........ 99

right angle: a 90-degree turn; quarter turn.. 94

rotation: turn........ 180

round number: a number that ends with one or more zeros 51

rounding off: a mathematical technique of adjusting the number's digits to make the number easier to use during calculations........ 51

row: a line of things, people, animals, etc. arranged next to each other – in a grid rows run horizontally........ 178

rule: a principle guiding the way things are done........ 192

S

sample: a portion of the whole population 111, 205

sample size: the measure of the number of individual samples used in an experiment........ 111

sample space: describes the total number of possible outcomes for any event 273

sampling: the way a researcher chooses the sample group........ 112

scale: the relation between the real size of something and its size on a map, model or diagram 64

sequence: a series of related things or events, or the order in which they follow each other........ 188

services: a particular type of help or work... 156

set: a group of objects with same characteristics 18

set notation: listing the members of a set in curly brackets and putting a comma between each member........ 19

shape: an arrangement that is formed by joining lines together in a particular way or by the line or lines around its outer edge 164

sides: the line segments that form the outline of the polygon 98

similar (shapes): shapes that have the same number of sides and angles, but their sizes are different........ 102

similarities: when people or things look or are the same........ 18

simplest form: the smallest possible equivalent fraction of the number 34

simplifying (a fraction): using division to make equivalent fractions 34

slide: when the source image moves horizontally and/or vertically to a new position; translation 180

solid: an object that has three dimensions: a height, width and length, and is not flat 242

solution: the answer to a problem 196, 258

sort: to put a number of things in an order or to separate them into groups........ 19

source object (in transformation of shapes): the shape you start with........ 180

square: a flat shape that has all its sides equal in length, and all four angles are right angles........ 99

square centimetre (cm²): an area equal to a square that is 1 centimetre on each side... 162

square corner: a corner that is a right angle 92

square kilometre (km²): an area equal to a square that is 1 kilometre on each side 162

square metre (m²): an area equal to a square that is 1 metre on each side 162

square unit: the metric unit used to measure area 164

standard form: the use of digits to show value of a number 9

standard unit: the unit usually used to measure the weight, length or capacity of objects 60

statistics: the area of mathematics concerned with gathering and understanding data 205

straight: continuing in one direction without bending or curving 241

strategies: plans or methods for achieving any specific goal 49

substitute: the replacement of the letter with the known number or value 260

substitution: putting numbers in place of letters to calculate the value of an expression 260

subtraction: the process of removing one number from another 46

sum: the whole number when two or more numbers have been added together 46

surface: top part or layer of something 162

survey: looking at something closely and collecting data 205

symmetrical (shape): when you can draw (or imagine) a line down the middle that divides it into two parts that match exactly 173

symmetry: the quality of having two parts that match exactly, either when one half is like an image of the other half in a mirror, or when one part can take the place of another if it is turned 90° or 180° 173

T

tabular form: when we express a pattern using numbers and set it out in a table 191

tally marks: a way for counting by drawing marks – the general way of writing tally marks is as a group or set of five lines 109

temperature: when we talk about how hot or cold something is 82

tenth: one of then equal parts of something 146

terms: the shapes (or groups of shapes) at each position in a pattern 188, 255

thermometer: the instrument used to measure temperature 82

thrust: a forward force 276

tonne: used for measuring heavy masses – 1 tonne = 1 000 kg 76

transformation: a change in the size, position or orientation of a shape 180

translation: when the source image moves horizontally and/or vertically to a new position; slide 180

trapezium: a flat shape that has one pair of parallel sides 99

trend: the general direction in which something is developing or changing over time 209, 215

triangle: a flat shape with three straight sides 98

turn: rotation 94, 180

two-step problem: a problem that requires somebody to work in more than one step 135

U

unit fraction: any fraction with 1 as its numerator and a whole number for the denominator 27, 29, 143

unknown: not known or familiar 196

unlikely: not probable or likely to happen 271

V

value: 1. magnitude; quantity 7
2. each number in a data set 203

variable: the letter that represents an unknown value 196, 253, 254